Lecture Notes in Mathematics

1658

Editors:
A. Dold, Heidelberg
F. Takens, Groningen

Subseries: Instituto de Matemática Pura e Aplicada
Rio de Janeiro, Brasil (vol. 49)

Adviser: C. Camacho

Springer
Berlin
Heidelberg
New York
Barcelona
Budapest
Hong Kong
London
Milan
Paris
Santa Clara
Singapore
Tokyo

Antonio Pumariño
J. Angel Rodríguez

Coexistence
and Persistence
of Strange Attractors

 Springer

Authors

Antonio Pumariño
Departamento de Matemáticas
Universidad de Oviedo
Calvo Sotelo s/n
33.007 Oviedo
Spain
e-mail: apv@pinon.ccu.uniovi.es

J. Angel Rodríguez
Departamento de Matemáticas
Universidad de Oviedo
Calvo Sotelo s/n
33.007 Oviedo
Spain
e-mail: chachi@pinon.ccu.uniovi.es

Cataloging-in-Publication Data applied for

Mathematics Subject Classification (1991): 58F12, 58F13, 58F14

ISSN 0075-8434
ISBN 3-540-62731-6 Springer-Verlag Berlin Heidelberg New York

© Springer-Verlag Berlin Heidelberg 1997
Printed in Germany

Typesetting: Camera-ready T$_E$X output by the author
SPIN: 10520379   46/3142-543210 - Printed on acid-free paper

Preface

For dissipative dynamics, chaos is defined as the existence of strange attractors. Chaotic behaviour was often numerically observed, but the first mathematical proof of the existence, with positive probability (persistence), of a strange attractor was given by Benedicks and Carleson for the Hénon family, at the begining of 1990's. A short time later, Mora and Viana extended the proof of Benedicks and Carleson to the Hénon-like families in order to demonstrate that a strange attractor is also persistent in generic one-parameter families of surface diffeomorphisms unfolding a homoclinic tangency, as conjectured by Palis. In the present book, we prove the coexistence and persistence of any number of strange attractors in a simple three-dimensional scenario. Moreover, infinitely many of them exist simultaneously.

Besides proving this new non-hyperbolic phenomenon, another goal of this book is to show how the Benedicks-Carleson proof can be extended to families different from the Hénon-like ones.

We would like to thank J. Palis and M. Viana for their constant support, their suggestions and guidance in the realization of this work, especially the iluminating discussions with M. Viana during our meetings at the Universities of Santiago de Compostela, Oporto and Oviedo and at the Instituto de Matemática Pura e Aplicada of Rio de Janeiro. We also wish to mention F. Costal, C. Masa and C. Simó for reading a first Spanish version of the book, J. Mateos for his help in the elaboration of the figures and our workmate S. Ibáñez for his friendship and unconditional help.

This research was partially supported by DGICYT grant number PS88-0054 and by the Projects DF-92/35 and DF-93/213-34 of the University of Oviedo.

The authors

Contents

Introduction   1

1 SADDLE-FOCUS CONNECTIONS   11

2 THE UNIMODAL FAMILY   21

   2.1 Introduction 21

   2.2 Preliminary results 24

   2.3 The binding period 27

   2.4 The inductive step 31

   2.5 Construction of the sets $(\Omega_n)_{n \in N}$ 32

   2.6 Estimates of the excluded set 43

3 CONTRACTIVE DIRECTIONS   53

4 CRITICAL POINTS OF THE BIDIMENSIONAL MAP   73

   4.1 Preliminary results 73

   4.2 Construction of critical points 79

     4.2.1 Critical approximations of generation zero: The algorithm A . 83

     4.2.2 Critical approximations of higher generations: The algorithm B   85

   4.3 The contractive fields 87

5 THE INDUCTIVE PROCESS   89

   5.1 The construction of C_n 91

   5.2 Returns, binding points and binding periods 94

     5.2.1 Comments and properties 97

   5.3 The folding period 100

     5.3.1 The splitting algorithm 102

   5.4 The remainder of the induction hypotheses 105

5.5   Appendix: The function $d_\mu(\xi_0)$ 115

6 THE BINDING POINT   119

7 THE BINDING PERIOD   135

8 THE EXCLUSION OF PARAMETERS   153

Appendix A: Numerical experiments   191

Bibliography   193

Index   195

INTRODUCTION

This book deals with the existence and persistence of any number of coexisting strange attractors in three-dimensional flows. More precisely, we shall define a one-parameter family X_μ of piecewise regular vector fields on $\mathbf{R}^3$ and we shall prove that for each natural number n, there exists a positive Lebesgue measure set of parameter values for which X_μ has, at least, n strange attractors. Moreover, X_μ exhibits an infinite number of strange attractors for some values of the parameter.

By an attractor we mean a compact invariant set Λ having a dense orbit (transitive) and whose stable set $W^s(\Lambda)$ has a non-empty interior. For different notions of attractor see [14]. We call an attractor strange if it contains a dense orbit with a positive Liapunov exponent (sensitive dependence on initial conditions).

The term strange attractor was first used by D. Ruelle and F. Takens [20] to suggest that turbulent behaviour in fluids might be caused by the presence of attractors which are locally the product of a Cantor set and a piece of two-dimensional manifold. The notion of strange attractor associated to the sensitive dependence on initial conditions was needed to explain asymptotic dynamics which numerically or empirically manifest this kind of unpredictable behaviour. One of the most relevant dynamics of this type was earlier observed by Lorenz [11] on analysing the quadratic vector field

$$\begin{cases} x' = -10x + 10y \\ y' = 28x - y - xz \\ z' = -\frac{8}{3}z + xy \end{cases}$$

which follows from a truncation of the Navier-Stokes equations. Surprisingly, under small perturbations of the system, he seemed to get a persistent but not stable attractor, i.e. small perturbations of the original system give rise to nearby attractors but, in general, they are not topologically equivalent.

From a physical point of view, a certain degree of persistence is as relevant as the unpredictability of the dynamics resulting from the afore-mentioned sensitivity with respect to initial conditions. So, if a family X_μ of vector fields exhibits a strange attractor for the value of the parameter $\mu = \mu_0$, the dynamics of the attractor should only be considered if for every $\delta > 0$, strange attractors still exist for values of the parameter belonging to a positive Lebesgue measure set $E \subset B(\mu_0, \delta)$. In this case, the attractor is said to be persistent for the family X_μ, and is said to be fully persistent if we can take $E = B(\mu_0, \delta)$ for some $\delta > 0$.

A non-periodic hyperbolic attractor is strange, fully persistent and even stable. From numerical analysis, Lorenz's attractor seems to be strange, fully persistent but not stable. M. Hénon [7] found a possible persistent (but not fully persistent) strange attractor for the family

$$H_{a,b}(x,y) = (1 - ax^2 + y, bx)$$

with $a = 1.4$ and $b = 0.3$. At the begining of 1990's, in a historical and very complex paper [3], M. Benedicks and L. Carleson proved mathematically that the Hénon family has persistent strange attractors for values of the parameters close to $a = 2$ and $b = 0$. A short time later, L. Mora and M. Viana [15] proved that, such as J. Palis had conjectured, generic one-parameter families of diffeomorphisms on a surface which unfold a homoclinic tangency have strange attractors or repellers (negative attractors) with positive probability in the parameter space. For a proof of this result in higher dimensions, see [25].

Homoclinic orbits were discovered by H. Poincaré a century ago. In his famous essay on the stability of the solar system, Poincaré showed that the invariant manifolds of a hyperbolic fixed point could cut each other at points, called homoclinics, which yield the existence of more and more points of this type and consequently, a very complicated configuration of the manifolds, [18]. Many years later, G. Birkhoff [4] showed that in general, near a homoclinic point there exists an extremely intrincated set of periodic orbits, mostly with a very high period. By the mid-1960's, S. Smale [21] placed his geometrical device, the Smale horseshoe, in a neighbourhood of a tranversal homoclinic orbit, thus explaining Birkhoff's result and arranged the complicated dynamics that occur near a homoclinic orbit by means of a conjugation to the shift of Bernouilli.

The strange attractors found in [15] arise from the creation or destruction of Smale horseshoes associated to the transversal homoclinic points which appear as the result of the bifurcation of a tangential homoclinic point. Roughly speaking, homoclinic bifurcations mean the creation of transversal homoclinic orbits resulting from small perturbations of the dynamical system. Though one case of homoclinic bifurcation is that of homoclinic tangency, there are, however, interesting examples of homoclinic bifurcations that do not correspond to homoclinic tangencies. For an extensive study of the phenomena which accompany homoclinic bifurcations, see the book by J. Palis and F. Takens, [17]. We also quote this reference as a suitable complement to this introduction. In Chapter 7 of [17] the authors propose homoclinic bifurcations as the doorways (the only ones in dimension two) to non-hyperbolic dynamics: coexistence of infinitely many sinks, persistence of Hénon-like attractors, etc. In the present book we place ourselves in one of these doorways from which we have access to an infinite number of strange attractors and to any finite number of persistent strange attractors, within a three-dimensional vector field framework.

In order to place ourselves within this framework, we evoke the following result proved by P. Sil'nikov [22]: In every neighbourhood of a homoclinic orbit of a hyperbolic fixed point of an analytical vector field on $\mathbf{R}^3$ with eigenvalues λ and $-\rho \pm i\omega$ such that $0 < \rho < \lambda$, there exists a countable set of periodic orbits. This result is similar to Birkhoff's for diffeomorphisms and thus, it should be completed by proceeding as Smale did. So, C. Tresser [24] proved that in every neighbourhood of such a homoclinic orbit, an infinity of linked horseshoes can be defined in such a way that the dynamics is conjugated to a subshift of finite type on an infinite number of symbols. If, on the contrary, $0 < \lambda < \rho$, then the dynamics is trivial: The ω-limit set of any point in a neighbourhood of the homoclinic orbit is contained in the closure of this orbit. In the case $\lambda = \rho$, we shall prove our main result:

Theorem A. *In the set of three-dimensional vector fields having a homoclinic orbit to a fixed point with eigenvalues $\lambda > 0$ and $-\lambda \pm i\omega$ satisfying $\left|\frac{\lambda}{\omega}\right| < 0.3319$, there exists a one-parameter family X_a of piecewise regular vector fields such that for every neighbourhood V of the homoclinic orbit, for each $k \in \mathbf{N}$ and for every value of the parameter a in a set of positive Lebesgue measure depending on k, at least k strange attractors coexist in V. Moreover, for some value of the parameter a, there exist infinitely many strange attractors contained in V.*

Piecewise regular vector fields with a Sil'nikov homoclinic orbit were constructed in [24]. In fact, these orbits arise in families of analytical vector fields as a codimension-one phenomenon, [19]. Recently, in [9] the authors proved the existence of vector fields verifying the hypotheses of Sil'nikov's theorem in generic unfoldings of codimensionfour singularities in $\mathbf{R}^3$ (Sil'nikov bifurcation). This homoclinic bifurcation occurs when the parameters take values on a manifold of codimension one. Just off this manifold the homoclinic orbit disappears and an infinite number of horseshoes given in [24] are destroyed. Then, as a consequence of [15], a method of constructing families of quadratic vector fields on $\mathbf{R}^3$ which display strange attractors is obtained.

Unlike the Sil'nikov bifurcation mentioned above, the homoclinic orbit in Theorem A endures for each vector field X_a. Therefore, in a neighbourhood of this orbit we can choose a suitable transversal section Π_0 and define the transformation $T : \Pi_0 \to \Pi_0$ associated to the flow. After splitting Π_0 into a countable union of rectangles, R_m, and carrying out adequate changes of variable, we get the following sequence of families of diffeomorphisms

$$T_{\lambda,a,b}(x,y) = \left(f_{\lambda,a}(x) + \frac{1}{\lambda} \log \left(1 + \sqrt{b}y \right), \sqrt{b} \left(1 + \sqrt{b}y \right) e^{\lambda x} \sin x \right),$$

with $b = e^{-2\pi\lambda m}$ and $m \in \mathbf{N}$. For a large enough m, each $T_{\lambda,a,b}$ is a small perturbation of $\Psi_{\lambda,a}(x,y) = (f_{\lambda,a}(x),0)$, where $f_{\lambda,a}(x) = \lambda^{-1} \log a + x + \lambda^{-1} \log \cos x$. Thus, Theorem A is an immediate consequence of the following one:

Theorem B. *Fixed* $0 < \lambda < 0.3319$, *for every* $m_0 \in \mathbf{N}$ *and for each* $k \in \mathbf{N}$ *there exists a positive Lebesgue measure set* $E = E(k)$ *of values of the parameter* a *such that, for every* $a \in E$, *there exist, at least,* k *transformations* $T_{\lambda,a,b}$, *with* $b < e^{-2\pi\lambda m_0}$, *having a strange attractor. Moreover, there exist values of the parameter* a *for which infinitely many* $T_{\lambda,a,b}$ *simultaneously have a strange attractor.*

Mora and Viana defined a renormalization in a neighbourhood of a homoclinic point to transform a generic family of diffeomorphisms unfolding a homoclinic bifurcation into a Hénon-like family. These families are defined in Proposition 2.1 of [15] so as to be suitable small perturbations of $H_a(x,y) = (1 - ax^2, 0)$ just as the Hénon family $H_{a,b}$ is for small values of b. The changes of variable which we have to carry out to obtain $T_{\lambda,a,b}$, play the same role as does the renormalization in [15]. We shall prove in Proposition 1.3 the conditions which make $T_{\lambda,a,b}$ a good perturbation of $\Psi_{\lambda,a}$. Then, we shall say that $T_{\lambda,a,b}$ is an adequate unfolding of $\Psi_{\lambda,a}$.

From this stage and in spite of $f_{\lambda,a}$ not being the quadratic map, the proof of Theorem B can be developed by means of a cautious adaptation of the ideas and the arguments in [3] and [15]. Nevertheless, since the density of these references makes them hard to read, we feel it is both useful and necessary to give a proof in detail in order to facilitate the understanding of the intricate inductive method and the control of the numerous estimates required. We also try, thereby, to show how the ideas for the Hénon-like families can be applied to adequate unfoldings of unimodal maps which are distinct from the quadratic one. Maybe these unfoldings take part in many other cases where possible strange attractors have also been observed numerically. See, for instance, [5].

This book is organized as follows:

In Chapter 1 we introduce the afore-mentioned changes of variable for defining the transformations $T_{\lambda,a,b}$. In fact, it is shown that $T_{\lambda,a,b}$ is an adequate unfolding of $f_{\lambda,a}$. Next, we prove that, for every positive λ, there exists a value of the parameter a, $a(\lambda)$, such that $f_{\lambda,a(\lambda)}$ has a homoclinic orbit and that, for a sufficiently small λ, for instance $\lambda < 0.3319$, the Schwarzian derivative of the map $f_{\lambda,a(\lambda)}$ is negative. This means that $f_{\lambda,a(\lambda)}$ has no periodic attractors, [23].

In Chapter 2 we study the unimodal family $f_{\lambda,a}$ for $0 < \lambda < 0.3319$. It is shown that there exists a constant $c_0 > 0$ such that, for every $0 < c < \min\{c_0, \log(1+\lambda)\}$, there is a value of the parameter $a_0 = a_0(\lambda, c) < a(\lambda)$ close to $a(\lambda)$ and a positive Lebesgue measure set $E = E(\lambda, c) \subset [a_0, a(\lambda)]$ such that every $a \in E$ satisfies the exponential growth condition for every $n \in \mathbf{N}$, i. e.,

$$|D_n(a)| = \left|\left(f_{\lambda,a}^n\right)'(f_{\lambda,a}(c_\lambda))\right| \geq e^{cn} \text{ for every } n \in \mathbf{N}.$$

This result, which is stated in Theorem 2.1, is a consequence of Theorem 6.1 in [13], that is, of the Benedicks and Carleson theorem for unimodal maps distinct from the quadratic one. However, since comprehension of the unidimensional case will be necessary to understand the bidimensional dynamics, which is studied in successive chapters, and since many of the specific ideas used in the study of $f_{\lambda,a}$ will be evoked in the study of $T_{\lambda,a,b}$, we have to develop a different proof from the one given in [13].

To construct a positive Lebesgue measure set E such that the exponential growth condition holds for every $a \in E$, we proceed by induction on the length n of the orbit of $f_{\lambda,a}(c_\lambda)$. Clearly, whenever this orbit remains far from the critical point (and this is easily obtained for a number N of initial iterates and for the values of the parameter belonging to an interval $[a_N(\lambda), a(\lambda)]$), the orbit of $f_{\lambda,a}(c_\lambda)$ will be e^c-expansive, where c depends on the distance between the initial orbit and c_λ. This remark allows us to start the inductive process, but, since the length of the interval $[a_N(\lambda), a(\lambda)]$ tends to zero as N tends to infinity, we have to let the orbit of $f_{\lambda,a}(c_\lambda)$

accede to any sufficiently small neighbourhood $(c_\lambda - \delta, c_\lambda + \delta)$ of the critical point, at iterates which will be called returns. In this case, since the derivative of the unimodal map tends to zero as the distance between the return and the critical point tends to zero, we have to control such distances. To this end, it seems to be natural to permit this distance to decrease as the return iterate increases, because the small derivative may be distributed in a larger exponent in the definition of expansiveness. Hence, if $\Omega_{n-1} \subset [a_N(\lambda), a(\lambda)]$ denotes the set of values of the parameter a for which $f_{\lambda,a}(c_\lambda)$ is e^c-expanding up to time $n - 1$, we remove, from Ω_{n-1}, those parameters for which the following basic assumption does not hold:

$$\left| f_{\lambda,a}^n(c_\lambda) - c_\lambda \right| \geq e^{-\alpha n},$$

where $\alpha > 0$ is a small positive constant. In this way, a set Ω_n' is constructed in a correct, but unfinished, posing of the problem. In fact, we also have to control the rate of previous iterates to the return whose expansiveness has been annihilated by the small derivative at the return. Here is where the reason for the inductive method becomes patent:

Since the orbit at the return is close to the critical point, their successive iterates, and consequently the derivatives at these iterates, are close each other. In this context, the binding period $[n + 1, n + p]$ is defined by taking the largest natural number p such that

$$\left| f_{\lambda,a}^{n+j}(c_\lambda) - f_{\lambda,a}^j(c_\lambda) \right| \leq e^{-\beta j} \text{ for } 1 \leq j \leq p,$$

where $\beta > 0$ is a small constant. By taking $\alpha < \beta$ small enough it is shown that the length of the binding period is smaller than n. Then, by using the inductive hypothesis for the orbit of the critical point and bearing in mind the closeness between its iterates and the respective iterates of the return, the small derivative at the return is proved to be compensated during the binding period. The remainder of the iterates outside the binding periods are called free iterates and they will be used to recover the exponential growth of the orbit. Therefore, the rate of these iterates has to be sufficiently large, for which we have to remove from Ω_n' the parameters not satisfying the following free assumption:

$$F_n(a) \geq (1 - \alpha)n,$$

where $F_n(a)$ denotes the number of free iterates in $[1, n]$.

In this way, the sets Ω_n are inductively constructed so that if, in each step, the measure of the excluded set exponentially decreases with respect to n, then the set E announced in Theorem 2.1 can be obtained by intersecting all the sets Ω_n. The detailed development of the whole process requires a large number of estimates. We finish this advance of Chapter 2 by calling attention to the relationship between the different constants taking part in the process and to their adequate and orderly selection:

First, we consider an arbitrary $\lambda < 0.3319$. Once λ is chosen, the constant c_0 of Theorem 2.1 depends on λ and is given in Proposition 2.2. Once c is fixed with $0 < c < \min\{c_0, \log(1 + \lambda)\}$, in the definition of binding period we take $\beta = \beta(\lambda, c)$ small enough. With respect to the constant α taking part in basic and free assumptions, this will depend on λ, c and β and will be taken sufficiently small with respect to them. In order to establish the concept of return, a constant $\delta = \delta(\lambda, c, \beta, \alpha)$ is chosen which is related to the natural number Δ ($\delta \approx e^{-\Delta}$) given in Proposition 2.2. In accordance with this proposition, Δ has to be large enough. Hence, δ will be taken small enough and, in particular, δ is always said to be sufficiently less than λ, c, β and α. Schematically, we write

$$\lambda \rightarrow c_0(\lambda) > c >> \beta(\lambda, c) >> \alpha(\lambda, c, \beta) >> \delta(\lambda, c, \beta, \alpha).$$

Finally, the inductive process will be started in an iterate $N = N(\lambda, c, \beta, \alpha, \delta) >> \Delta$. Then, for fixed N, a set $\Omega_0 = [a_0, a(\lambda)]$ is constructed, where the inductive process starts. Lastly, a_0 only depends on λ and c.

In the remaining chapters we prove Theorem B. From Chapter 1, we know that the closure of the unstable manifold of the saddle-point $P_{a,m}$ is an attracting set. Therefore, this set will be a strange attractor whenever the existence of a dense orbit with a positive Liapunov exponent is stated.

Though $T_{\lambda,a,b}$ is close to $\Psi_{\lambda,a}$ for small values of b, the expansiveness, with positive probability, along the orbit of the critical point of $\Psi_{\lambda,a}$ does not easily extend to $T_{\lambda,a,b}$. The hardness of this extension begins to appear in the definition of critical points for the bidimensional map. In fact, the role of critical points is now played by points on $W^u(P_{a,m})$ such that the differential map of $T_{\lambda,a,b}$ sends the tangent vector to $W^u(P_{a,m})$ at these points into a contractive direction, that is, into a direction which is exponentially contracted by all the iterates of the differential map. These concepts will be accurately stated in the inductive process framework which takes part in the proof of Theorem B.

In Chapter 3 we study, for each $n \in \mathbf{N}$, the maximally contracting and maximally expanding directions for the n-th iterate of the differential map. Under inductive hypotheses of expansiveness, remarkable properties of these directions are established. Of course, the known expansiveness on the first iterates allows the inductive process to start.

The algorithms used for constructing critical approximations of order n from critical approximations of order $n-1$ are introduced in Chapter 4. A point z belonging to $W^u(P_{a,m})$ is said to be a critical approximation of order n if the image under the differential map of the tangent vector to $W^u(P_{a,m})$ at z lies on the maximally contracting direction for the n-th iterate of the differential map. Critical approximations of order n play, in the respective step of the inductive process, the same role as that of the critical point in the unimodal case. In order to prove expansiveness on the orbit of every critical approximation z, we have to control, at its returns, the distance between the respective iterate and the critical approximations of order equal to or lower than the order of z. In fact, it will be sufficient to control the distance to a certain critical approximation placed in a determined situation (tangential position). As in the unidimensional case, it will be possible to compare the exponential growth in the successive iterates of returns with the growth in the respective iterates of a critical approximation in tangential position (binding point) during a period of time which will also be called binding period.

To guarantee the existence of binding points, the afore-mentioned algorithms will need to adduce sufficient critical approximations and these approximations will have to be distributed in a suitable way, as their orders increase, on the different branches that $W^u(P_{a,m})$ defines in its continuous folding process. This adequate distribution is obtained by ordering the branches of $W^u(P_{a,m})$ by means of the concept of generation. In the fourth chapter it is also proved that from old critical approximations close new ones are constructed in such a way that each critical approximation generates a convergent sequence of critical approximations. By definition, critical points are the limits of these sequences. Expansiveness along the orbits of the image of every critical approximation yields expansiveness of the orbit of the respective critical point.

We argue by induction in Chapter 5 in order to rigorously define the recurrent process for constructing the critical set C_n (set of critical approximations of order

$n - 1$). In this chapter the concepts of returns, binding points and binding periods associated to each point bound to C_n are introduced. Next, we deal with a new difficulty which arises in the treatment of the bidimensional problem: the folding phenomenon. When a return μ of a critical approximation z_0 takes place, that is, when $z_\mu = T_{\lambda,a,b}^\mu(z_0)$ is bound to C_μ, the slope of the vector $\omega_\mu = DT_{\lambda,a,b}^\mu(z_1)(1,0)$ may be high. In this case, the study of the behaviour of ω_μ, which coincides with the study of the expansiveness of the orbit of z_1, is far from being an unidimensional problem. Nevertheless, it will be proved that, after a number l of iterates, the slope of $\omega_{\mu+l} = DT_{\lambda,a,b}^{\mu+l}(z_1)(1,0)$ is very small again. The period $[\mu + 1, \mu + l]$ will be called the folding period associated to the return μ. During this period we only have knowledge about the evolution of the vector h_μ, which corresponds to the horizontal component of ω_μ. The choice of iterates, on which the study of the behaviour of the vectors ω_j is replaced by the study of the behaviour of h_j, is called the splitting algorithm. Chapter 5 ends by establishing the set of inductive hypotheses which allow us to state the expansiveness, up to time n, of every point bound to C_n. In Chapters 6 and 7 these inductive hypotheses are proved at time n.

The main objective of Chapter 6 is to find, for every free return n of a critical approximation $z_0 \in C_n$, a binding point $\zeta_0 \in C_n$ in tangential position. The loss of exponential growth at each return is estimated throughout the chapter. In Chapter 7 it is shown that these losses are compensated by the exponential growth, in the first iterates, of the vectors $h_j(\zeta_1)$, where ζ_0 is the binding point associated to the considered return, $\zeta_1 = T_{\lambda,a,b}(\zeta_0)$ and $h_j(\zeta_1)$ is the respective vector given by the splitting algorithm related to the orbit of ζ_1. Finally, an upper bound for the binding period associated to each return of every critical approximation is obtained.

Chapter 8 is the longest of this book and many references to previous chapters, especially to Chapter 2, are made there. The process of exclusion of parameters needed to deduce the e^c-expansiveness of the image of every critical point for a positive Lebesgue measure set is developed. The starting point of this chapter is the existence, with certain properties, of analytic continuations of the critical approximations. These properties are also inductively proved and they permit us to assume that the binding point is independent of a (for small changes of a) as occurs in dimension one. To procede as in Chapter 2, it is necessary to redefine the sets C_n, taking new critical sets with, perhaps, less elements but still sufficient ones so as to ensure the existence of binding points in every return. On the other hand, the cardinal of C_n is small enough so that, after removing the parameters for which the orbit of some critical approximation fails to be expansive, a positive Lebesgue measure set E remains.

The global interpretation of the proof of Theorem B, developed throughout the final six chapters of the book, is not simple but a much simpler treatment does not seem to exist. The inductive method is used so frequently that the reader will have to control which step is applied each time. Furthermore, some concepts have to be redefined and, therefore, the validity of many arguments already proved have to be supervised later on. As in Chapter 2, special attention has to be paid to the relationship and order of choice of the different constants. Here, two new constants are needed: The constant K introduced in Chapter 1 when the adequate unfolding of $f_{\lambda,a}$ is stated and the constant b. K only depends on λ and almost every constant arising from Chapter 2 depends on it. The constant b depends on the remaining constants, is the last one to be selected and is chosen sufficiently small in each argument.

Once the expansiveness of the orbit of the critical points is achieved, which corresponds to the longest part of the proof of Theorem B, the density of the orbit of the critical point of generation zero is demonstrated for a set of parameters with positive Lebesgue measure, say ϵ. Since ϵ does not depend on m, provided that m is sufficiently large, we deduce the coexistence and persitence of any number of strange attractors.

The book ends with the exposition of some numerical experiments.

Chapter 1

SADDLE-FOCUS CONNECTIONS

In this chapter, we consider autonomous differential equations in $\mathbf{R}^3$

$$\begin{cases} x' = -\rho x + \omega y + P(x, y, z) \\ y' = -\omega x - \rho y + Q(x, y, z) \\ z' = \lambda z + R(x, y, z) \end{cases} \tag{1.1}$$

where ρ, λ and ω are positive real numbers and P, Q and R are sufficiently smooth maps, vanishing together with their first order derivatives at the origin. Then, the origin θ is a fixed point of the saddle-focus type, with eigenvalues λ and $-\rho \pm \omega i$. Under linearizing assumptions, the flow in a neighbourhood of θ is given by

$$\begin{cases} x(t) = e^{-\rho t}(x_0 \cos \omega t + y_0 \sin \omega t) \\ y(t) = e^{-\rho t}(-x_0 \sin \omega t + y_0 \cos \omega t) \\ z(t) = z_0 e^{\lambda t} \end{cases} \tag{1.2}$$

In a more general framework, we now consider a sufficiently smooth family of vector fields

$$f : (\mu, \mathbf{x}) \in I \times \mathbf{R}^3 \longrightarrow f(\mu, \mathbf{x}) \in \mathbf{R}^3,$$

where I is an interval of parameters and, for each $\mu \in I$, $f(\mu, \mathbf{x})$ is a vector field of type (1.1). We also assume that for every $\mu \in I$, $f(\mu, \mathbf{x})$ is topologically conjugated to its linear part in a neighbourhood U of θ, [6].

These families unfold interesting dynamic behaviours when there exists a homoclinic orbit to θ for some value of the parameter μ. Thus, we choose P, Q and R in such a way that $f(0, \mathbf{x})$ has a solution $\mathbf{p} : t \in \mathbf{R} \rightarrow \mathbf{p}(t) \in \mathbf{R}^3$ satisfying that $\mathbf{p}(t) \rightarrow \theta$ as $t \rightarrow \pm\infty$. This solution defines a homoclinic orbit $\Gamma_0 = \{\mathbf{p}(t) : t \in \mathbf{R}\}$.

In order to describe the dynamics near Γ_0, a return map can be defined on a certain rectangle Π_0 contained in the set $\{(x, y, z) \in U : x = 0, y > 0, z > 0\}$. This map T_μ will be given by the composition of two maps L_μ and R_μ that we define below.

After a rescaling of time we may suppose $\omega = 1$. From (1.2), every $(x, y, z) \in \Pi_0$ takes a time $t = \lambda^{-1} \log z^{-1}$ to reach the plane $\Pi_1 = \{(x, y, z) \in \mathbf{R}^3 : z = 1\}$. So, we can set a map L_μ from Π_0 to Π_1 given by

$$L_\mu(y, z) = \left(y z^{\frac{\varrho}{\lambda}} \sin\left(\frac{1}{\lambda} \log \frac{1}{z}\right), y z^{\frac{\varrho}{\lambda}} \cos\left(\frac{1}{\lambda} \log \frac{1}{z}\right) \right).$$

In this expression we are neglecting the coordinates $x = 0$ and $z = 1$ of the points in Π_0 and Π_1, respectively. We also assume that U contains the ball with center θ and radius $r = 2$.

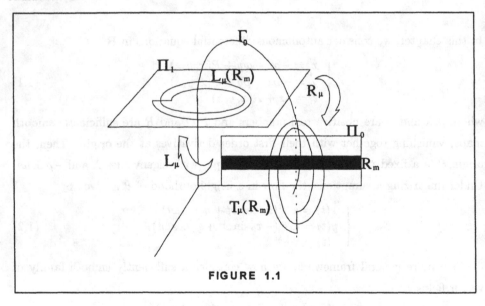

FIGURE 1.1

Now, we choose Π_0 so that L_μ is a homeomorphism. For this, it is enough to take $d > 0$ and

$$\Pi_0 = \left\{ (x, y, z) \in \mathbf{R}^3 : x = 0, d e^{-2\pi\rho} \leq y \leq d, 0 < z \leq e^{-\pi\lambda} \right\}.$$

In particular, we take $d = 2 \left(1 + e^{-2\pi\rho} \right)^{-1}$ so that $(0, 1, 0)$ is the mid-point of the base of Π_0. If we define the rectangle

$$R_m = \left\{ (x, y, z) \in \Pi_0 : e^{-(2m+1)\pi\lambda} \leq z \leq e^{-(2m-1)\pi\lambda} \right\},$$

for each $m \in \mathbf{N}$, then $\Pi_0 = \underset{m \in \mathbf{N}}{\cup} R_m$.

The shape of $L_\mu(R_m)$ suggests studying its image under a new transformation R_μ, associated to the flow in a neighbourhood of Γ_0 and defined between the tranversal planes Π_1 and $\tilde{\Pi}_0 = \{(x,y,z) : x = 0\}$. This map R_μ is a diffeomorphism that, neglecting once more the coordinates $z = 1$ and $x = 0$, may be expressed by

$$R_\mu(x,y) = (1 + c_1 x + c_2 y, \mu + c_3 x + c_4 y) + h.o.t.$$

where c_1, c_2, c_3 and c_4 are real constants and μ is a parameter which generically unfolds a homoclinic bifurcation.

Up to higher order terms, we may define $T_\mu = R_\mu \circ L_\mu : \Pi_0 \longrightarrow \tilde{\Pi}_0$ in the following way

$$T_\mu(y,z) = \begin{pmatrix} 1 + c_1 y z^{\frac{\rho}{\lambda}} \sin\left(\lambda^{-1} \log z^{-1}\right) + c_2 y z^{\frac{\rho}{\lambda}} \cos\left(\lambda^{-1} \log z^{-1}\right) \\ \mu + c_3 y z^{\frac{\rho}{\lambda}} \sin\left(\lambda^{-1} \log z^{-1}\right) + c_4 y z^{\frac{\rho}{\lambda}} \cos\left(\lambda^{-1} \log z^{-1}\right) \end{pmatrix}.$$

The dynamics of these maps, and hence those of $x' = f(\mu, x)$, were studied when $\lambda \neq \rho$:

If $\rho > \lambda$, there exists for $\mu > 0$ a limit cycle which approaches Γ_0 as μ tends to 0. Of course, this limit cycle is the unique attractor in a neighbourhood of Γ_0, [1].

If $\rho < \lambda$, there exists an infinite number of periodic orbits in a neighbourhood of Γ_0, [22]. In fact, infinitely many horseshoes appear when $\mu \to 0$ and, in an arbitrarily small neighbourhood of Γ_0, there exist trajectories in one to one correspondence with a subshift on infinitely many symbols, [24]. This creation of horseshoes involves generic unfoldings of homoclinic tangencies and, consequently, the existence of strange attractors for a positive Lebesgue measure set of parameters, [15].

In this book, we deal with the case $\rho = \lambda$ and, unlike the above mentioned cases, where the results do not depend on R_μ, now R_μ will play an essential role. We take $\mu = c_2 = c_3 = 0$, $c_1 = -1$ and $c_4 = a$, and we shall prove that the family of maps from Π_0 to $\tilde{\Pi}_0$ given by

$$T_{\lambda,a}(y,z) = \left(1 + yz \sin\left(\lambda^{-1} \log z\right), ayz \cos\left(\lambda^{-1} \log z\right)\right),$$

exhibits strange attractors in a persistent way in the sense of measure theory. More precisely, we show that for a set of values of the parameter a with positive Lebesgue measure, for every neighbourhood V of Γ_0 and for every $k \in \mathbb{N}$, there exist at the same time k strange attractors contained in V. Moreover, there are values of the parameter a for which there exist simultaneously infinitely many strange attractors contained in V. Notice that these strange attractors should not necessarily be related to those that can appear in the case $\lambda > \rho$ as $\mu \to 0$. In this case, if $\mu = 0$, there are no attractors near Γ_0.

We would also like to point out that $\lambda = \rho$ is a resonant case and that we do not know of any result of linearization in this situation. However, since we focus on the coexistence of strange attractors, let us suppose that the maps P, Q and R given in (1.1) are null in a neighbourhood of the origin. In fact, this assumption is not a handicap to get a homoclinic orbit Γ_0.

Next, we shall introduce some changes of variable which will allow us to prove that for every sufficiently large m, the respective restriction of $T_{\lambda,a}$ to R_m is C^3 close to a unidimensional map.

By means of new coordinates $\eta = y$ and $\xi = \lambda^{-1} \log z$, Π_0 becomes a vertical strip

$$B = \left\{ (\eta, \xi) \in \mathbf{R}^2 : \frac{2e^{-2\pi\lambda}}{1 + e^{-2\pi\lambda}} \leq \eta \leq \frac{2}{1 + e^{-2\pi\lambda}}, \xi < -\pi \right\},$$

and each rectangle R_m is transformed into a new rectangle

$$R'_m = \{ (\eta, \xi) \in B : -(2m+1)\pi \leq \xi \leq -(2m-1)\pi \}.$$

Now, the map $T_{\lambda,a}$ is given by

$$T_{\lambda,a}(\eta, \xi) = \left(1 + \eta e^{\lambda\xi} \sin \xi, \xi + \frac{1}{\lambda} \left(\log a + \log \eta + \log \cos \xi \right) \right), \qquad (1.3)$$

and it is only defined on

$$C_m = \left\{ (\eta, \xi) \in R'_m : -\left(2m + \frac{1}{2} \right) \pi < \xi < -\left(2m - \frac{1}{2} \right) \pi \right\}.$$

Proposition 1.1. *For every $\lambda > 0$ there exist $a_1(\lambda) > 1$ and $m_0 = m_0(\lambda, a_1(\lambda)) \in \mathbf{N}$ such that, if $a \in (1, a_1(\lambda))$ and $m > m_0$, then there exists a rectangle $S_m \subset C_m$ such that $T_{\lambda,a}(S_m) \subset S_m$.*

Before proving this proposition we need to quote some properties of the following family of unimodal maps

$$f_{\lambda,a}(x) = \frac{1}{\lambda} \log a + x + \frac{1}{\lambda} \log \cos x, \qquad (1.4)$$

defined on the intervals

$$J_m = \left(-\left(2m + \tfrac{1}{2}\right)\pi, -\left(2m - \tfrac{1}{2}\right)\pi \right).$$

Since these maps satisfy $f_{\lambda,a}(x + 2\pi) = f_{\lambda,a}(x) + 2\pi$, it is only necessary to consider these maps on the interval $\left(-\tfrac{\pi}{2}, \tfrac{\pi}{2}\right)$.

For $a > 1$ each map $f_{\lambda,a}$ has two fixed points $P_a = \arccos a^{-1}$, $Q_a = -\arccos a^{-1}$ and a critical point $c_\lambda = \operatorname{arctg} \lambda$. If $a = 1$, both fixed points are the same and a saddle-node bifurcation takes place there. If a is close to 1, then P_a is an attractor and Q_a is a repeller. If the eigenvalue in P_a is equal to -1, which happens when a takes the value $\bar{a}(\lambda) = \sqrt{1 + 4\lambda^2}$, then a flip bifurcation occurs.

Let us first state the following result:

Lemma 1.2. *If* $f_{\lambda,a}^2(c_\lambda) > Q_a$, *then there exists an interval* $I \subset \left(-\tfrac{\pi}{2}, \tfrac{\pi}{2}\right)$ *such that* $f_{\lambda,a}(I) \subsetneq I$.

Proof. Let $\overline{Q}_a \in \left(0, \tfrac{\pi}{2}\right)$ be such that $f_{\lambda,a}(\overline{Q}_a) = Q_a$. By choosing $I = (x_1, x_2)$ with $x_2 \in \left(f_{\lambda,a}(c_\lambda), \overline{Q}_a\right)$ and $x_1 \in (Q_a, f_{\lambda,a}(x_2))$ the lemma holds. $\square$

Proof of Proposition 1.1. Let us write

$$T_{\lambda,a}(\eta, \xi) = \left(1 + \eta e^{\lambda\xi}\sin\xi, \tfrac{1}{\lambda}\log\eta + f_{\lambda,a}(\xi)\right).$$

We fix $\lambda > 0$ and take $a_1(\lambda) > 1$ to get $f_{\lambda,a}^2(c_\lambda) > Q_a$ for every $a \in [1, a_1(\lambda))$. From Lemma 1.2, there exists an interval $I \subset \left(-\tfrac{\pi}{2}, \tfrac{\pi}{2}\right)$ such that $f_{\lambda,a}(I) \subsetneq I$. Let l be the smallest of the lengths of the connected components of $I - f_{\lambda,a}(I)$. For each m sufficiently large we define $S_m = \left\{(\eta, \xi) \in C_m : e^{-\lambda l} < \eta < e^{\lambda l}, \xi + 2\pi m \in I\right\}$.

Since $\xi + 2\pi m \in I$, the image under $T_{\lambda,a}$ of the segment $\eta = 1$ of S_m does not go beyond the horizontal boundaries of S_m. Furthermore, since $e^{-\lambda l} < \eta < e^{\lambda l}$, the same follows for every point of S_m.

To show that $T_{\lambda,a}(S_m)$ does not go beyond the vertical boundaries of S_m, we shall check that the first coordinates of $T_{\lambda,a}\left(e^{\lambda l}, \left(-2m - \tfrac{1}{2}\right)\pi\right)$ and $T_{\lambda,a}\left(e^{\lambda l}, \left(-2m + \tfrac{1}{2}\right)\pi\right)$ belong to the interval $\left(e^{-\lambda l}, e^{\lambda l}\right)$, that is to say, $1 - e^{\lambda l}e^{-\frac{1}{2}\pi\lambda}e^{-2\pi\lambda m} > e^{-\lambda l}$ and $1 + e^{\lambda l}e^{\frac{1}{2}\pi\lambda}e^{-2\pi\lambda m} < e^{\lambda l}$. Now, since $1 - e^{-\lambda l} < e^{\lambda l} - 1$, it is sufficient to take $e^{\lambda l}e^{\frac{1}{2}\pi\lambda}e^{-2\pi\lambda m} < 1 - e^{-\lambda l}$ or equivalently,

$$m \geq \frac{1}{2\pi\lambda}\log\left(\frac{e^{\lambda l}e^{\frac{1}{2}\pi\lambda}}{1 - e^{-\lambda l}}\right). \quad \square$$

It is interesting to remark that, from the proof of Proposition 1.1, it follows that the rectangle

$$S'_m = \left\{ (\eta, \xi) \in C_m : 1 - be^{\lambda l}e^{-\frac{1}{2}\pi\lambda} < \eta < 1 + be^{\lambda l}e^{\frac{1}{2}\pi\lambda}, \xi + 2\pi m \in I \right\} \subset S_m,$$

with $b = e^{-2\pi\lambda m}$, is also invariant. So, we redefine S_m and, henceforth, we shall take $S_m = S'_m$.

If we introduce the following new coordinates $x = \xi + 2\pi m$ and $y = \left(\sqrt{b}\right)^{-1}(\eta - 1)$ inside each S_m, then the maps $T_{\lambda,a}$ can be expressed in the following way

$$T_{\lambda,a,b}(x, y) = \left(f_{\lambda,a}(x) + \frac{1}{\lambda}\log\left(1 + \sqrt{b}y\right), \sqrt{b}\left(1 + \sqrt{b}y\right)e^{\lambda x}\sin x \right) \qquad (1.5)$$

and each S_m becomes a new rectangle

$$U_m = \left\{ (x, y) \in \mathbf{R}^2 : x \in I, -\sqrt{b}e^{\lambda l}e^{-\frac{1}{2}\pi\lambda} \le y \le \sqrt{b}e^{\lambda l}e^{\frac{1}{2}\pi\lambda} \right\}, \qquad (1.6)$$

such that $T_{\lambda,a,b}(U_m) \subset U_m$.

Next, we shall prove some crucial properties of $T_{\lambda,a,b}$, for which we denote

$$DT_{\lambda,a,b}(x, y) = \begin{pmatrix} A & B \\ C & D \end{pmatrix}.$$

Proposition 1.3. *There exists a constant* $K = K(\lambda) > 0$ *such that, for every* $(a, x, y) \in \widetilde{D} = (1, \infty) \times U_m$, *it follows that:*

(a) $|A| \le K$, $|B| \le K\sqrt{b}$, $|C| \le K\sqrt{b}$, $|D| \le Kb$, $|detDT_{\lambda,a,b}| \le Kb$.

(b) $\left\|D_{(a,x,y)}A\right\| \le K$, $\left\|D_{(a,x,y)}B\right\| \le K\sqrt{b}$, $\left\|D_{(a,x,y)}C\right\| \le K\sqrt{b}$, $\left\|D_{(a,x,y)}D\right\| \le K\sqrt{b}$.

(c) $\left\|D^2_{(a,x,y)}A\right\| \le K$, $\left\|D^2_{(a,x,y)}B\right\| \le K\sqrt{b}$, $\left\|D^2_{(a,x,y)}C\right\| \le K\sqrt{b}$, $\left\|D^2_{(a,x,y)}D\right\| \le K\sqrt{b}$.

(d) *Finally, setting* $T_{\lambda,b}(a, x, y) = T_{\lambda,a,b}(x, y)$ *and* $\Psi_\lambda(a, x, y) = (f_{\lambda,a}(x), 0)$, *then*

$$\|T_{\lambda,b} - \Psi_\lambda\|_{C^3(\widetilde{D})} \le K\sqrt{b}.$$

Proof. Deriving (1.5), we obtain that

$$A = A(x) = \lambda^{-1}\left(\lambda - \text{tg } x\right), \quad B = B(y) = \lambda^{-1}\left(1 + \sqrt{b}y\right)^{-1}\sqrt{b},$$

$$C = C(x, y) = \sqrt{b}\left(1 + \sqrt{b}y\right)e^{\lambda x}(\lambda\sin x + \cos x), \quad D = D(x) = be^{\lambda x}\sin x.$$

Hence, neither $DT_{\lambda,a,b}$ nor the partial derivatives depend on a. Furthermore, there exists a constant $K = K(\lambda)$ such that, for every $x \in U_m$, it follows that $|A| \leq K$, $|B| \leq K\sqrt{b}$, $|C| \leq K\sqrt{b}$ and $|D| \leq Kb$. On the other hand, since

$$\det DT_{\lambda,a,b}(x,y) = \frac{-be^{\lambda x}}{\lambda \cos x},$$

it follows that $|\det DT_{\lambda,a,b}| \leq Kb$ and the first statement is proved.

Now, let us write the non-trivial second order derivatives of $T_{\lambda,a,b}$:

$$\partial_x A = -\lambda^{-1}(1 + \text{tg}\,^2 x), \quad \partial_y B = -\lambda^{-1}\left(1 + \sqrt{b}y\right)^{-2} b,$$

$$\partial_x C = \sqrt{b}\left(1 + \sqrt{b}y\right)e^{\lambda x}\left((\lambda^2 - 1)\sin x + 2\lambda \cos x\right),$$

$$\partial_y C = \partial_x D = be^{\lambda x}\left(\lambda \sin x + \cos x\right).$$

Thus, $\left\|D_{(a,x,y)}A\right\| \leq K$, $\left\|D_{(a,x,y)}B\right\| \leq Kb$, $\left\|D_{(a,x,y)}C\right\| \leq K\sqrt{b}$ and $\left\|D_{(a,x,y)}D\right\| \leq Kb$ and the second statement holds.

To prove statement (c), let us bound the non-trivial third order partial derivatives of $T_{\lambda,a,b}$:

$$\partial_x^2 A = -2\lambda^{-1}\text{tg}\,x\,(1 + \text{tg}\,^2 x), \quad \partial_y^2 B = 2\lambda^{-1}\left(1 + \sqrt{b}y\right)^{-3} b\sqrt{b},$$

$$\partial_x^2 C = \sqrt{b}\left(1 + \sqrt{b}y\right)e^{\lambda x}\left(\lambda(\lambda^2 - 3)\sin x + (3\lambda^2 - 1)\cos x\right),$$

$$\partial_{xy}^2 C = \partial_x^2 D = be^{\lambda x}\left((\lambda^2 - 1)\sin x + 2\lambda \cos x\right),$$

so as to obtain $\left\|D_{(a,x,y)}^2 A\right\| \leq K$, $\left\|D_{(a,x,y)}^2 B\right\| \leq K\sqrt{b}$, $\left\|D_{(a,x,y)}^2 C\right\| \leq K\sqrt{b}$ and $\left\|D_{(a,x,y)}^2 D\right\| \leq Kb$.

Finally, the proposition will be proved if we show that $T_{\lambda,b}$ is C^0 close to Ψ_λ. To this end, notice that $T_{\lambda,b} - \Psi_\lambda$ does not depend on the parameter a and, for every $(x,y) \in U_m$, $\left|\lambda^{-1}\log\left(1 + \sqrt{b}y\right)\right| < K\sqrt{b}$, and $\left|\sqrt{b}\left(1 + \sqrt{b}y\right)e^{\lambda x}\sin x\right| < K\sqrt{b}$. $\square$

For fixed $\lambda > 0$, let us consider the interval $(\overline{a}(\lambda), a^*(\lambda))$ with $\overline{a}(\lambda) = \sqrt{1 + 4\lambda^2}$ and $a^*(\lambda)$ being the value of a for which $f_{\lambda,a}(c_\lambda) = \frac{1}{2}\pi$, that is

$$a^*(\lambda) = \sqrt{1 + \lambda^2}\exp\left(\lambda\left(\frac{\pi}{2} - \text{arctg}\,\lambda\right)\right).$$

We shall now study some properties of the unimodal maps $f_{\lambda,a}$.

Proposition 1.4. *For every $\lambda > 0$ there exists a unique $a = a(\lambda) \in (\overline{a}(\lambda), a^*(\lambda))$ such that $f^2_{\lambda,a}(c_\lambda) = Q_a$. Moreover, $\lambda \to a(\lambda)$ is an increasing C^1 map and $\lim_{\lambda \to 0} a(\lambda) = 1$.*

Proof. Let us fix $\lambda > 0$ and write $F(\lambda, a) = f^2_{\lambda,a}(c_\lambda) + \arccos a^{-1}$. Since $F(\lambda, \overline{a}(\lambda)) > 0$ and $\lim_{a \to a^*(\lambda)} F(\lambda, a) = -\infty$, the existence of $a(\lambda)$ holds.

On the other hand, if we derive the function

$$F(\lambda, a) = \frac{2}{\lambda} \log a + \text{arctg } \lambda - \frac{1}{2\lambda} \log(1 + \lambda^2) + \frac{1}{\lambda} \log \cos(f_{\lambda,a}(c_\lambda)) + \arccos \frac{1}{a},$$

with respect to a, then we obtain

$$\partial_a F(\lambda, a) = \frac{n(\lambda, a)\sqrt{a^2 - 1} + \lambda^2}{\lambda^2 a \sqrt{a^2 - 1}},$$

where $n(\lambda, a) = 2\lambda - \text{tg } (f_{\lambda,a}(c_\lambda))$.

We shall show that $n(\lambda, a) < 0$ for every $\lambda > 0$ and for every $a \in (\overline{a}(\lambda), a^*(\lambda))$. In fact, $\partial_a n(\lambda, a) = -(1 + \text{tg }^2 f_{\lambda,a}(c_\lambda)) (\lambda a)^{-1} < 0$. Furthermore, since $f_{\lambda,\overline{a}(\lambda)}(c_\lambda) > P_{\overline{a}(\lambda)}$ and $f''_{\lambda,a}(x) < 0$, we get $-1 > f'_{\lambda,\overline{a}(\lambda)}\left(f_{\lambda,\overline{a}(\lambda)}(c_\lambda)\right) = \lambda^{-1}\left(\lambda - \text{tg } f_{\lambda,\overline{a}(\lambda)}(c_\lambda)\right)$. Hence $n(\lambda, \overline{a}(\lambda)) < 0$. Now, let us define

$$m(\lambda, a) = n(\lambda, a)\sqrt{a^2 - 1} + \lambda^2. \tag{1.7}$$

Since $\partial_a m(\lambda, a) < 0$, $\partial_a F(\lambda, a)$ vanishes, at the most, at one point of $(\overline{a}(\lambda), a^*(\lambda))$, where $F(\lambda, a)$ should reach its maximum value. Therefore, for each $\lambda > 0$, there exists a unique solution $a(\lambda) \in (\overline{a}(\lambda), a^*(\lambda))$ of $F(\lambda, a) = 0$, with $\partial_a F(\lambda, a(\lambda)) < 0$. In order to prove that

$$a'(\lambda) = -\frac{\partial_\lambda F(\lambda, a(\lambda))}{\partial_a F(\lambda, a(\lambda))} > 0,$$

we shall see that $\partial_\lambda F(\lambda, a) \geq 0$ for every $a \in (\overline{a}(\lambda), a^*(\lambda))$. In fact, for $0 < \lambda_1 < \lambda_2$, it is easy to check that $f_{\lambda_2,a}(c_{\lambda_2}) < f_{\lambda_1,a}(c_{\lambda_1})$ and, since $f_{\lambda_2,a}$ is a decreasing map on the right of c_{λ_2}, it follows that $f^2_{\lambda_2,a}(c_{\lambda_2}) > f_{\lambda_2,a}(f_{\lambda_1,a}(c_{\lambda_1}))$. In addition, since $f_{\lambda_1,a}(c_{\lambda_1}) > P_a$, we obtain $f^2_{\lambda_1,a}(c_{\lambda_1}) < f_{\lambda_2,a}(f_{\lambda_1,a}(c_{\lambda_1}))$ and consequently, $f^2_{\lambda_1,a}(c_{\lambda_1}) < f^2_{\lambda_2,a}(c_{\lambda_2})$. Thus, $F(\lambda, a)$ increases with respect to λ and then $\partial_\lambda F(\lambda, a) \geq 0$. The proposition is proved. $\square$

From Propositions 1.3 and 1.4, we obtain some properties for the maps $T_{\lambda,a}$. For instance, provided that $\lambda > 0$, for each $a \in (\overline{a}(\lambda), a(\lambda))$ the map $\Psi_{\lambda,a}(x, y) = (f_{\lambda,a}(x), 0)$ has a unique fixed point $(P_a, 0)$ inside the domain U_m given by (1.6). This fixed point is hyperbolic, therefore, it has an analytic continuation $P_{a,m}$ when we consider any map $T_{\lambda,a,b}$ with $b << 1$ (that is m large enough). $P_{a,m}$ is a saddle-point and it is the unique fixed point of $T_{\lambda,a,b}$ inside U_m.

Since $T_{\lambda,a,b}(U_m) \subset U_m$, it follows that $W^u(P_{a,m}) \subset U_m$. So, $\Lambda_{a,m} = cl\,(W^u(P_{a,m}))$ is a compact set contained in U_m.

On the other hand, the existence of homoclinic points associated to $P_{a,m}$ implies the existence of open sets $\Omega_{a,m} \subset U_m$ such that $\lim_{n \to \infty} d\left(T^n_{\lambda,a,b}(x,y), \Lambda_{a,m}\right) = 0$ for every $(x,y) \in \Omega_{a,m}$, which means that $\Lambda_{a,m}$ is an attracting set for $T_{\lambda,a,b}$ whenever m is large enough. We shall prove throughout this book that $\Lambda_{a,m}$ is, in fact, a strange attractor with positive probability.

Next, we shall study some properties of the map $f_{\lambda,a}$ which will be used in the forthcoming chapters:

Proposition 1.5. *Given* $\lambda > 0$ *let* $\overline{Q}_a \in \left(0, \frac{\pi}{2}\right)$ *be such that* $f_{\lambda,a}(\overline{Q}_a) = Q_a$. *Then*

$$\frac{d}{da}\left(\overline{Q}_a - f_{\lambda,a}(c_\lambda)\right)\mid_{a=a(\lambda)} \neq 0.$$

Proof. Let $G_\lambda(a,x) = f_{\lambda,a}(x) - Q_a$. Then $G_\lambda(a, \overline{Q}_a) = 0$, $\partial_x G_\lambda(a, \overline{Q}_a) \neq 0$ and, by applying the implicit function theorem, we have

$$\frac{d}{da}\overline{Q}_a = -\frac{\partial_a G_\lambda(a, \overline{Q}_a)}{\partial_x G_\lambda(a, \overline{Q}_a)} = -\frac{\sqrt{a^2 - 1} + \lambda}{\left(\lambda - tg\,\overline{Q}_a\right)a\sqrt{a^2 - 1}}.$$

Finally,

$$\frac{d}{da}(\overline{Q}_a - f_{\lambda,a}(c_\lambda))\mid_{a=a(\lambda)} = -\frac{m(\lambda, a(\lambda))}{\left(\lambda - tg\,f_{\lambda,a(\lambda)}(c_\lambda)\right)\lambda a(\lambda)\sqrt{a^2(\lambda) - 1}},$$

where $m(\lambda, a)$ is the map given by (1.7). Now, from the proof of Proposition 1.4, it follows that $m(\lambda, a(\lambda)) \neq 0$. $\square$

Lemma 1.6. *For each* λ *sufficiently small the set* $\left[Q_{a(\lambda)}, f_{\lambda,a(\lambda)}(c_\lambda)\right] = W^u(P_{a(\lambda)})$ *is contained in the region where the Schwarzian derivative of* $f_{\lambda,a(\lambda)}$ *is negative.*

Proof. Let us compute the Schwarzian derivative of $f_{\lambda,a(\lambda)}$:

$$Sf_{\lambda,a(\lambda)}(x) = \frac{f'''_{\lambda,a(\lambda)}(x)}{f'_{\lambda,a(\lambda)}(x)} - \frac{3}{2}\left(\frac{f''_{\lambda,a(\lambda)}(x)}{f'_{\lambda,a(\lambda)}(x)}\right)^2 =$$

$$= -\frac{1 + \operatorname{tg}^2 x}{2 \left(\lambda - \operatorname{tg} x \right)^2} \left(4 \left(\lambda - \operatorname{tg} x \right) \operatorname{tg} x + 3 \left(1 + \operatorname{tg}^2 x \right) \right).$$

Since $3 + 4\lambda \operatorname{tg} x - \operatorname{tg}^2 x > 0$ when

$$x \in \left(\operatorname{arctg} \left(2\lambda - \sqrt{4\lambda^2 + 3} \right), \operatorname{arctg} \left(2\lambda + \sqrt{4\lambda^2 + 3} \right) \right),$$

we have $Sf_{\lambda,a(\lambda)}(x) < 0$ in the above interval. To finish the proof of the lemma we may check that, for every λ sufficiently small, it follows that

$$\operatorname{arctg} \left(2\lambda - \sqrt{4\lambda^2 + 3} \right) < Q_{a(\lambda)} \text{ and } f_{\lambda,a(\lambda)}(c_\lambda) < \operatorname{arctg} \left(2\lambda + \sqrt{4\lambda^2 + 3} \right). \quad \square$$

Lemma 1.7. *If $\lambda > 0$ is small enough, then $a(\lambda) > \tilde{a}(\lambda) = \sqrt{(1 + 4\lambda^2)(1 + \lambda^2)}$.*
Proof. Let $F(\lambda, a)$ be the function defined in the proof of Proposition 1.4. Clearly, the lemma holds if we prove that $F(\lambda, a) > 0$ for every $a \in (\bar{a}(\lambda), \tilde{a}(\lambda))$. Since $f_{\lambda,a}(c_\lambda) = \operatorname{arctg} \lambda + \lambda^{-1} \log \left(a \left(\sqrt{1 + \lambda^2} \right)^{-1} \right) < \operatorname{arctg} \lambda + \frac{1}{2}\lambda^{-1} \log (1 + 4\lambda^2)$, it is sufficient to check that $f_{\lambda,a} \left(\operatorname{arctg} \lambda + \frac{1}{2}\lambda^{-1} \log (1 + 4\lambda^2) \right) > - \arccos a^{-1} = -\operatorname{arctg} \sqrt{a^2 - 1}$. Furthermore, since $a > \sqrt{1 + 4\lambda^2}$, it suffices to verify that

$$\operatorname{arctg} \lambda + \operatorname{arctg} 2\lambda + \frac{1}{\lambda} \log \left(1 + 4\lambda^2 \right) +$$

$$+ \frac{1}{\lambda} \log \cos \left(\operatorname{arctg} \lambda + \frac{1}{2\lambda} \log(1 + 4\lambda^2) \right) > 0.$$

Since $\log \cos x$ is decreasing and $\operatorname{arctg} \lambda + \frac{1}{2}\lambda^{-1} \log (1 + 4\lambda^2) < 3\lambda$, the proof finishes by showing that $\operatorname{arctg} \lambda + \operatorname{arctg} 2\lambda + \lambda^{-1} \log (1 + 4\lambda^2) + \lambda^{-1} \log \cos(3\lambda) > 0$. To this end, use the power series expansion around $\lambda = 0$. $\quad \square$

Lemma 1.8. *For each λ sufficiently small and for every $a \in (\tilde{a}(\lambda), a(\lambda))$, it follows that*

$$\left| f'_{\lambda,a} \left(f_{\lambda,a}(c_\lambda) \right) \right| > \frac{3}{2}.$$

Proof. Since $f'_{\lambda,a} \left(f_{\lambda,a}(c_\lambda) \right) = \lambda^{-1} \left(\lambda - \operatorname{tg} f_{\lambda,a}(c_\lambda) \right)$ and $a > \tilde{a}(\lambda)$, the proof ends by using once again the power series expansion around $\lambda = 0$ to prove that

$$\frac{1}{2\lambda} \log \left(1 + 4\lambda^2 \right) > \operatorname{arctg} \frac{5}{2}\lambda - \operatorname{arctg} \lambda. \quad \square$$

Remark 1.9. *It is easy to verify that the last three lemmas hold for $0 < \lambda < 0.3319...$ Under this condition, we may extend Proposition 1.3 to obtain $|C| \geq K^{-1}\sqrt{b}$. In fact, since $C = \sqrt{b} \left(1 + \sqrt{b}y \right) e^{\lambda x} \left(\lambda \sin x + \cos x \right)$, it suffices to note that $\lambda \sin x + \cos x = 0$ only if $x < Q_a$, whenever $\lambda > 0$ and $a \in (\bar{a}(\lambda), a(\lambda))$.*

Chapter 2

THE UNIMODAL FAMILY

2.1 Introduction

In this chapter we shall study the dynamics of the unimodal maps

$$f_{\lambda,a}(x) = \frac{1}{\lambda} \log a + x + \frac{1}{\lambda} \log \cos x,$$

defined on $\left(-\frac{\pi}{2}, \frac{\pi}{2}\right)$ for $\lambda > 0$ and $a \in [\tilde{a}(\lambda), a(\lambda)]$, where $\tilde{a}(\lambda) = \sqrt{1 + 5\lambda^2}$. Recall that $a(\lambda)$ is such that $f_{\lambda,a(\lambda)}^2(c_\lambda) = Q_a$. From Lemma 1.7, it is clear that $a(\lambda) > \tilde{a}(\lambda)$ for every small enough λ. Let us denote by $\{\xi_n(a)\}_{n=1}^{\infty}$ the orbit of the point $\xi_1(a) = f_{\lambda,a}(c_\lambda)$ and write $D_n(a) = \left(f_{\lambda,a}^n\right)'(\xi_1(a))$. From now on, we shall omit the subscript λ whenever it does not lead into error. We shall prove the following statement:

Theorem 2.1. *For each $\lambda > 0$ sufficiently small there exists a constant $c_0 > 0$ such that, whenever $0 < c < \min\{c_0, \log(1+\lambda)\}$, there exists a value $a_0 = a_0(\lambda, c) < a(\lambda)$ of the parameter a close to $a(\lambda)$ and a positive Lebesgue measure set $E = E(\lambda, c) \subset [a_0, a(\lambda)]$ such that if $a \in E$, then $|D_n(a)| \geq e^{cn}$ for every $n \in \mathbf{N}$.*

This theorem was originally proved for the quadratic family by Benedicks and Carleson, [3]. A consequence of this theorem, even of a weaker result proved by Benedicks and Carleson themselves [2], is the theorem demonstrated years earlier by Jacobson [10]. The relationship between both theorems can be found in [13] where the authors also prove Theorem 2.1 for families $f_a : I \longrightarrow I$ of sufficiently smooth unimodal maps such that:

i. f_a has a quadratic critical point c.

ii. f_a has a fixed point Q_a in the boundary of I which is repelling.

iii. The map $(x, a) \longrightarrow (f_a(x), Df_a(x), D^2 f_a(x))$ is C^1.

iv. There exists a value a_* of the parameter a for which f_{a_*} is a Misiurewicz map, that is, the forward iterates of $f_{a_*}(c)$ remain outside a neighbourhood U of c.

v. f_{a_*} has no periodic attractors.

vi. The transversality condition $\frac{d}{da}(x_a - f_a(c)) \neq 0$ holds for $a = a_*$, where, in our case, $x_a \neq Q_a$ is the point such that $f_a(x_a) = Q_a$.

From Chapter 1, it follows that for any sufficiently small λ our family $f_{\lambda,a}$ satisfies the above assumptions: The value a_* of the parameter a is given by $a(\lambda)$ and $I^*_\lambda = [Q_{a_*}, f_{a_*}(c_\lambda)]$ is the interval where f_{a_*} is defined on. Each interval $I_{\lambda,a} = [Q_a, x_a(\lambda)]$ where the respective $f_{\lambda,a}$ works, can be transformed into a new interval I^*_λ by means of a C^∞ diffeomorphism leaving c_λ invariant. The new family $\tilde{f}_{\lambda,a}$, which is C^∞ conjugated to $f_{\lambda,a}$, satisfies all the above assumptions. The negative Schwarzian derivative of f_{a_*}, just as proved in Lemma 1.6 and Singer's Theorem [23], yield f_{λ,a_*} has no attracting periodic orbits.

Though Theorem 2.1 follows from the more general one stated in [13], we shall next develop a proof by following the nice one given in [16] for the quadratic family. However, unlike the quadratic case, we do not know a conjugation between $f_{a(\lambda)}$ and the tent map. This and other facts make some estimates a bit more complicated. We think that a sketch of the proof will be helpful to understand the forthcoming chapters.

As in the case of the quadratic family, the proof of Theorem 2.1 lies basically in inductively constructing a decreasing sequence of sets $(\Omega_n)_{n \in \mathbb{N}}$, with $\Omega_1 = [a_0, a(\lambda)]$ and such that every $a \in \Omega_n$ verifies the condition

$$|D_j(a)| \geq e^{cj} \text{ for } 1 \leq j \leq n, \quad (\text{EG}_n)$$

which is called the exponential growth hypothesis up to time n. Hence, it is sufficient to show that $E = \bigcap_{n \in \mathbb{N}} \Omega_n$ has positive Lebesgue measure.

For any fixed $n \in \mathbb{N}$ the existence of a positive Lebesgue measure set Ω_n is clear: The condition (EG_n) is satisfied whenever the orbit of the critical point remains outside some of its neighbourhoods. Since this is true for $f_{\lambda,a(\lambda)}$, the claim follows for every $a \in [a_n(\lambda), a(\lambda)]$, with $a_n(\lambda)$ close to $a(\lambda)$. Nevertheless, on increasing n, the length of the interval $[a_n(\lambda), a(\lambda)]$ tends to zero. So, we have to let the orbit of the critical point return arbitrarily close to this point. Thus, in order to ensure exponential growth for a positive Lebesgue measure set of parameters, we shall have to assume that each $a \in \Omega_n$ satisfies the basic assumption

$$|\xi_j(a) - c_\lambda| \geq e^{-\alpha j} \text{ for } N \leq j \leq n, \quad \text{(BA}_n\text{)}$$

where $\alpha > 0$ is a sufficiently small number and N is an arbitrarily large number of iterates fixed at the beginning. The right side in (BA$_n$) sets a lower bound of the derivatives on the iterates of the critical point and it makes each of these derivatives be non-zero. Thus, since the derivative along an orbit is the product of the derivatives on each iterate, one expects to compensate the small derivatives near the critical point with those on the points away from it. Nevertheless, this compensation will only be possible if the orbit has too many points far from the critical point. This fact holds if we assume a third hypothesis called free assumption. To set this hypothesis we shall introduce some notions.

Let us split the orbit $\{\xi_k(a)\}_{k=1}^\infty$ into different pieces, each of them given by the union of three different kinds of consecutive iterates:

$$\{\mu_i\} \cup \{\mu_i + 1, \mu_i + 2, ..., \mu_i + p_i\} \cup \{\mu_i + p_i + 1, ..., \mu_{i+1} - 1\},$$

that will be called returns, binding periods and free periods, respectively. Heuristically, returns μ_i correspond to iterates of the orbit $\{\xi_k(a)\}_{k=1}^\infty$ returning to near the critical point. Binding periods consist of the iterates $\mu_i + s$ that follow μ_i and for which $\xi_{\mu_i+s}(a)$ remains close enough to $\xi_s(a)$. This closeness will be established in terms of s. Between two consecutives returns, free and binding periods are complementary.

The length of a binding period depends on the distance $|\xi_{\mu_i}(a) - c_\lambda|$ and so, by means of (BA$_n$), we shall show that p_i is less than μ_i. Then, and this is crucial, we may use the closeness between $\xi_{\mu_i+s}(a)$ and $\xi_s(a)$ up to $s = p_i$ and the inductive method to prove the exponential growth needed during the binding period in order to compensate the small derivative on $\xi_{\mu_i}(a)$.

Finally, if the number $F_n(a)$ of free iterates is large enough for each $n \in \mathbf{N}$, then we shall obtain exponential growth along the whole orbit. Therefore, we shall remove, from Ω_n, the parameters which do not verify the following free assumption

$$F_n(a) \geq (1 - \alpha)n. \quad \text{(FA}_n\text{)}$$

In this way, the number of iterates belonging to binding periods is rather inferior to the number of free iterates and this fact will be essential to inductively prove (EG$_n$)

from (EG_{n-1}). More precisely, it is shown that

$$(EG_n{}_{-1}) + (FA_n) + (BA_n) \Rightarrow (EG_n).$$

In fact, this is a suitable way of going from Ω_{n-1} to Ω_n.

2.2   Preliminary results

To begin with, we need some knowledge of the expansiveness of f_a in points far enough from c_λ and for a close to $a(\lambda)$. To set the statements, let us define the following neighbourhoods of c_λ:

$$U_m = \left(c_\lambda - e^{-m}, c_\lambda + e^{-m}\right) \text{ for } m \geq \Delta - 1 \text{ and } \Delta \in \mathbf{N} \text{ large enough.}$$

Proposition 2.2. *For each $\lambda > 0$ sufficiently small there exist positive constants C_0, c_0 and a neighbourhood W of c_λ such that, for any $\Delta \in \mathbf{N}$ with $U_{\Delta+1} \subset W$, there exists $a_1 = a_1(\Delta, c_0) < a(\lambda)$ such that: If $a \in [a_1, a(\lambda)]$ and $f_a^j(x) \notin U_{\Delta+1}$ for $0 \leq j \leq k-1$, then it follows that:*

(a) $\left|\left(f_a^k\right)'(x)\right| \geq C_0 e^{c_0 k} \displaystyle\min_{0 \leq i \leq k-1} |f_a'(f_a^i(x))| \geq C_0 e^{-\Delta} e^{c_0 k}.$

(b) If $f_a^k(x) \in W$, then $\left|\left(f_a^k\right)'(x)\right| \geq C_0 e^{c_0 k}.$

(c) If $x \in f_a(U_{\Delta+1})$, then $\left|\left(f_a^k\right)'(x)\right| \geq C_0 e^{c_0 k}.$

Proof. Statements (a) and (b) follow from Theorem III 6.4 in [13]. To prove (c) take $\overline{x}$ such that $Q_{a(\lambda)} < \overline{x} < 0$. Fixed $\Delta \in \mathbf{N}$, there exists $n_1 = n_1(\Delta)$ such that $\left[Q_{a(\lambda)}, \overline{x}\right] \subset f_{a(\lambda)}^{n_1}\left(f_{a(\lambda)}(U_{\Delta+1})\right)$ and $f_{a(\lambda)}^j\left(f_{a(\lambda)}(U_{\Delta+1})\right) \subsetneq \left[Q_{a(\lambda)}, \overline{x}\right]$ for $j = 1, ..., n_1 - 1$. Fix $a_1(\lambda)$ sufficiently close to $a(\lambda)$. Regardless of $a_1(\lambda)$ and for every $a \in [a_1(\lambda), a(\lambda)]$, it follows that $|f_a^{n_1}\left(f_a(U_{\Delta+1})\right)| > \frac{1}{2}\left(\overline{x} - Q_a\right) = L$ and $f_a^j\left(f_a(U_{\Delta+1})\right) \subsetneq [Q_a, \overline{x}]$ for $1 \leq j \leq n_1 - 1$. Since for every $x \in [Q_a, \overline{x}]$ there exists a positive constant $\widehat{c}_0$ such that $f_a'(x) > e^{\widehat{c}_0} > 1$, we obtain (c) whenever $k \leq n_1$. Otherwise, notice that $f_a'(x) < e^{G_0}$ for each $x \in [Q_a, \overline{x}]$, where $\overline{x}$ is chosen near $Q_{a(\lambda)}$ and consequently, G_0 is a constant close to $\widehat{c}_0$. Furthermore, since the length of $f_a(U_{\Delta+1})$ is of order $e^{-2\Delta}$, we have $e^{\frac{4}{3}\widehat{c}_0 n_1} e^{-2\Delta} > e^{G_0 n_1} e^{-2\Delta} \geq const > 0$. Then, from statement (a), we obtain

$$\left|\left(f_a^k\right)'(x)\right| = \left|\left(f_a^{k-n_1}\right)'(f_a^{n_1}(x))\right| \left|\left(f_a^{n_1}\right)'(x)\right| \geq C_0 e^{-\Delta} e^{c_0(k-n_1)} e^{\widehat{c}_0 n_1} \geq$$

$$\geq \widehat{C}_0 e^{c_0(k-n_1)} e^{\frac{1}{3}\widehat{c}_0 n_1} \geq \widehat{C}_0 e^{c_0 k},$$

where $c_0' = \min\left\{\frac{1}{3}\hat{c}_0, c_0\right\}$. By redefining the constants, statement (c) holds. $\square$

Proposition 2.3. *For $0 < c < \log(1 + \lambda)$ there exist a constant $A = A(\lambda) > 1$, a natural number $N_0 = N_0(\lambda, c)$ and a value of the parameter $a_1 = a_1(\lambda, N_0) < a(\lambda)$ such that if $a \in [a_1, a(\lambda)]$ satisfies*

$$|D_k(a)| \geq e^{ck} \text{ for } k = N_0, ..., n - 1, \tag{2.1}$$

then

$$\frac{1}{A} \leq \frac{|\xi_n'(a)|}{|D_{n-1}(a)|} \leq A.$$

Proof. Let $a = a(\lambda)$ and $x_a \neq Q_a$ such that $f_a(x_a) = Q_a$. Since $f_a'(Q_a) > 1 + \sqrt{5}$ and Lemma 1.8 implies $|f_a'(x_a)| > \frac{3}{2}$, we have $\sum\limits_{n=1}^{\infty} |D_n(a)|^{-1} < \frac{2}{3} \sum\limits_{n=0}^{\infty} \left(1 + \sqrt{5}\right)^{-n} < 1 - \epsilon$. Choose $N_0 \in \mathbf{N}$ such that $\sum\limits_{n=N_0+1}^{\infty} e^{-cn} < \frac{1}{4}\epsilon$. From the continuity of $a \to f_a$, there exists $a_1 = a_1(\lambda, N_0)$ such that, for each $a \in [a_1, a(\lambda)]$, it follows that $\sum\limits_{n=1}^{N_0} |D_n(a)|^{-1} < 1 - \frac{1}{2}\epsilon$.

From the recurrent expressions $D_k(a) = \lambda^{-1} (\lambda - \operatorname{tg} \xi_k(a)) D_{k-1}(a)$ with $D_0(a) = 1$ and $\xi_{k+1}'(a) = \xi_1'(a) + \lambda^{-1} (\lambda - \operatorname{tg} \xi_k(a)) \xi_k'(a)$, we obtain that

$$\frac{\xi_{k+1}'(a)}{D_k(a)} = \frac{\xi_1'(a)}{D_k(a)} + \frac{\xi_k'(a)}{D_{k-1}(a)}$$

and hence,

$$\left|\frac{\xi_n'(a)}{D_{n-1}(a)} - \frac{\xi_1'(a)}{D_0(a)}\right| \leq \frac{\xi_1'(a)}{D_0(a)} \sum\limits_{k=1}^{n-1} \frac{1}{|D_k(a)|} < \frac{\xi_1'(a)}{D_0(a)} \left(1 - \frac{\epsilon}{4}\right) < \frac{1}{\lambda} \left(1 - \frac{\epsilon}{4}\right).$$

The proof ends by taking $A(\lambda) = \max\left\{4\epsilon^{-1}\lambda, \left(2 - \frac{1}{4}\epsilon\right)\lambda^{-1}\right\}$. $\square$

As a consequence of the above proposition, if $\omega \subset [a_1(\lambda), a(\lambda)]$ is an interval such that each $a \in \omega$ satisfies (2.1), then ξ_k' is different from zero in ω for $k = N_0, ..., n$. Therefore, ξ_k is a homeomorphism on ω and hence, for any integers i, j such that $N_0 \leq i \leq j \leq n$, we may define the map

$$\Psi : x \in \xi_i(\omega) \longrightarrow \left(\xi_j \circ \xi_i^{-1}\right)(x) \in \xi_j(\omega),$$

whose derivative is given by

$$\Psi'(\xi_i(a)) = \frac{\xi_j'(a)}{\xi_i'(a)}.$$

By applying the mean value theorem we have the following statement

Proposition 2.4. *Let $\omega \subset [a_1(\lambda), a(\lambda)]$ be an interval such that (2.1) holds for every $a \in \omega$. Then, for any i, j such that $N_0 \leq i \leq j \leq n$, it follows that*

(a) For every $a, b \in \omega$ there exists $t \in \omega$ such that

$$\frac{1}{A^2}\left|\left(f_i^{j-i}\right)'(\xi_i(t))\right| \leq \frac{|\xi_j(a) - \xi_j(b)|}{|\xi_i(a) - \xi_i(b)|} \leq A^2\left|\left(f_i^{j-i}\right)'(\xi_i(t))\right|.$$

(b) There exists $t \in \omega$ such that

$$\frac{1}{A^2}\left|\left(f_i^{j-i}\right)'(\xi_i(t))\right| \leq \frac{|\xi_j(\omega)|}{|\xi_i(\omega)|} \leq A^2\left|\left(f_i^{j-i}\right)'(\xi_i(t))\right|.$$

The next result provides a sequence of parameter intervals which will be used later for starting the construction of the sets $(\Omega_n)_{n\in\mathbb{N}}$.

Proposition 2.5. *Let $a_1(\lambda)$ and N_0 be as in Proposition 2.3 and let $U_\lambda = \left(\frac{3}{4}c_\lambda, \frac{5}{4}c_\lambda\right)$. There exists an unbounded sequence of natural numbers $N \geq N_0$ and intervals $\Omega_N \subset [a_1(\lambda), a(\lambda)]$ such that*

(a) $\xi_j(\Omega_N) \cap U_\lambda = \emptyset$ for $1 \leq j \leq N - 1$.

(b) $\xi_N(\Omega_N) \supset U_\lambda$.

(c) $|D_j(a)| \geq (1 + \lambda)^j$ for every $a \in \Omega_N$ and $1 \leq j \leq N - 1$.

Proof. Let $I(\lambda) = \left[Q_{a(\lambda)}, -\text{arctg } \lambda^2\right]$. If $\xi_j(a) \in I(\lambda)$ for $j = 2, ..., N_1$, then

$$|D_k(a)| = \prod_{j=1}^{k} \frac{1}{\lambda}|\lambda - \text{tg } \xi_j(a)| \geq (1 + \lambda)^k.$$

Let us take $N_1 > N_0$ and $V_1 \subset [a_1(\lambda), a(\lambda)]$ such that $\xi_j(V_1) \subset I$ for $j = 2, ..., N_1$. From Proposition 2.3 it follows that $|\xi_j(V_1)| = \left|\xi_j'(t)\right||V_1| \geq A^{-1}|D_{j-1}(t)||V_1| \geq A^{-1}(1 + \lambda)^{j-1}|V_1|$.

So, we obtain exponential growth of $|\xi_n(V_1)|$ up to a first iterate $n = N_2 > N_1$ for which $\xi_{N_2}(V_1) \not\subset I$. Since $\xi_{N_2}(a(\lambda)) = Q_{a(\lambda)}$, there exists $t_* \in V_1$ such that $\xi_{N_2}(t_*) = -\text{arctg } \lambda^2$. Let $N = N_2 + 1$ and $\Omega_N = [t_*, a(\lambda)]$. Notice that statements (a) and (c) are proved. Statement (b) follows by using Lemma 1.7 to check that $\xi_{N_2+1}(t_*) > \frac{5}{4}\text{arctg } \lambda$ for $\lambda < \frac{1}{3}$. $\square$

2.3 The binding period

In the former epigraph we introduced the basis of neighbourhoods of c_λ, $U_m = (c_\lambda - e^{-m}, c_\lambda + e^{-m})$ for $m \geq \Delta - 1$ and $\Delta \in \mathbf{N}$ arbitrarily large. Now, we need new sets and terminology to define the concept of binding period.

Let us consider the maps $T_\lambda(x) = x + c_\lambda$ and the intervals $A_m = \left[e^{-(m+1)}, e^{-m}\right]$, $A_m^+ = A_{m+1} \cup A_m \cup A_{m-1}$. Denote by $-A$ the symmetrical interval of A with respect to 0. For $m \geq \Delta - 1$ write $I_m = T_\lambda(A_m)$ and $I_m^+ = T_\lambda(A_m^+)$ and for $m \leq -(\Delta - 1)$, $I_m = T_\lambda(-A_{-m})$ and $I_m^+ = T_\lambda(-A_{-m}^+)$. Finally, define $U_m^+ = U_{m-1}$.

Definition 2.6. *Fix $0 < \beta << 1$. The binding period of the interval U_m^+ is the largest element $p(a, m)$ of $\mathbf{N} \cup \{\infty\}$ such that $|f_a^j(x) - \xi_j(a)| \leq e^{-\beta j}$ for every $x \in U_m^+$ and for every $j = 1, ..., p(a, m)$.*

As an immediate consequence of this definition we obtain that

$$\left| f_a^{p(a,m)+1}(U_m^+) \right| \geq e^{-\beta(p(a,m)+1)} \text{ and } \left| f_a^j(U_m^+) \right| \leq 2e^{-\beta j} \text{ for } 1 \leq j \leq p(a, m). \quad (2.2)$$

Remark 2.7. *The assumption (BA_n), as given before, has not been used yet. It was posed for $n \geq N$ as long as we always get expansiveness up to time N. However, once α is fixed, we may choose N to set the assumption in the following equivalent way:*

$$|\xi_j(a) - c_\lambda| \geq \Lambda e^{-\alpha j} \text{ for } 1 \leq j \leq n, \quad (BA_n)$$

where $\Lambda = \Lambda(\lambda) = f_{a(\lambda)}(c_\lambda) - c_\lambda$ is a constant smaller than one. Henceforth, we shall use this formulation.

Recall that Ω_k denotes a set of parameter values satisfying (BA_k) and (EG_k). Let us write $D_k(a, x) = \left(f_a^k\right)'(f_a(x))$ and $D_k(a) = D_k(a, c_\lambda)$.

Lemma 2.8. *Let $a \in \Omega_{n-1}$, $\xi_n(a) \in I_m^+$ and $p = p(a, m)$. There exists $\Delta = \Delta(\lambda, c, \alpha, \beta)$ sufficiently large such that if $\Delta \leq |m| \leq [\alpha n] - 1$, then:*

(a) There exists a constant $B_1 = B_1(\lambda, \alpha, \beta)$ such that for every $x \in U_m^+$ and for every $k = 1, ..., p$ it follows that

$$\frac{1}{B_1} \leq \frac{|D_k(a, x)|}{|D_k(a)|} \leq B_1.$$

(b)
$$\frac{|m|}{\beta + \log \frac{10}{\lambda}} \le p \le \frac{3|m|}{\beta + c} < \frac{n}{2}.$$

(c) For every $x \in I_m^+$,
$$\left|\left(f_a^{p+1}\right)'(x)\right| \ge e^{\left(1 - \frac{4\beta}{\beta+c}\right)|m|}.$$

Proof. We obtain
$$\frac{|D_k(a,x)|}{|D_k(a)|} \le \exp\left\{\sum_{j=1}^{k} \frac{|\text{tg } f_a^j(x) - \text{tg } \xi_j(a)|}{|\lambda - \text{tg } \xi_j(a)|}\right\}$$

for $k = 1, ..., \min\{p, n\}$. Now, we use Definition 2.6 and the basic assumption to prove that $|\text{tg } f_a^j(x) - \text{tg } \xi_j(a)| < 10e^{-\beta j}$ and $|\lambda - \text{tg } \xi_j(a)| > \Lambda e^{-\alpha j}$, respectively. Notice here that, since $a \in \Omega_{n-1}$, then (BA$_j$) does not work when $j = n$. For such case we have to use the condition $\xi_n(a) \in I_m^+$. By taking $\alpha < \beta$ we obtain

$$\frac{|D_k(a,x)|}{|D_k(a)|} \le \exp\left\{\frac{10}{\Lambda} \sum_{j=1}^{\infty} e^{(\alpha-\beta)j}\right\} = B_1'$$

and in a similar way we conclude that

$$\frac{|D_k(a)|}{|D_k(a,x)|} \le \frac{1}{B_1''}.$$

In summary, statement (a) holds by taking $B_1 = \max\{B_1', B_1''\}$ and considering that statement (b) implies $p < n$. Next, we shall prove statement (b) by only using statement (a) for $k = 1, ..., \min\{p, n\}$.

Let $j = \min\{p, n\} - 1$. Then, there exists a constant $\overline{B}_1 = \overline{B}_1(\lambda, c, \alpha, \beta)$ such that $|\xi_{j+1}(a) - f_a^{j+1}(x)| \ge B_1 |D_j(a)| |f_a(x) - f_a(c_\lambda)| \ge \overline{B}_1 e^{c(j+1)} e^{-2|m|}$ for every $x \in I_m^+$. Now, since $\overline{B}_1 e^{c(j+1)} e^{-2|m|} > e^{-\beta(j+1)}$ if and only if $(c + \beta)(j + 1) > 2|m| - \log \overline{B}_1$, we have

$$\min\{n, p\} \le \frac{2|m| - \log \overline{B}_1}{c + \beta} < \frac{n}{2}$$

whenever $\Delta > -\log \overline{B}_1$. On the other hand, from $|f_a'(x)| < 3\lambda^{-1}$, it follows that $|\xi_j(a) - f_a^j(x)| \le (10\lambda^{-1})^j e^{-|m|}$ for every $x \in U_m^+$. Hence, $p \ge |m|(\beta + \log 10\lambda^{-1})^{-1}$.

To prove statement (c) notice that from (2.2), $|f_a^{p+1}(U_m^+)| \ge e^{-\beta(p+1)}$. Then, there exists $y \in U_m^+$ such that

$$\frac{e^{-\beta(p+1)}}{|U_m^+|} < \left|\left(f_a^p\right)'(f_a(y))\right| |f_a'(y)|.$$

On the other hand, according to (a), each $x \in U_m^+$ satisfies

$$\left|\left(f_a^{p+1}\right)'(x)\right| \geq \frac{1}{B_1^2}\left|\left(f_a^p\right)'(f_a(y))\right|\left|f_a'(x)\right| \geq \frac{1}{B_1^2}\frac{e^{-\beta(p+1)}}{|U_m^+|}\frac{|f_a'(x)|}{|f_a'(y)|}.$$

Moreover, since $x \in I_m^+$ and $y \in U_m^+$, we have $|f_a'(x)| \geq const\, |f_a'(y)|$. In short, by using (b) we obtain $\left|\left(f_a^{p+1}\right)'(x)\right| \geq const\, e^{-\beta(p+1)}e^{|m|} \geq \tau e^{|m|\left(1-\frac{3\beta}{\beta+c}\right)}$, where $\tau = \tau(\lambda, \alpha, \beta)$ is a positive constant. This statement holds by taking $\Delta > (\beta + c)\beta^{-1}\log(\tau^{-1})$. $\square$

Remark 2.9. *Under the assumptions of Lemma 2.8, p is finite,* $\left|\left(f_a^k\right)'(\xi_{n+1}(a))\right| \geq B_1^{-1}|D_k(a)| \geq B_1^{-1}e^{ck}$ *for* $k = 1, ..., p$ *and* $\left|\left(f_a^{p+1}\right)'(\xi_n(a))\right| \geq e^{\left(1-\frac{4\beta}{\beta+c}\right)|m|} \geq 1$.

The estimates above are helpful to compensate the loss of expansiveness on returns. We need to obtain similar properties when defining p to be constant on small intervals of the parameter. This fact will be fundamental to get bounded distortion for D_n in such intervals. Bounded distortion will play an essential role in the exclusion of the parameters.

Let ω be an interval of parameters such that $\xi_n(\omega) \subset I_m^+$ with $|m| \geq \Delta$. Define $p(\omega, m) = \min_{a \in \omega} p(a, m)$. Statements (a) and (b) of Lemma 2.8 immediately follow for $p = p(\omega, m)$, but statement (c) does not. For instance, the inequality $|f_a^{p+1}(U_m^+)| \geq e^{-\beta(p+1)}$ of (2.2), used to prove (c), is not true for $p(\omega, m)$. The following results enable us to prove (c).

Lemma 2.10. *Let* $\omega \subset \Omega_{n-1}$. *Then, for each* $a, b \in \omega$, *it follows that* $|a - b| \leq 3\pi A e^{-cn}$, *where A is the constant given in Proposition 2.3.*

Proof. This follows directly from Proposition 2.3. $\square$

Lemma 2.11. *Let* $\Delta \in \mathbf{N}$ *be large enough. If* $\omega \subset \Omega_{n-1}$ *and* $\xi_n(\omega) \subset I_m^+$, *with* $\Delta \leq |m| \leq [\alpha n] - 1$, *then* $|\xi_j(a) - \xi_j(b)| \leq e^{-\beta j}$ *for each* $a, b \in \omega$ *and each* $j = 1, ..., p(\omega, m)$.

Proof. Since $\omega \subset \Omega_{n-1}$, $j \leq p(\omega, m) \leq n - 1$ and $\xi_1'(a) = (\lambda a)^{-1}$ we have, from the proof of Proposition 2.3, that

$$\left|\frac{\xi_j'(a)}{D_{j-1}(a)}\right| \leq \frac{1}{\lambda}\left(1 + \sum_{k=1}^{j-1}e^{-ck}\right) = L_1(\lambda, c).$$

According to the mean value theorem, there exists $t \in \omega$ such that $|\xi_j(a) - \xi_j(b)| \leq L_1|D_{j-1}(t)||a - b|$ and $\left|f_t^j(U_m^+)\right| \geq B_1^{-1}|D_{j-1}(t)||f_t(U_m^+)|$, where $B_1 = B_1(\lambda, \alpha, \beta)$ is the constant given in Lemma 2.8. So, we have that

$$|\xi_j(a) - \xi_j(b)| \leq L_1 B_1 \frac{|a-b|}{|f_t(U_m^+)|} \left| f_t^j(U_m^+) \right|.$$

On the other hand, by using a bound for the second order derivative, we obtain $|f_t(U_m^+)| > \lambda^{-1} e^{-2\alpha n}$ and from Lemma 2.10 it follows that $|\xi_j(a) - \xi_j(b)| \leq L_2 \left| f_t^j(U_m^+) \right| e^{(2\alpha-c)n}$ with $L_2 = L_2(\lambda, \alpha, \beta) = 3\pi \lambda A L_1 B_1$.

Finally, from (2.2), we conclude that $\left| f_t^j(U_m^+) \right| \leq 2e^{-\beta j}$ and the proof ends by taking $2\alpha < c$ and $\Delta \in \mathbf{N}$ such that $2L_2 e^{(2\alpha-c)\Delta} \leq 1$. $\square$

The next lemma is an approach to Proposition 2.20 which will be crucial for the proof of Theorem 2.1.

Lemma 2.12. *Let $\Delta \in \mathbf{N}$ be large enough. There exists a constant $B_2 = B_2(\lambda, \alpha, \beta)$ such that if $\omega \subset \Omega_{n-1}$ and $\xi_n(\omega) \subset I_m^+$ with $\Delta \leq |m| \leq [\alpha n] - 1$, then, for each $a, b \in \omega$ and for each $x, y \in U_m^+$,*

$$\frac{|D_j(a,x)|}{|D_j(b,y)|} \leq B_2 \quad for \ \ j = 1, ..., p = p(\omega, m).$$

Proof. From Lemma 2.11 and (BA_{n-1}) we have $|\mathrm{tg}\, \xi_i(a) - \mathrm{tg}\, \xi_i(b)| \leq 10 e^{-\beta i}$ and $|\lambda - \mathrm{tg}\, \xi_i(b)| \geq \Lambda e^{-\alpha i}$, respectively. Hence

$$\frac{|D_j(a)|}{|D_j(b)|} \leq \exp \left\{ \sum_{i=1}^{j} \frac{|\mathrm{tg}\, \xi_i(a) - \mathrm{tg}\, \xi_i(b)|}{|\lambda - \mathrm{tg}\, \xi_i(b)|} \right\} \leq \exp \left\{ \frac{10}{\Lambda} \sum_{i=1}^{\infty} e^{(\alpha-\beta)i} \right\} = const.$$

Now, from Lemma 2.8(a) it follows that

$$\frac{|D_j(a,x)|}{|D_j(b,y)|} \leq B_1^2 \frac{|D_j(a)|}{|D_j(b)|}$$

and the lemma holds. $\square$

We shall prove below the main result in this section, which, as we said before, will be an enlargement of Lemma 2.8 to the interval $\omega \subset \Omega_{n-1}$.

Proposition 2.13. *Let $\Delta \in \mathbf{N}$ be large enough. If $\omega \subset \Omega_{n-1}$ and $\xi_n(\omega) \subset I_m^+$ with $\Delta \leq |m| \leq [\alpha n] - 1$ then*

(a) There exists a constant $B = B(\lambda, \alpha, \beta)$ such that for each $a \in \omega$ and $x \in U_m^+$ it follows that

$$\frac{1}{B} \leq \frac{|D_k(a,x)|}{|D_k(a)|} \leq B \quad for \ \ k = 1, ..., p = p(\omega, m).$$

(b)

$$\frac{|m|}{\beta + \log \frac{10}{\lambda}} \leq p \leq \frac{3|m|}{\beta + c} < \frac{n}{2}.$$

(c) For each $a \in \omega$ and each $x \in I_m^+$,

$$\left| \left(f_a^{p+1} \right)' (x) \right| \geq e^{\left(1 - \frac{5\beta}{\beta+c}\right)|m|}.$$

Proof. Let $a_* \in \omega$ be such that $p(a_*, m) = p(\omega, m)$. From Lemmas 2.8 and 2.12 it follows, respectively, that $\left| \left(f_{a_*}^{p+1} \right)' (x) \right| \geq e^{\left(1 - \frac{4\beta}{\beta+c}\right)|m|}$ for each $x \in I_m^+$ and

$$\frac{|D_p(a_*, x)|}{|D_p(a, x)|} \leq B_2$$

for every $a \in \omega$. Therefore,

$$\frac{\left| \left(f_{a_*}^{p+1} \right)' (x) \right|}{\left| \left(f_a^{p+1} \right)' (x) \right|} = \frac{|D_p(a_*, x)|}{|D_p(a, x)|} \frac{|f_{a_*}'(x)|}{|f_a'(x)|} \leq B_2$$

and

$$\left| \left(f_a^{p+1} \right)' (x) \right| \geq \frac{1}{B_2} e^{\left(1 - \frac{4\beta}{\beta+c}\right)|m|} \geq e^{\left(1 - \frac{5\beta}{\beta+c}\right)|m|}$$

whenever Δ is large enough. So, statement (c) is proved. The remaining ones follow as in Lemma 2.8. □

2.4 The inductive step

As we have already mentioned, for every $a \in \Omega_{n-1}$, we shall define two finite sequences $(\mu_i)_{i=0,\ldots,s+1}$ and $(p_i)_{i=0,\ldots,s}$ with $\mu_0 = 1$, $p_0 = -1$, $\mu_{s+1} = n$ such that $\mu_i \leq \mu_i + p_i + 1 \leq \mu_{i+1} \leq n$ for $i = 0, \ldots, s-1$. Furthermore, we shall proceed in such a way that if we define

$$q_i = \mu_{i+1} - (\mu_i + p_i + 1) \text{ for } i = 0, \ldots, s-1,$$

$$q_s = \begin{cases} 0, & \text{if } n \leq \mu_s + p_s \\ n - (\mu_s + p_s + 1), & \text{if } n \geq \mu_s + p_s + 1 \end{cases}$$

then the following properties will be satisfied:

(a) $\left|(f_a^{q_i})'(\xi_{\mu_i+p_i+1}(a))\right| \geq C_0 e^{c_0 q_i}$ for $i = 0, ..., s-1$.

(b) $\left|(f_a^{q_s})'(\xi_{\mu_s+p_s+1}(a))\right| \geq C_0 e^{-\Delta} e^{c_0 q_s}$ if $q_s \neq 0$.

(2.3)

(c) $\left|(f_a^{p_i+1})'(\xi_{\mu_i}(a))\right| \geq 1$ for $i = 1, ..., s$.

(d) $\left|(f_a^k)'(\xi_{\mu_i+1}(a))\right| \geq B^{-1} e^{ck}$ for $k = 1, ..., p_i$ and $i = 1, ..., s$.

where $B > 1$ is given in Proposition 2.13.

Proposition 2.14. *Let $N \in \mathbf{N}$ be large enough. If $n > N$, then*

$$(\mathrm{EG_{n-1}}) + (\mathrm{BA_n}) + (\mathrm{FA_n}) \Rightarrow (\mathrm{EG_n}).$$

Proof. First, suppose that $n \geq \mu_s + p_s + 1$. Then, from (2.3),

$$|D_n(a)| = \prod_{i=0}^{s} \left|(f_a^{q_i})'(\xi_{\mu_i+p_i+1}(a))\right| \left|(f_a^{p_i+1})'(\xi_{\mu_i}(a))\right| \geq C_0^{s+1} e^{-\Delta} e^{c_0 F_n(a)},$$

where $F_n(a) = q_0 + q_1 + ... + q_s \geq (1-\alpha)n$ according to (FA$_n$). Once $c < c_0$ is fixed, take $\alpha \ll 1$ such that $c_0(1-\alpha) > c + \alpha$. Then, if $\alpha N > 2\Delta$, we have $C_0^{s+1} e^{-\Delta} e^{c_0 F_n(a)} \geq C_0^{s+1} e^{\frac{1}{2}\alpha n} e^{cn}$. Furthermore, since

$$\mu_{i+1} - \mu_i > p_i > \frac{|m_i|}{\beta + \log \frac{10}{\lambda}} > \frac{\Delta}{\beta + \log \frac{10}{\lambda}}$$

implies $(s+1)\Delta \leq n(\beta + \log 10\lambda^{-1})^{-1}$, we derive

$$|D_n(a)| \geq C_0^{s+1} e^{\frac{1}{2}\alpha n} e^{cn} \geq \left(C_0^{\Delta^{-1}(\beta+\log(10\lambda^{-1}))} e^{\frac{1}{2}\alpha}\right)^n e^{cn} > e^{cn},$$

by taking Δ sufficiently large so that $(\beta + \log(10\lambda^{-1}))\Delta^{-1} \log C_0 + \frac{1}{2}\alpha > 0$.

Now, assume that $n \leq \mu_s + p_s$. Apply again (2.3) and use (BA$_n$) to obtain $|D_n(a)| \geq C e^{-2\alpha\mu_s} B^{-1} e^{c(n-\mu_s-1)} C_0^s e^{c_0 F_n(a)}$, where $C > 1$ is a bound of the second derivative of f_a in a neighbourhood of the critical point. Once $c < c_0$ is fixed, take α such that $c + \frac{5}{2}\alpha < c_0(1-\alpha)$. As in the case above, we obtain

$$|D_n(a)| \geq \frac{C_0^{s+1}}{B} e^{-2\alpha n} e^{c_0(1-\alpha)n} > e^{cn}. \qquad \square$$

2.5   Construction of the sets $(\Omega_n)_{n \in \mathbf{N}}$

In this section we shall accurately define the sets Ω_n and, for each $a \in \Omega_n$, the afore-mentioned sequences $(\mu_i)_i$, $(p_i)_i$, verifying (2.3). First, we shall give a partition defined on I_m in the following way:

For $m \geq \Delta$, let $r_m = m^{-2}|I_m|$ and $A_{m,k} = [e^{-m} - kr_m, e^{-m} - (k-1)r_m)$, with $k = 1, ..., m^2$. For $m = \Delta - 1$ and $k \geq 1$, let $A_{\Delta-1,k} = [e^{-\Delta}, e^{-\Delta} + kr_{\Delta-1})$, with $r_{\Delta-1} = (\Delta - 1)^{-2}|I_{\Delta-1}|$. We extend the definitions for $m \leq -(\Delta - 1)$ setting $A_{m,k} = -A_{-m,k}$. For $|m| \geq \Delta - 1$, write $I_{m,k} = T_\lambda(A_{m,k})$.

In short, for $|m| \geq \Delta$ we have a partition of I_m into intervals of equal length, $I_m = I_{m,1} \cup ... \cup I_{m,m^2}$. Denote $I_{m,k}^+ = I_{m_1,k_1} \cup I_{m,k} \cup I_{m_2,k_2}$, where I_{m_1,k_1} and I_{m_2,k_2} are the adjacent intervals of $I_{m,k}$. Note that $I_{m,k} \subset I_m$, $I_{m,k}^+ \subset I_m^+$ and that $\left|I_{m,k}^+\right| \leq 5m^{-2}|I_m|$ whenever Δ is large enough. Henceforth, it will be useful to take $I_{\Delta-1,k}^+ = \left(c_\lambda, f_{a(\lambda)}(c_\lambda)\right]$ and $I_{1-\Delta,k}^+ = \left[Q_{a(\lambda)}, c_\lambda\right)$ regardless of k. These partitions defined on U_Δ enable us to inductively define partitions P_n into the space of the parameters in order to get bounded distortion between ξ_n' and D_{n-1} on every $\omega \in P_{n-1}$. The sets Ω_n will be given by $\Omega_n = \bigcup_{\omega \in P_n} \omega$.

To begin the inductive process, fix Δ sufficiently large to get all the previous estimates. Then, from Proposition 2.2, there exists $a_1(\lambda) < a(\lambda)$ such that if $a \in [a_1(\lambda), a(\lambda)]$, then we have exponential growth on all the free iterates before the first return. On the other hand, taking $N > N_0$ to be large enough, Proposition 2.5 furnishes an interval $\Omega_{N-1} = [a_N(\lambda), a(\lambda)]$ with $a_N(\lambda) > a_1(\lambda)$, such that, for every $a \in \Omega_{N-1}$, (BA$_n$), (FA$_n$) and (EG$_{n-1}$) are satisfied up to $n = N - 1$. Therefore, we take $\Omega = \Omega_{N-1}$ to begin the inductive process. That is, $\Omega = \Omega_i$ and $P_i = \{\Omega\}$ for $i = 1, ..., N - 1$.

Now, let $n \geq N$. Let us assume the following conditions for every element ω of the partition P_{n-1}:

H.1. There exists a sequence of parameter intervals $\Omega = \omega_1 \supset \omega_2 \supset ... \supset \omega_{n-1} = \omega$, such that $\omega_k \in P_k$ for $k = 1, ..., n - 1$.

H.2. There exists a set $R_{n-1} = R_{n-1}(\omega) = \{\mu_0, ..., \mu_s\}$ for some $s = s(n - 1, \omega)$ such that $\mu_0 = 1$ and $\mu_i \leq n - 1$ for $i = 1, ..., s$. For $i = 1, ..., s$ we shall say that μ_i is a return of ω. Furthermore, if $k \leq n - 1$, then $R_k(\omega_k) = R_{n-1}(\omega) \cap \{m \in \mathbf{N} : m \leq k\}$.

H.3. For each return $\mu_i \in R_{n-1}$ there exists an associated interval I_{m_i,k_i}^+ with $|m_i| \geq \Delta$, which is called the host interval of ω at the return μ_i, such that $\xi_{\mu_i}(\omega_{\mu_i}) \subset I_{m_i,k_i}^+$. If

$p_i = p(\omega_{\mu_i}, m_i)$, then $\{\mu_i + 1, ..., \mu_i + p_i\}$ will be called the binding period associated to the return μ_i. For a suitable notation we write $p_0 = -1$.

H.4. For each $k = 1, ..., n - 1$, ω_k satisfies (BA$_k$) and (EG$_k$).

So, in order to check (2.3), notice that for each return $\mu_i \in R_{n-1}$, ω_{μ_i} verifies (BA$_{\mu_i}$) and (EG$_{\mu_i - 1}$) and, from Proposition 2.13, it follows that $p_i < \frac{3}{\beta + c}|m_i|$. That is, the binding periods are finite and, according to the same proposition, for each $a \in \omega \subset \omega_{\mu_i}$ we obtain $\left|(f_a^{p_i + 1})'(\xi_{\mu_i}(a))\right| \geq e^{\left(1 - \frac{5\beta}{\beta + c}\right)|m_i|} \geq 1$ and $\left|(f_a^j)'(\xi_{\mu_i + 1}(a))\right| \geq$ $B^{-1}|D_j(a)|$ for $j = 1, ..., p_i$. Now, from $\xi_{\mu_i}(\omega_{\mu_i}) \subset I_{m_i, k_i}^+ \subset I_{m_i}^+$ and (BA$_{\mu_i}$), it follows that $2\alpha\mu_i > |m_i|$. Therefore, $j < \frac{3}{\beta + c}|m_i| < \frac{6\alpha}{\beta + c}\mu_i < \mu_i$ for $j \leq p_i$ and thus, $|D_j(a)| \geq e^{cj}$ and $\left|(f_a^j)'(\xi_{\mu_i + 1}(a))\right| \geq B^{-1}e^{cj}$.

H.5. The sets $\{\mu_i + p_i + 1, ..., \mu_{i+1} - 1\}$, for $i < s$, and $\{\mu_s + p_s + 1, ..., n - 1\}$ if $n > \mu_s + p_s + 1$, are called free periods. During these periods we have $\xi_{\mu_i + p_i + j}(\omega) \cap U_{\Delta + 1} = \emptyset$ for $j = 1, ..., q_i$, with $q_i = \mu_{i+1} - (\mu_i + p_i + 1)$ for $i < s$ and $q_s = n - (\mu_s + p_s + 1)$.

Notice that, by applying Proposition 2.2, we obtain, for every $a \in \omega$, the remaining inequalities in (2.3).

Remark 2.15. *Whenever we define partitions P_n verifying hypotheses H.1 to H.5, we have the properties stated in (2.3) and assumptions (BA$_n$) and (EG$_{n-1}$). So, we only need assumption (FA$_n$) to apply the inductive process given in Proposition 2.14. Now, assumption (EG$_n$), which follows from this proposition, is clearly given in hypothesis H.4. However, although Proposition 2.14 seems unnecessary, it will turn out to be crucial to construct P_n from P_{n-1}.*

Next, we shall show how to choose the elements $\omega \in P_n$ so that the above five hypotheses hold. Since $R_{n-1}(\Omega) = \{1\}$, these hypotheses are trivially satisfied by $\Omega = P_{n-1}$ for $n \leq N$. For $n > N$, we inductively get a sequence of partitions satisfying hypotheses H.1 to H.5. So, let us suppose that we have defined P_{n-1} satisfying these hypotheses. To construct P_n we introduce an auxiliary partition P_n' which is obtained from the elements of P_{n-1} in the following way:

(a) If n belongs to the binding period associated to a previous return, then we do not partition ω any further and we let $\omega \in P_n'$ and $R_n(\omega) = R_{n-1}(\omega)$. Notice that, although the orbit intersects $U_{\Delta + 1}$, n is not a return of ω in this case.

(b) If $\xi_n(\omega) \cap U_\Delta \subset I_{\Delta, 1} \cup I_{-\Delta, 1}$, then we take again $\omega \in P_n'$ and $R_n(\omega) = R_{n-1}(\omega)$.

(c) In other cases, we say that n is a return situation and distinguish between two possibilities:

(c.1) $\xi_n(\omega)$ does not completely contain an interval $I_{m,k}$. We say that the return situation is inessential and $\xi_n(\omega) \subset I_{m,k}^+$ necessarily holds. In this case, we let $\omega \in P_n'$ and define $R_n(\omega) = R_{n-1}(\omega) \cup \{n\}$. The return n is called inessential and $I_{m,k}^+$ is its host interval.

(c.2) $\xi_n(\omega)$ contains at least one interval $I_{m,k}$ with $|m| \geq \Delta$. We say that the return situation is essential and define $\omega_{1-\Delta,1}' = \xi_n^{-1}\left(\left[Q_{a(\lambda)}, c_\lambda\right] \setminus U_\Delta\right) \cap \omega$, $\omega_{\Delta-1,1}' = \xi_n^{-1}\left(\left[c_\lambda, f_{a(\lambda)}(c_\lambda)\right] \setminus U_\Delta\right) \cap \omega$ and $\omega_{m,k}' = \xi_n^{-1}(I_{m,k}) \cap \omega$ for $|m| \geq \Delta$. If A denotes the set of pairs (i,j) such that $\omega_{i,j}' \neq \emptyset$, then we may split

$$\omega - \left\{\xi_n^{-1}(c_\lambda)\right\} = \bigcup_{(i,j)\in A} \omega_{i,j}'. \tag{2.4}$$

Now, from Proposition 2.3, we deduce that $\xi_n |_\omega$ is a homeomorphism and thus every $\omega_{m,k}'$ is an interval. So, $\xi_n(\omega_{m,k}')$ completely contains one $I_{m,k}$ except, maybe, when $I_{m,k}$ is one of the extreme intervals of $\xi_n(\omega)$. In this case, we join $\omega_{m,k}'$ to its adjacent interval in the splitting (2.4). The same can be done for $\omega_{\Delta-1,1}'$ and $\omega_{1-\Delta,1}'$ when their images do not completely contain $I_{\Delta-1,1}$ or $I_{1-\Delta,1}$, respectively. In this way, we obtain from (2.4) a new partition of $\omega - \{\xi_n^{-1}(c_\lambda)\}$ into intervals $\omega_{m,k}$ such that $I_{m,k} \subset \xi_n(\omega_{m,k}) \subset I_{m,k}^+$.

At last, we have to rule out the intervals $\omega_{m,k}$ which do not satisfy (BA$_n$). So, $\omega_{m,k} \in P_n'$ if and only if $|m| \leq [\alpha n] - 1$. Note that the excluded portion of ω is an interval ω_{exc} such that $\xi_n(\omega_{exc}) \subset U_{[\alpha n]-1}$. If $|m| \geq \Delta$, then $R_n(\omega_{m,k}) = R_{n-1}(\omega) \cup \{n\}$ and n is called an essential return of $\omega_{m,k}$. If $|m| = \Delta - 1$, then $R_n(\omega_{m,k}) = R_{n-1}(\omega)$. The essential return situations for which $|\xi_n(\omega)| \geq e^{-\frac{1}{2}\Delta}$ are called escape situations of ω. In this case, note that at least one of the components $\omega_{m,k}$, for $|m| = \Delta - 1$, verifies $|\xi_n(\omega_{m,k})| \geq \frac{1}{3}e^{-\frac{1}{2}\Delta}$ and we say that this $\omega_{m,k}$ is an escape component.

From the above construction it follows that P_n' satisfies H.1, H.2, H.3 and H.5 in any of the cases (a), (b) and (c). Next, we shall prove that condition (BA$_n$) in hypothesis H.4 holds.

Proposition 2.16. *Each $\omega \in P_n'$ verifies* (BA$_n$).

Proof. We shall carry out the proof for each case in the above construction of P_n':

(a) If $n = \mu + j$ belongs to the binding period associated to a previous return μ, then we have $|\xi_{\mu+j}(a) - \xi_j(a)| \leq e^{-\beta j}$ for every $a \in \omega$. On the other hand, since a satisfies (BA$_{n-1}$) and $j \leq p \leq \frac{1}{2}\mu < n-1$, it follows that $|\xi_{\mu+j}(a) - c_\lambda| \geq e^{-\alpha j} - e^{-\beta j} \geq e^{-\alpha n}$, provided that N is large enough.

(b) If $\xi_n(\omega) \cap U_\Delta \subset I_{\Delta,1} \cup I_{-\Delta,1}$, then we have $|\xi_n(a) - c_\lambda| > e^{-(\Delta+1)}$ for every $a \in \omega$. Hence $|\xi_n(a) - c_\lambda| > e^{-\alpha n}$ holds by taking $\alpha N \geq \Delta + 1$.

(c.1) If n is an inessential return, then $\xi_n(\omega)$ does not contain any $I_{m,k}$. On the other hand, if ω does not satisfy (BA$_n$), then there exists $x \in \xi_n(\omega) - \{c_\lambda\}$ with $|x - c_\lambda| < e^{-\alpha n}$. Therefore, if $I_{m,k}^+$ is the host interval of ω at time n, then $|m| > [\alpha n] - 1$. Both facts lead to $|\xi_n(\omega)| \leq |I_{m,k}^+| \leq 5m^{-2}e^{-|m|} < e^{-(|m|+2)} < e^{-(1+[\alpha n])} < e^{-\alpha n}$, provided that Δ is large enough. Nevertheless, in Lemma 2.19 we shall prove that $|\xi_n(\omega)| \geq e^{-\alpha n}$.

(c.2). In this case, hypothesis (BA$_n$) follows from the construction of P_n'. $\square$

At this stage, if we define $\Omega_n' = \bigcup_{\omega \in P_n'} \omega$, we know that hypotheses (BA$_n$) and (EG$_{n-1}$) are fulfilled for every $a \in \Omega_n'$. As for free assumption (FA$_n$), note that if $F_n(a)$ is the number of free iterates in $\{\xi_j(a)\}_{j=1,...,n}$, then $F_n(a)=F_n(b)$ for each $a, b \in \omega$ and for every $\omega \in P_n'$. Therefore, it makes sense to refer to $F_n(\omega)$ and define $P_n = \{\omega \in P_n' : F_n(\omega) \geq (1 - \alpha)n\}$. Then P_n satisfies (BA$_n$), (FA$_n$) and (EG$_{n-1}$). According to Remark 2.15, we use Proposition 2.14 for concluding that P_n verifies (EG$_n$).

In fact, this section should be finished by proving Lemma 2.19, which was called on the last proof. However, besides this and other preliminary results, we also prove here Proposition 2.20, which is one of the key statements in this chapter.

According to our definition of $\{\Omega_n\}_{n\in\mathbb{N}}$, it follows that if $a \in \Omega_n$, then a belongs to a unique $\omega_k \in P_k$ for each $k = 1, ..., n$. These intervals are defined in the following way: $\omega_1 = ... = \omega_{N-1} = \Omega$ and, since $\xi_N(\Omega) \supset U_\lambda$, $\nu_1(a) = N$ is an essential return situation of ω_{N-1}. Then Ω is split into intervals $\Omega_{(m,k)}$ with $\Omega_{(m,k)} \in P_N$ for $\Delta - 1 \leq |m| \leq [\alpha N] - 1$. Since $a \in \Omega_n$, there exists (m_1, k_1) such that $a \in \omega_N = \Omega_{(m_1,k_1)}$. Now, let $\nu_2(a)$ be the next essential return situation of $\Omega_{(m_1,k_1)}$. Then $\omega_k = \Omega_{(m_1,k_1)}$ for $k = \nu_1(a) + 1, ..., \nu_2(a) - 1$. For $\nu_2(a)$, let us again split the interval $\Omega_{(m_1,k_1)}$ to obtain a new component $\Omega_{(m_1,k_1)(m_2,k_2)}$ of $P_{\nu_2(a)}$ such that $\omega_{\nu_2(a)} = \Omega_{(m_1,k_1)(m_2,k_2)}$. Iterating the process we obtain sequences $\nu_1, ..., \nu_s$ and $(m_1, k_1), ..., (m_s, k_s)$, with s depending on a and n, such that $\omega_{\nu_i} = \Omega_{(m_1,k_1)...(m_i,k_i)}$, $\omega_k = \omega_{\nu_i}$ for $k = \nu_i + 1, ..., \nu_{i+1} - 1$, $\omega_{\nu_i} \subset \omega_{\nu_{i-1}}$ and $I_{m_i,k_i} \subset \xi_{\nu_i}(\omega_{\nu_i}) \subset I_{m_i,k_i}^+$. Moreover, since the restriction $\xi_{\nu_i} |_{\omega_{\nu_i}}$ is a homeomorphism, each $\omega \in P_n$ is an interval $\Omega_{(m_1,k_1)...(m_s,k_s)}$ for a unique sequence $(m_1, k_1), ..., (m_s, k_s)$, with $|m_i| \geq \Delta - 1$ and $1 \leq k_i \leq m_i^2$, $1 \leq i \leq s$.

Lemma 2.17. *Assume that* $\omega \in P_{\nu_0}$ *is an escape component in the construction of* P_{n-1}. *If* ν *is the next return situation of* ω, *then* $|\xi_\nu(\omega)| \geq e^{-\frac{1}{2}\Delta}$, *that is*, ν *is an escape situation of* ω.

Proof. From Proposition 2.4, there exists $x = \xi_{\nu_0}(t)$, with $t \in \omega$, such that $|\xi_\nu(\omega)| \geq A^{-2}\left|\left(f_t^{\nu-\nu_0}\right)'(x)\right||\xi_{\nu_0}(\omega)|$. Now, $x, f_t(x), ..., f_t^{\nu-\nu_0-1}(x)$ is an orbit outside $U_{\Delta+1}$ and we may suppose that $f_t^{\nu-\nu_0}(x) \in W$, where W is the neighbourhood given in Proposition 2.2. Otherwise, the lemma follows easily by taking a sufficiently large Δ so that $U_{\frac{\Delta}{4}} \subset W$. Then, we apply Proposition 2.2 to get

$$|\xi_\nu(\omega)| \geq \frac{1}{A^2}C_0 e^{c_0(\nu-\nu_0)}|\xi_{\nu_0}(\omega)| \geq \frac{1}{3A^2}C_0 e^{c_0(\nu-\nu_0)}e^{-\frac{1}{2}\Delta}.$$

The proof ends if $\nu - \nu_0 \geq c_0^{-1}\log\left(3C_0^{-1}A^2\right)$. So, let us assume that $\nu - \nu_0 < c_0^{-1}\log\left(3C_0^{-1}A^2\right)$. Since ν_0 is an escape situation of ω, there exists $a_* \in \omega$ such that $|\xi_{\nu_0}(a_*) - c_\lambda| = e^{-\Delta}$. Therefore, we may take a sufficiently large Δ and $a_1(\lambda)$ close enough to $a(\lambda)$ to obtain $f_{a_*}^{\nu-\nu_0}(\xi_{\nu_0}(a_*)) < 0$. The proof finishes by taking Δ such that $c_\lambda > e^{-\Delta} + e^{-\frac{1}{2}\Delta}$. $\square$

Lemma 2.18. *Suppose that* $\mu < n$ *is a return of* $\omega \in P_{n-1}$, *with host interval* $I_{m,k}^+$. *Let* p *be the length of its binding period and* $\omega_\mu \in P_\mu$ *such that* $\omega \subset \omega_\mu$.

(a) If $\mu' \leq n$ *is the next return after* μ *and* $q = \mu' - (\mu + p + 1)$, *then*

(a.1) $|\xi_{\mu'}(\omega)| \geq e^{cq}e^{\left(1-\frac{6\beta}{\beta+c}\right)|m|}|\xi_\mu(\omega)| \geq 3|\xi_\mu(\omega)|.$

(a.2) If μ *is an essential return and* μ' *is the first return situation of* ω_μ *after* μ, *then* $|\xi_{\mu'}(\omega_\mu)| \geq e^{cq}e^{-\frac{6\beta}{\beta+c}|m|}.$

(b) If μ *is the last return before* n *and* n *is a free iterate, then for* $q = n - (\mu+p+1)$ *it follows that*

(b.1) $|\xi_n(\omega)| \geq e^{cq-\Delta}e^{\left(1-\frac{6\beta}{\beta+c}\right)|m|}|\xi_\mu(\omega)|.$

(b.2) If μ *is an essential return, and there exist no return situations of* ω_μ *in the piece of orbit* (μ, n), *then* $|\xi_n(\omega)| \geq e^{cq-\Delta}e^{-\frac{6\beta}{\beta+c}|m|}.$

Proof. From Proposition 2.4, there exist $x = \xi_{\mu+p+1}(t)$ and $x' = \xi_\mu(t')$, with $t, t' \in \omega$, such that

$$\frac{|\xi_{\mu'}(\omega)|}{|\xi_{\mu+p+1}(\omega)|} \geq \frac{1}{A^2}\left|(f_t^q)'(x)\right|, \quad \frac{|\xi_{\mu+p+1}(\omega)|}{|\xi_\mu(\omega)|} \geq \frac{1}{A^2}\left|\left(f_{t'}^{p+1}\right)'(x')\right|.$$

Then, according to Proposition 2.13, we obtain

$$\frac{|\xi_{\mu'}(\omega)|}{|\xi_\mu(\omega)|} \geq \frac{1}{A^4}\left|(f_t^q)'(x)\right|e^{\left(1-\frac{5\beta}{\beta+c}\right)|m|}.$$

To prove statement (a) notice that, from Proposition 2.2(b), $\left|(f_t^q)'(x)\right| \geq C_0 e^{coq}$. For $\Delta > (\beta + c)\beta^{-1}\log\left(C_0^{-1}A^4\right)$, we have

$$\frac{|\xi_{\mu'}(\omega)|}{|\xi_\mu(\omega)|} \geq e^{coq} e^{\left(1 - \frac{6\beta}{\beta+c}\right)|m|}$$

and statement (a.1) follows by taking $6\beta < \beta + c$ and a large enough Δ.

Now, if μ is an essential return, then $I_{m,k} \subset \xi_\mu(\omega)$. Furthermore, since μ' is the first return situation of ω_μ, it necessarily follows that $\omega_\mu \subset P_{\mu'-1}$, otherwise $\omega = \emptyset$. Then, we may apply statement (a) to get

$$|\xi_{\mu'}(\omega_\mu)| \geq e^{coq} e^{\left(1 - \frac{5\beta}{\beta+c}\right)|m|} \frac{e^{-|m|}}{2m^2} \frac{C_0}{A^4} \geq e^{coq} e^{-\frac{6\beta}{\beta+c}|m|},$$

for $|m| > \Delta > (\beta + c)\beta^{-1}\log\left(2C_0^{-1}\Delta^2 A^4\right)$. So, statement (a.2) is proved.

To prove (b) we proceed in the same way, but we must bear in mind that only statement (a) of Proposition 2.2 can be applied. Notice that in (b.2) we have $\omega = \omega_\mu$.
□

Lemma 2.19. *If n is a return situation of $\omega \in P_{n-1}$, then $|\xi_n(\omega)| \geq e^{-\frac{1}{2}\alpha n}$.*

Proof. Since n is a return situation, it does not belong to any binding period associated to previous returns. Let $\nu < n-1$ be the first natural number for which $\omega_\nu = \omega$. Then, ν is a return situation and there exists $|m| \geq \Delta - 1$ such that $I_{m,k} \subset \xi_\nu(\omega) \subset I_{m,k}^+$.

If $|m| = \Delta - 1$, then we apply Propositions 2.2 and 2.4 as we did in Lemma 2.17. So, we have $|\xi_n(\omega)| \geq A^{-2}C_0 e^{-\Delta}(\Delta - 1)^{-2} \geq e^{-\frac{1}{2}\alpha n}$, provided that $n > N > 2\alpha^{-1}\left\{\Delta + \log\left(C_0^{-1}A^2(\Delta - 1)^2\right)\right\}$.

If $[\alpha\nu] - 1 \geq |m| \geq \Delta$, then ν is an essential return. Let $\mu_0 = \nu$, $\mu_{s+1} = n$ and $(\mu_i)_{i=1,...,s}$ be all the returns of ω between ν and n. If $s = 0$, then there exist no return situations of $\omega = \omega_\nu$ in (ν, n). So, from Lemma 2.18(b.2), we have $|\xi_n(\omega)| \geq e^{-\frac{6\beta}{\beta+c}|m|-\Delta}$. On the other hand, if $s > 0$, then, from Lemma 2.18, we deduce

$$|\xi_n(\omega)| = |\xi_{\mu_1}(\omega)| \left(\prod_{i=1}^{s-1} \frac{|\xi_{\mu_{i+1}}(\omega)|}{|\xi_{\mu_i}(\omega)|}\right) \frac{|\xi_{\mu_{s+1}}(\omega)|}{|\xi_{\mu_s}(\omega)|} \geq e^{-\frac{6\beta}{\beta+c}|m|-\Delta}.$$

In short, whether $s = 0$ or whether $s > 0$, we obtain

$$|\xi_n(\omega)| \geq e^{-\frac{1}{4}|m|-\Delta} \geq e^{-\frac{1}{4}\alpha\nu-\Delta} \geq e^{-\frac{1}{2}\alpha n}$$

for $n \geq N \geq 4\alpha^{-1}\Delta$. □

Proposition 2.20. *There exists a constant* $C = C(\lambda, \alpha, \beta)$ *such that if* $\omega \in P_{n-1}$ *with* $n \geq N$ *and* $\xi_n(\omega) \subset U_{\frac{\Delta}{4}}$, *then for every* $a, b \in \omega$ *it follows that*

$$(a) \quad \frac{|D_{n-1}(a)|}{|D_{n-1}(b)|} \leq C \qquad (b) \quad \frac{|\xi'_n(a)|}{|\xi'_n(b)|} \leq C.$$

Proof. Let $(\mu_j)_{j=1,\dots,s}$ be the returns of ω up to $n-1$, $(p_j)_{j=1,\dots,s}$ be the lengths of the respective binding periods and $\left\{ I^+_{m_j, k_j} \right\}_{j=1,\dots,s}$ the corresponding host intervals. Let us denote $\mu_0 = 1$, $p_0 = -1$ and write $\sigma_i = \xi_{\mu_i}(\omega)$ for $i = 1, \dots, s$. Notice that, according to Proposition 2.3, it suffices to prove statement (a).

As in Lemma 2.8, we obtain

$$\frac{|D_{n-1}(a)|}{|D_{n-1}(b)|} \leq \exp \left\{ 10 \sum_{i=1}^{n-1} \frac{|\xi_i(a) - \xi_i(b)|}{|c_\lambda - \xi_i(b)|} \right\}$$

and hence, we only need to bound

$$S = \sum_{i=1}^{n-1} \frac{|\xi_i(a) - \xi_i(b)|}{|c_\lambda - \xi_i(b)|}.$$

To this end, we distinguish between two cases:

1. Let $n \leq \mu_s + p_s$. Then S is the addition of the following terms:

$$R_j = \frac{\left| \xi_{\mu_j}(a) - \xi_{\mu_j}(b) \right|}{\left| c_\lambda - \xi_{\mu_j}(b) \right|} \quad \text{for } j = 1, \dots s$$

$$S'_j = \sum_{i=\mu_j+1}^{\mu_j+p_j} \frac{|\xi_i(a) - \xi_i(b)|}{|c_\lambda - \xi_i(b)|} \quad \text{for } j = 1, \dots, s-1$$

$$S'_s = \sum_{i=\mu_s+1}^{n-1} \frac{|\xi_i(a) - \xi_i(b)|}{|c_\lambda - \xi_i(b)|} \quad \text{and } S''_j = \sum_{i=\mu_j+p_j+1}^{\mu_{j+1}-1} \frac{|\xi_i(a) - \xi_i(b)|}{|c_\lambda - \xi_i(b)|} \quad \text{for } j = 0, \dots, s-1.$$

First let us bound each S''_j. Recall that, from Proposition 2.4, there exist $t_i \in \omega$ such that

$$\frac{\left| \xi_{\mu_{j+1}}(a) - \xi_{\mu_{j+1}}(b) \right|}{|\xi_i(a) - \xi_i(b)|} \geq \frac{1}{A^2} \left| \left(f_{t_i}^{\mu_{j+1}-i} \right)' (\xi_i(t_i)) \right|.$$

Now, from Proposition 2.2(b),

$$\left| \xi_{\mu_{j+1}}(a) - \xi_{\mu_{j+1}}(b) \right| \geq C_0 A^{-2} e^{c_0(\mu_{j+1}-i)} |\xi_i(a) - \xi_i(b)|$$

and, since i is a free iterate, it follows that $|c_\lambda - \xi_i(b)| \geq e^{-(\Delta+1)}$. Then

$$\frac{|\xi_i(a) - \xi_i(b)|}{|c_\lambda - \xi_i(b)|} \leq \frac{A^2 e^{\Delta+1}}{C_0} e^{c_0(i-\mu_{j+1})} \left| \xi_{\mu_{j+1}}(\omega) \right|.$$

Finally,

$$S_j'' \leq L_1 e^\Delta \left|I_{m_j+1}\right| \frac{\left|\xi_{\mu_j+1}(\omega)\right|}{\left|I_{m_j+1}\right|} \leq L_1 \frac{\left|\sigma_{j+1}(\omega)\right|}{\left|I_{m_j+1}\right|}$$

where the constant L_1 only depends on λ and c_0.

To bound each R_j notice that $\xi_{\mu_j}(b) \in I_{m_j,k_j}^+ \subset I_{m_j}^+$ and consequently, we obtain $\left|c_\lambda - \xi_{\mu_j}(b)\right| > e^{-(|m_j|+2)}$. From $\left|I_{m_j}\right| < e^{-|m_j|}$, it follows that

$$R_j = \frac{\left|\xi_{\mu_j}(a) - \xi_{\mu_j}(b)\right|}{\left|c_\lambda - \xi_{\mu_j}(b)\right|} \leq \frac{|\sigma_j|}{e^{-(|m_j|+2)}} \leq 9 \frac{|\sigma_j|}{\left|I_{m_j}\right|}. \tag{2.5}$$

Finally, let us bound S_j' for every $j = 1, \ldots, s$. To this end, write

$$\frac{\left|\xi_i(a) - \xi_i(b)\right|}{\left|c_\lambda - \xi_i(b)\right|} = \frac{\left|\xi_i(a) - \xi_i(b)\right|}{\left|\xi_i(b) - \xi_{i-\mu_j}(b)\right|} \frac{\left|\xi_i(b) - \xi_{i-\mu_j}(b)\right|}{\left|c_\lambda - \xi_i(b)\right|}.$$

With regard to the second factor, $\left|\xi_i(b) - \xi_{i-\mu_j}(b)\right| \leq e^{-\beta(i-\mu_j)}$. So, provided that $\alpha << \beta$, it follows that $\left|c_\lambda - \xi_i(b)\right| \geq e^{-\alpha(i-\mu_j)} - e^{-\beta(i-\mu_j)} \geq \left(1 - e^{\alpha-\beta}\right) e^{-\alpha(i-\mu_j)}$. Hence, there exists a constant $L_2 = L_2(\alpha, \beta)$ such that

$$\frac{\left|\xi_i(b) - \xi_{i-\mu_j}(b)\right|}{\left|c_\lambda - \xi_i(b)\right|} \leq L_2 e^{(\alpha-\beta)(i-\mu_j)}.$$

In relation to the first factor, from Proposition 2.4, there exist $x_i = \xi_{\mu_j}(t_i)$, with $t_i \in \omega$, such that

$$\frac{\left|\xi_i(a) - \xi_i(b)\right|}{\left|\xi_{\mu_j}(a) - \xi_{\mu_j}(b)\right|} \leq A^2 \left|\left(f_{t_i}^{i-\mu_j-1}\right)'(f_{t_i}(x_i))\right| \left|(f_{t_i})'(x_i)\right|.$$

Furthermore, there exists $x \in U_{m_j}^+$ such that

$$\left|\xi_i(b) - \xi_{i-\mu_j}(b)\right| \geq L_3 \left(\xi_{\mu_j}(b) - c_\lambda\right)^2 \left|\left(f_b^{i-\mu_j-1}\right)'(f_b(x))\right|$$

and so,

$$\frac{\left|\xi_i(a) - \xi_i(b)\right|}{\left|\xi_i(b) - \xi_{i-\mu_j}(b)\right|} \leq \frac{A^2}{L_3} \frac{\left|\left(f_{t_i}^{i-\mu_j-1}\right)'(f_{t_i}(x_i))\right|}{\left|\left(f_b^{i-\mu_j-1}\right)'(f_b(x))\right|} \frac{\left|(f_{t_i})'(x_i)\right|}{\left|\xi_{\mu_j}(b) - c_\lambda\right|} \frac{\left|\xi_{\mu_j}(a) - \xi_{\mu_j}(b)\right|}{\left|\xi_{\mu_j}(b) - c_\lambda\right|}.$$

The last factor was bounded in (2.5). From Lemma 2.12, there exists a constant B_2 such that

$$\frac{\left|\left(f_{t_i}^{i-\mu_j-1}\right)'(f_{t_i}(x_i))\right|}{\left|\left(f_b^{i-\mu_j-1}\right)'(f_b(x))\right|} \leq B_2.$$

Since $\xi_{\mu_j}(\omega) \subset I^+_{m_j,k_j}$ leads to $\left|\xi_{\mu_j}(t_i) - c_\lambda\right| < e^{-(|m_j|-1)}$ and $\left|\xi_{\mu_j}(b) - c_\lambda\right| > e^{-(|m_j|+2)}$, it follows that

$$\frac{\left|(f_{t_i})'(x_i)\right|}{\left|\xi_{\mu_j}(b) - c_\lambda\right|} \leq K \frac{\left|\xi_{\mu_j}(t_i) - c_\lambda\right|}{\left|\xi_{\mu_j}(b) - c_\lambda\right|} < 25K.$$

Finally, we get

$$\frac{|\xi_i(a) - \xi_i(b)|}{|c_\lambda - \xi_i(b)|} \leq L_4 \frac{|\sigma_j|}{|I_{m_j}|} e^{(\alpha - \beta)(i - \mu_j)}$$

and consequently, $S'_j \leq L_5 |\sigma_j| \left|I_{m_j}\right|^{-1}$ both for $j = 1, ..., s - 1$ and for $j = s$. In short, we get $S''_{j-1} + R_j + S'_j \leq L_6 |\sigma_j| \left|I_{m_j}\right|^{-1}$ for $j = 1, ..., s$.

Let us define $N_m = \{i \in \{1, ..., s\} : m_i = m\}$ and let $r_m = \max N_m$, provided that $N_m \neq \emptyset$. Since Lemma 2.18 yields $|\sigma_{j+1}| > 2 |\sigma_j| > ...$, we get

$$\sum_{j \in N_m} \frac{|\sigma_j|}{|I_{m_j}|} \leq 2 \frac{|\sigma_m|}{|I_m|} \leq 2 \frac{\left|I^+_{m,k}\right|}{|I_m|} \leq \frac{10}{m^2}$$

and eventually,

$$S \leq L_6 \sum_{m : N_m \neq \emptyset} \sum_{j \in N_m} \frac{|\sigma_j|}{|I_{m_j}|} \leq 10 L_6 \sum_{m \geq \Delta} \frac{1}{m^2} = L_7(\lambda, \alpha, \beta).$$

2. Let us assume that $n \geq \mu_s + p_s + 1$. From the previous estimates, we only need to bound

$$S''_s = \sum_{i = \mu_s + p_s + 1}^{n-1} \frac{|\xi_i(a) - \xi_i(b)|}{|c_\lambda - \xi_i(b)|}.$$

With this end in mind, we consider again two cases:

(2.a) Let $|\xi_{n-1}(\omega)| \leq e^{-2\Delta}$. Then, from Proposition 2.4, there exist $t_i \in \omega$ such that

$$\frac{|\xi_{n-1}(a) - \xi_{n-1}(b)|}{|\xi_i(a) - \xi_i(b)|} \geq \frac{1}{A^2} \left|\left(f_{t_i}^{n-i-1}\right)'(\xi_i(t_i))\right| \geq \frac{C_0}{A^2} e^{-\Delta} e^{c_0(n-i-1)},$$

whenever $\mu_s + p_s + 1 \leq i \leq n - 1$. On the other hand, $|c_\lambda - \xi_i(b)| \geq e^{-(\Delta+1)}$ and so

$$\frac{|\xi_i(a) - \xi_i(b)|}{|c_\lambda - \xi_i(b)|} \leq \frac{3A^2}{C_0} e^{c_0(i+1-n)}.$$

Hence, $S''_s \leq 3A^2 C_0^{-1} \sum_{i=\mu_s+p_s+1}^{n-1} e^{c_0(i+1-n)} \leq L_8$.

(2.b) Let now $|\xi_{n-1}(\omega)| > e^{-2\Delta}$. Let $k_0 \geq \mu_s + p_s + 1$ be the first natural number such that $|\xi_{k_0}(\omega)| > e^{-2\Delta}$. Then, according to the case (2.a), we get

$$\frac{|D_{k_0-1}(a)|}{|D_{k_0-1}(b)|} \leq L_9.$$

Therefore, since

$$\frac{|D_{n-1}(a)|}{|D_{n-1}(b)|} = \frac{\left|\left(f_a^{n-k_0}\right)'(\xi_{k_0}(a))\right|}{\left|\left(f_b^{n-k_0}\right)'(\xi_{k_0}(b))\right|} \frac{|D_{k_0-1}(a)|}{|D_{k_0-1}(b)|},$$

it suffices to bound the first factor.

Recall that I is called a homterval for f if $f^n|_I$ is monotonous for every $n \in \mathbf{N}$. Furthermore, it is known that I is a homterval if and only if I is either a wandering interval or every $x \in I$ is contained in the basin of a periodic attractor. In our case, since $f_{a(\lambda)}$ is a C^2 map without flat critical points, it has no wandering intervals, see [12]. Furthermore, since its Schwarzian derivative is negative, $f_{a(\lambda)}$ has no periodic attractors either. Therefore, $f_{a(\lambda)}$ has no homtervals and, by means of a usual compactness argument, it follows that, for each $l > 0$, there exists $n_0(l) \in \mathbf{N}$ such that each interval of length higher than or equal to l contains a point x satisfying $f_{a(\lambda)}^j(x) = c_\lambda$ for some $j \leq n_0(l)$. Because of the continuity of f_a with respect to the parameter a, we derive the same conclusion for intervals of length higher than or equal to $2l$ and for every $a \in \Omega = [a_0(\lambda), a(\lambda)]$, with $a_0(\lambda)$ close enough to $a(\lambda)$. Take $l = \frac{1}{2}e^{-2\Delta}$, then, regardless of $\omega \in P_{n-1}$ verifying $|\xi_{n-1}(\omega)| > e^{-2\Delta}$, we may choose $\Omega = [a_0(\lambda), a(\lambda)]$ so that $f_a^j(\xi_{k_0}(a)) = \xi_{k_0+j}(a) = c_\lambda \in U_{\Delta+1}$ for $a \in \omega \subset \Omega$ and $j \leq n_0(\Delta)$.

On the other hand, since $\xi_i(\omega) \cap U_{\Delta+1} = \emptyset$ for $i = k_0, ..., n-1$, it follows that $n - k_0 < n_0(\Delta)$ for every $\omega \in P_{n-1}^M = \left\{\omega \subset \Omega : |\xi_{n-1}(\omega)| > e^{-2\Delta}\right\}$. Since $n_0(\Delta)$ only depends on Δ, we may state that, fixed $\Delta \in \mathbf{N}$ and $\delta > 0$, there exists $\epsilon > 0$ sufficiently small such that, if $|\Omega| = |a(\lambda) - a_0(\lambda)| < \epsilon$, then $\sup_x \left|f_a^j(x) - f_b^j(x)\right| < \delta$ for each $j = 1, ..., n - k_0$ and for each $a, b \in \omega$. Hence, for any $\omega \in P_{n-1}^M$, $f_{a(\lambda)}^{n-k_0}(\xi_{k_0}(\omega))$ is completely contained in a small neighbourhood of $\xi_n(\omega)$. Thus, once the neighbourhood W given in Proposition 2.2 is fixed and taking into account the fact that $\xi_n(\omega) \subset U_{\frac{\Delta}{4}}$, we get a neighbourhood V of c_λ such that $f_{a(\lambda)}^{n-k_0}(\xi_{k_0}(\omega)) \subset V \subset \overline{V} \subset W$, provided that Δ is large enough. Now, Proposition V.6.1 in [13] furnishes a constant K such that

$$\frac{\left|\left(f_{a(\lambda)}^{n-k_0}\right)'(\xi_{k_0}(a))\right|}{\left|\left(f_{a(\lambda)}^{n-k_0}\right)'(\xi_{k_0}(b))\right|} \leq K$$

for any $a, b \in \omega$. Finally, we get

$$\frac{\left|\left(f_a^{n-k_0}\right)'(\xi_{k_0}(a))\right|}{\left|\left(f_b^{n-k_0}\right)'(\xi_{k_0}(b))\right|} \leq L_{10}. \quad \square$$

2.6 Estimates of the excluded set

Let us denote by $m(A)$ the Lebesgue measure of A. In this section, we shall show that, for each $n \in \mathbf{N}$, the set Ω_n satisfies $m(\Omega_{n-1} \setminus \Omega'_n) \leq e^{-\frac{1}{3}\alpha n} m(\Omega_{n-1})$ and $m(\Omega'_n \setminus \Omega_n) \leq e^{-\frac{1}{100}c\alpha n} |\Omega|$. Therefore, since $m(\Omega_{n-1}) - m(\Omega_n) = m(\Omega_{n-1} \setminus \Omega'_n) + m(\Omega'_n \setminus \Omega_n)$, we shall obtain $m(\Omega_n) \geq \left(1 - e^{-\frac{1}{3}\alpha n}\right) m(\Omega_{n-1}) - e^{-\frac{1}{100}c\alpha n} |\Omega|$ for each $n \in \mathbf{N}$. Then

$$m(\Omega_n) \geq \left[\prod_{j=N}^{n}\left(1 - e^{-\frac{1}{3}\alpha j}\right) - \sum_{j=N}^{n} e^{-\frac{1}{100}c\alpha j}\right] |\Omega| > \frac{1}{2}|\Omega|,$$

provided that N is large enough.

Proposition 2.21. *There exists $N \in \mathbf{N}$ such that for every $n \geq N$*

$$m(\Omega_{n-1} \setminus \Omega'_n) \leq e^{-\frac{1}{3}\alpha n} m(\Omega_{n-1}).$$

Proof. Let $\omega \in P_{n-1}$ and $\omega_{exc} = \{a \in \omega : a \notin \Omega'_n\}$. Whether n is not a return situation or whether n is an inessential return, we have $\omega_{exc} = \emptyset$. Then, suppose that n is an essential return of ω. In this case, ω_{exc} is an interval contained in $U_{[\alpha n]-1}$ and hence, $|\xi_n(\omega_{exc})| \leq 2e^{-\alpha n+2}$.

Let us define $\tilde{\omega} = \xi_n^{-1}(U_{\frac{\Delta}{4}}) \cap \omega$. Then $\omega_{exc} \subset \tilde{\omega}$ and, from the main value theorem and Proposition 2.20, there exist $t_1 \in \tilde{\omega}$ and $t_2 \in \omega_{exc}$ such that

$$\frac{|\omega_{exc}|}{|\omega|} \leq \frac{|\omega_{exc}|}{|\tilde{\omega}|} = \frac{|\xi_n'(t_1)|}{|\xi_n'(t_2)|} \frac{|\xi_n(\omega_{exc})|}{|\xi_n(\tilde{\omega})|} \leq C \frac{|\xi_n(\omega_{exc})|}{|\xi_n(\tilde{\omega})|}.$$

Let us now consider the following cases:

If $\tilde{\omega} = \omega$, then, from Lemma 2.19, $|\xi_n(\tilde{\omega})| = |\xi_n(\omega)| \geq e^{-\frac{1}{2}\alpha n}$.

If $\tilde{\omega} \neq \omega$, then $|\xi_n(\tilde{\omega})| \geq e^{-\frac{1}{4}\Delta} - e^{-\Delta} \geq e^{-\frac{1}{2}\alpha n}$, whenever N is large enough.

In either case, we have proved that there exists $N \in \mathbf{N}$ such that if $n \geq N$, then

$$\frac{|\omega_{exc}|}{|\omega|} \leq 2Ce^{-\frac{1}{2}\alpha n+2} \leq e^{-\frac{1}{3}\alpha n}.$$

Since P_{n-1} is a partition of Ω_{n-1}, it is clear that

$$m\left(\Omega_{n-1} \setminus \Omega_n'\right) = \sum_{\omega \in P_{n-1}} |\omega_{exc}| \leq \sum_{\omega \in P_{n-1}} e^{-\frac{1}{3}\alpha n} |\omega| = e^{-\frac{1}{3}\alpha n} m\left(\Omega_{n-1}\right). \quad \square$$

Next, we shall confirm that $m\left(\Omega_n' \setminus \Omega_n\right) \leq e^{-\frac{1}{100}c\alpha n} |\Omega|$, that is,

$$m\left(\{a \in \Omega_n' : F_n(a) \leq (1-\alpha)n\}\right) \leq e^{-\frac{1}{100}c\alpha n} |\Omega|.$$

We begin by proving a result that shows how often the essential returns take place.

Lemma 2.22. Let ν_0 be an essential return of $\omega \in P_{\nu_0}$, with $\xi_{\nu_0}(\omega) \supset I_{m_0,k_0}$ for $|m_0| \geq \Delta$. If ν_1 is the next essential return situation of ω, then $\nu_1 - \nu_0 \leq 5c^{-1}|m_0|$.

Proof. Let $\mu_1 < ... < \mu_s$ be the inessential returns in (ν_0, ν_1) and $I_{m_1,k_1}^+, ..., I_{m_s,k_s}^+$ be the corresponding host intervals. Set $\mu_0 = \nu_0$, $\mu_{s+1} = \nu_1$ and $\sigma_i = \xi_{\mu_i}(\omega)$ for $i = 0, ..., s+1$. Let $q_i = \mu_{i+1} - (\mu_i + p_i + 1)$ be the length of the free periods associated to the returns μ_i for $i = 0, ..., s$. We consider the two possible cases:

1. Let $s > 0$. Then, from Lemma 2.18(a), it follows that $|\sigma_1| \geq e^{c_0 q_0} e^{-\frac{6\beta}{\beta+c}|m_0|}$ and

$$\frac{|\sigma_{i+1}|}{|\sigma_i|} \geq e^{c_0 q_i} e^{\left(1-\frac{6\beta}{\beta+c}\right)|m_i|}$$

for $1 \leq i \leq s-1$. On the other hand, from $\sigma_s \subset I_{m_s,k_s}^+$ and $|m_s| \geq \Delta$, it follows that $|\sigma_s| \leq 5\Delta^{-2} e^{-\Delta} < e^{-(\Delta+1)}$. In short, writing

$$|\sigma_s| = |\sigma_1| \prod_{i=1}^{s-1} \frac{|\sigma_{i+1}|}{|\sigma_i|},$$

we obtain

$$\exp\left(-(\Delta+1)\right) \geq \exp\left\{-\frac{6\beta}{\beta+c}|m_0| + \sum_{i=0}^{s-1} c_0 q_i + \sum_{i=1}^{s-1}\left(1-\frac{6\beta}{\beta+c}\right)|m_i|\right\},$$

or, what is the same,

$$\frac{6\beta}{\beta+c}|m_0| - (\Delta+1) \geq \sum_{i=0}^{s-1} c_0 q_i + \sum_{i=1}^{s-1}\left(1-\frac{6\beta}{\beta+c}\right)|m_i|. \tag{2.6}$$

Now, we apply Lemma 2.18(b) to get

$$\frac{|\sigma_{s+1}|}{|\sigma_s|} \geq e^{c_0 q_s - \Delta} e^{\left(1-\frac{6\beta}{\beta+c}\right)|m_s|}.$$

Since this lemma also leads to $|\sigma_s| \geq 3^{s-1}|\sigma_1| \geq |\sigma_1|$, we obtain

$$\exp\left\{-\frac{6\beta}{\beta+c}|m_0| + c_0 q_s - \Delta + \left(1 - \frac{6\beta}{\beta+c}\right)|m_s|\right\} \leq \frac{|\sigma_{s+1}|}{|\sigma_s|}|\sigma_1| \leq |\sigma_{s+1}| < e.$$

Then we may write

$$\frac{6\beta}{\beta+c}|m_0| + \Delta + 1 \geq c_0 q_s + \left(1 - \frac{6\beta}{\beta+c}\right)|m_s| \tag{2.7}$$

and, adding (2.6) to (2.7), it follows that

$$\sum_{i=0}^{s} c_0 q_i + \sum_{i=1}^{s}\left(1 - \frac{6\beta}{\beta+c}\right)|m_i| \leq \frac{12\beta}{\beta+c}|m_0|.$$

On the other hand, since $c_0 \leq \log(1.3319) < \frac{5}{17}$, Proposition 2.13 yields

$$p_i + 1 \leq \frac{3}{\beta+c}\frac{|m_i|}{} + \frac{1}{5}\frac{|m_i|}{\beta+c} \leq \frac{1}{c_0}\left(\frac{1 - 6\beta}{\beta+c}\right)|m_i|.$$

Hence, for any $\beta < c$ small enough, we have

$$\nu_1 - \nu_0 \leq \frac{4|m_0|}{\beta+c} + \frac{1}{c_0(\beta+c)^2}\sum_{i=1}^{s}\left(1 - \frac{6\beta}{\beta+c}\right)|m_i| + \frac{1}{c_0(\beta+c)^2}\sum_{i=0}^{s} c_0 q_i \leq$$

$$\leq \frac{4|m_0|}{\beta+c} + \frac{12\beta}{c_0(\beta+c)^3}|m_0| \leq \frac{5}{c}|m_0|.$$

2. Let $s = 0$. Then, once again from Lemma 2.18(b), $|\sigma_1| \geq e^{c_0 q_0 - \Delta}e^{-\frac{6\beta}{\beta+c}|m_0|}$. Now, since $|\sigma_1| < e$, we have $c_0 q_0 \leq 6\beta(\beta+c)^{-1}|m_0| + \Delta + 1$. Then $\nu_1 - \nu_0 = p_0 + q_0 + 1 < 5c^{-1}|m_0|$. $\square$

In order to estimate $m(\Omega_n' \setminus \Omega_n)$ let us go back to the construction of the sets Ω_n and introduce the notion of escape period.

Given $a \in \Omega_n'$, take the sets $\omega_k \in P_k$ such that $a \in \omega_k$ for $k = 1, ..., n - 1$. Let $1 < e_1 < ... < e_s < n - 1$ be all the escape situations of a, that is to say, such that $I_{m,k} \subset \xi_{e_i}(\omega_{e_i}) \subset I_{m,k+1}$ for some interval $I_{m,k}$ with $|m| = \Delta - 1$. Set $e_0 = 1$, $e_{s+1} = n$ and, for $i = 1, ..., s + 1$, $E_i(a) = e_i - \hat{\mu}_i$, where $\hat{\mu}_i$ is the first return situation after e_{i-1}.

Definition 2.23. *Each set of iterates $\{e_{i-1}, ..., \hat{\mu}_i - 1\}$ is called an escape period.*

Notice that each escape period consists only of free iterates. So, if we denote $T_n(a) = E_1(a) + ... + E_{s+1}(a)$, then $n - T_n(a) \leq F_n(a)$. Thus, to prove that

$$m(\{a \in \Omega_n' : F_n(a) \leq (1 - \alpha)n\}) \leq e^{-\frac{1}{100}c\alpha n}|\Omega|$$

it suffices to demonstrate that

$$m\left(\{a \in \Omega'_n : T_n(a) \ge \alpha n\}\right) \le e^{-\frac{1}{100}c\alpha n} |\Omega|.$$

In order to prove this last inequality notice that no return situation $\hat{\mu}_i$ following the escape situation e_{i-1} depends on $a \in \omega^{i-1}$. Hence, for each $\omega \in P'_n$ we may define $E_i(\omega)$ and $T(\omega) = E_1(\omega^1) + ... + E_{s+1}(\omega^{s+1})$, where $e_{s+1} = n$ and $\omega^{s+1} = \omega$. In short, for each $\omega \in P'_n$ we have defined the sequences $1 = e_0 < e_1 < ... < e_s < e_{s+1} = n$ with $e_i = e_i(\omega)$ and $s = s(\omega, n) < n$, $N = \hat{\mu}_1 < \hat{\mu}_2 < ... < \hat{\mu}_s < \hat{\mu}_{s+1} \le n$ with $\hat{\mu}_i = \hat{\mu}_i(\omega)$, $\Omega = \omega^0 \supset \omega^1 \supset ... \supset \omega^s \supset \omega^{s+1} = \omega$ with $\omega^i = \omega_{e_i}(\omega)$ and the function $T_n(\omega) = \sum\limits_{i=1}^{s+1} E_i(\omega^i)$. Here, $e_i < ... < e_s$ are the escape situations, $\omega^1, ..., \omega^s$ are the respective escape components and $\hat{\mu}_i$ is the first return situation of ω^{i-1} after e_{i-1}.

Let us complete the previous sequences for $s+2, ..., n$ by defining $e_i = \hat{\mu}_i = n$, $\omega^i = \omega$ and $E_i(\omega^i) = 0$. So, if we do $T_j(\omega) = \sum\limits_{i=1}^{j} E_i(\omega^i)$ for $j = 1, ..., n$, then the value of $T_n(\omega)$ coincides with the one given just above. It will be also helpful to take the partition $\tilde{P}_i$ being made up of escape components ω^i in the i-th escape situation of all $\omega \in P'_n$. Then, we define $\tilde{\Omega}_i = \bigcup\limits_{\omega^i \in \tilde{P}_i} \omega^i$. In order to extend this definition to all $i \le n$, we consider $\omega^i = \omega$ for $i \ge s+1$.

Let ω^{i-1} be the element of $\tilde{P}_{i-1}$ which contains $\omega \in P'_n$. Let us define the set $\langle \omega^{i-1} \rangle = \{a \in \omega^{i-1} : a \in \tilde{\omega}^i, \text{ with } \tilde{\omega} \subset \tilde{\omega}^i \text{ for some } \tilde{\omega} \in P'_n\} = \omega^{i-1} \cap \tilde{\Omega}_i$. We shall see that the function $E_i(a)$ (not necessarily constant on $\langle \omega^{i-1} \rangle$) satisfies the following capital property:

Lemma 2.24. For $i = 1, ..., n$ it follows that

$$\int_{\langle \omega^{i-1} \rangle} e^{\frac{1}{40}cE_i(a)} da \le e^{\frac{1}{100}c\alpha} |\omega^{i-1}|.$$

Before proving this lemma, we deduce the following proposition, which completes the proof of Theorem 2.1:

Proposition 2.25.

$$m\left(\{a \in \Omega'_n : T_n(a) \ge \alpha n\}\right) \le e^{-\frac{1}{100}c\alpha n} |\Omega|.$$

Proof. Since

$$e^{\frac{1}{40}c\alpha n}m\left(\{a \in \Omega'_n : T_n(a) \geq \alpha n\}\right) \leq \int_{\Omega'_n} e^{\frac{1}{40}cT_n(a)}da,$$

it suffices to prove that

$$\int_{\Omega'_n} e^{\frac{1}{40}cT_n(a)}da = \int_{\widetilde{\Omega}_n} e^{\frac{1}{40}cT_n(a)}da \leq e^{\frac{1}{100}c\alpha n}|\Omega|.$$

Let us proceed by induction: If $n = 1$, then $\widetilde{\Omega}_1 = \langle\Omega\rangle = \langle\omega^0\rangle$ and $T_1(a) = E_1(a)$. Therefore, from Lemma 2.24, we obtain

$$\int_{<\omega^0>} e^{\frac{1}{40}cE_1(a)}da \leq e^{\frac{1}{100}c\alpha}|\omega^0|.$$

Assume now that the result is true for $j - 1$. Then, since $\widetilde{\Omega}_j = \bigcup_{\omega' \in \widetilde{\Omega}_{j-1}} \langle\omega'\rangle$ and

$$T_j(a) = T_{j-1}(a) + E_j(a) = T_{j-1}(\omega') + E_j(a),$$

we apply Lemma 2.24 to obtain

$$\int_{\widetilde{\Omega}_j} e^{\frac{1}{40}cT_j(a)}da = \sum_{\omega' \in \widetilde{\Omega}_{j-1}} \int_{<\omega'>} e^{\frac{1}{40}cT_j(a)}da - \sum_{\omega' \in \widetilde{\Omega}_{j-1}} e^{\frac{1}{40}cT_{j-1}(\omega')} \int_{<\omega'>} e^{\frac{1}{40}cE_j(a)}da \leq$$

$$\leq e^{\frac{1}{100}c\alpha} \int_{\Omega'_{j-1}} e^{\frac{1}{40}cT_{j-1}(a)}da \leq e^{\frac{1}{100}c\alpha j}|\Omega|. \quad \square$$

Proof of Lemma 2.24. Let e_{i-1} be an escape situation of $\omega \in P'_n$, $\omega^{i-1} = \omega_{e_{i-1}}$ be the respective escape component and $\hat{\mu}_i$ the first return situation of ω^{i-1} after e_{i-1}. If $\hat{\mu}_i = n$, then $E_i(a) = 0$ for each $a \in \omega^{i-1}$ and the lemma is obvious because $\langle\omega^{i-1}\rangle \subset \omega^{i-1}$. So, assume that $\hat{\mu}_i < n$. Then, $\hat{\mu}_i$ is necessarily an essential return situation. Indeed, if $i = 1$ remember the choice of N in Proposition 2.5, otherwise see Lemma 2.17.

Therefore, on constructing the set $\Omega_{\hat{\mu}_i}$, the interval ω^{i-1} is split into disjoint intervals $\omega_{m,k}$. Now, each $\omega_{m,k}$ belonging to $P'_{\hat{\mu}_i}$ verifies the next lemma, whose proof will be postponed to the end of this chapter.

Lemma 2.26. Let $\langle\omega_{m,k}\rangle = \omega_{m,k} \cap \widetilde{\Omega}_i$. If $t > 10c^{-1}|m|$, then

$$m\left(\{a \in \langle\omega_{m,k}\rangle : E_i(a) = t\}\right) \leq e^{-\frac{1}{20}ct}|\omega_{m,k}|.$$

Now we continue with the proof of Lemma 2.24. Denote by Υ the set of subscripts (m, k) such that $\omega_{m,k} \in P_{\widehat{\mu}_i}$. Then $\langle \omega^{i-1} \rangle = \bigcup_{(m,k) \in \Upsilon} \langle \omega_{m,k} \rangle$ and we have

$$\int_{<\omega^{i-1}>} e^{\frac{1}{40}cE_i(a)} da \leq \sum_{(m,k) \in \Upsilon} \int_{<\omega_{m,k}>} e^{\frac{1}{40}cE_i(a)} da = S_1 + S_2 + S_3,$$

where

$$S_1 = \sum_{(m,k) \in \Upsilon} m\left(\{a \in \langle \omega_{m,k} \rangle : E_i(a) = 0\}\right)$$

$$S_2 = \sum_{(m,k) \in \Upsilon} \sum_{t=1}^{10c^{-1}|m|} e^{\frac{1}{40}ct} m\left(\{a \in \langle \omega_{m,k} \rangle : E_i(a) = t\}\right),$$

$$S_3 = \sum_{(m,k) \in \Upsilon} \sum_{t > 10c^{-1}|m|} e^{\frac{1}{40}ct} m\left(\{a \in \langle \omega_{m,k} \rangle : E_i(a) = t\}\right).$$

Next, we shall bound each S_i. On the one hand, we have

$$S_1 \leq \sum_{(m,k) \in \Upsilon} m\left(\langle \omega_{m,k} \rangle\right) \leq \sum_{(m,k) \in \Upsilon} |\omega_{m,k}| \leq \left|\omega^{i-1}\right|.$$

On the other, for $a \in \langle \omega_{m,k} \rangle$, $E_i(a) > 0$ if and only if $\omega_{m,k}$ is not an escape component. Hence, $|m| \geq \Delta$ and therefore

$$S_2 \leq \sum_{\substack{(m,k) \in \Upsilon \\ |m| \geq \Delta}} \sum_{t=1}^{10c^{-1}|m|} e^{\frac{1}{4}|m|} m\left(\langle \omega_{m,k} \rangle\right) \leq \sum_{\substack{(m,k) \in \Upsilon \\ |m| \geq \Delta}} 10c^{-1} |m| e^{\frac{1}{4}|m|} |\omega_{m,k}|.$$

Let us define $\widetilde{\omega}^{i-1} = \xi_{\widehat{\mu}_i}^{-1}(U_{\frac{\Delta}{4}}) \cap \omega^{i-1}$. Then, there exist $a \in \widetilde{\omega}^{i-1}$ and $b \in \omega_{m,k}$ such that

$$\frac{|\omega_{m,k}|}{|\omega^{i-1}|} \leq \frac{|\omega_{m,k}|}{|\widetilde{\omega}^{i-1}|} = \frac{\left|\xi'_{\widehat{\mu}_i}(a)\right|}{\left|\xi'_{\widehat{\mu}_i}(b)\right|} \frac{\left|\xi_{\widehat{\mu}_i}(\omega_{m,k})\right|}{\left|\xi_{\widehat{\mu}_i}(\widetilde{\omega}^{i-1})\right|}.$$

Notice that, since $|m| \geq \Delta$, $\omega_{m,k} \subset \widetilde{\omega}^{i-1} \subset \omega^{i-1}$. Furthermore, since $\widehat{\mu}_i$ is the first return situation after e_{i-1}, it follows that $\omega^{i-1} \in P_{\widehat{\mu}_i - 1}$. Therefore, we may apply Proposition 2.20 to get

$$\frac{|\omega_{m,k}|}{|\omega^{i-1}|} \leq C \frac{\left|\xi_{\widehat{\mu}_i}(\omega_{m,k})\right|}{\left|\xi_{\widehat{\mu}_i}(\widetilde{\omega}^{i-1})\right|} \leq \frac{5Ce^{-|m|}}{m^2 \left|\xi_{\widehat{\mu}_i}(\widetilde{\omega}^{i-1})\right|},$$

where we are taking into account the fact that

$$\left|\xi_{\widehat{\mu}_i}(\omega_{m,k})\right| \leq \left|I_{m,k}^+\right| \leq 5m^{-2}e^{-|m|}.$$

If $\widetilde{\omega}^{i-1} \neq \omega^{i-1}$, then

$$\left|\xi_{\widehat{\mu}_i}(\widetilde{\omega}^{i-1})\right| \geq e^{-\frac{1}{4}\Delta} - e^{-\Delta} > e^{-\frac{1}{2}\Delta}.$$

Otherwise, from Lemma 2.17, we obtain

$$\left|\xi_{\widehat{\mu}_i}(\widetilde{\omega}^{i-1})\right| = \left|\xi_{\widehat{\mu}_i}(\omega^{i-1})\right| \geq e^{-\frac{1}{2}\Delta}.$$

Therefore, in either case,

$$|\omega_{m,k}| \leq 5Cm^{-2}e^{-|m|+\frac{1}{2}\Delta}\left|\omega^{i-1}\right|$$

holds and we conclude that

$$S_2 \leq \left|\omega^{i-1}\right| \sum_{|m|=\Delta}^{\infty} \sum_{k=1}^{m^2} \frac{50C\,|\,m\,|}{cm^2}e^{\frac{1}{2}\Delta-\frac{3}{4}|m|} \leq L(\Delta)\left|\omega^{i-1}\right|,$$

where $L(\Delta)$ is a constant which approaches zero as $\Delta \to \infty$.

Finally, from Lemma 2.26,

$$S_3 \leq \sum_{(m,k)\in\Upsilon} \sum_{t>10c^{-1}|m|} e^{-\frac{1}{40}ct}|\omega_{m,k}| \leq \sum_{(m,k)\in\Upsilon} e^{-\frac{1}{4}|m|}|\omega_{m,k}| \leq e^{-\frac{1}{4}\Delta}\left|\omega^{i-1}\right|$$

and so

$$\int_{\langle\omega^{i-1}\rangle} e^{\frac{1}{40}cE_i(a)}da \leq \left|\omega^{i-1}\right| + L(\Delta)\left|\omega^{i-1}\right| + e^{-\frac{1}{4}\Delta}\left|\omega^{i-1}\right| \leq e^{\frac{1}{100}c\alpha}\left|\omega^{i-1}\right|,$$

provided that Δ is large enough. □

Proof of Lemma 2.26. Let $a \in \langle\omega_{m,k}\rangle$ be such that $E_i(a) = t$. By the definitions of $\langle\omega_{m,k}\rangle$ and $E_i(a)$, there exists $\omega \in P_n'$ such that $a \in \omega \subset \omega_{m,k}$ and ω has its i-th escape situation at $e_i = \widehat{\mu}_i + t$. Now, regardless of whether ω has its i-th escape situation e_i exactly at $\widehat{\mu}_i + t$ or not, we define for each $\omega \subset \omega_{m,k}$ the sequence $\widehat{\mu}_1 = \nu_0(\omega) < \nu_1(\omega) < ... < \nu_s(\omega) < e_i$ of all the essential return situations which are not escape situations. Denote by $\omega_j = \omega_{(m_0,k_0)...(m_j,k_j)}$ the component in $P_{\nu_j(\omega)}$ which contains ω and by I_{m_j,k_j}^+ its host interval. Of course, s depends on $\omega \in P_n'$. Given $s \in \mathbb{N}$ and $M \in \mathbb{N}$, let $\eta_s(M)$ be the number of all the components ω_s for all $\omega \in P_n'$, such that $|m_1| + ... + |m_s| = M$. Let $\eta(M) = \sum_{s\geq 0} \eta_s(M)$. We shall check that

$$\eta(M) \leq e^{\frac{1}{8}M}, \tag{2.8}$$

$$|\omega_s| \leq |\omega_{m,k}| \exp\left\{\frac{6\beta}{\beta+c}|m| - \frac{7}{8}M\right\}. \tag{2.9}$$

Then, for each set

$$X_M = \left\{a \in \omega_{m,k} : a \in \omega_s = \omega_{(m_0,k_0)...(m_s,k_s)}, \text{with } |m_1| + ... + |m_s| = M\right\},$$

it follows that

$$m\left(X_M\right) = \sum_{s \geq 0} \eta_s(M) \left|\omega_s\right| \leq \eta(M) \left|\omega_{m,k}\right| \exp\left\{\frac{6\beta}{\beta+c} \left|m\right| - \frac{7}{8}M\right\} \leq$$

$$\leq \left|\omega_{m,k}\right| \exp\left\{-\frac{3}{4}M + \frac{6\beta}{\beta+c} \left|m\right|\right\}.$$

Now, let us define $Y_{M,t} = X_M \cap \{a \in \langle\omega_{m,k}\rangle : E_i(a) = t\}$. Notice that if $t \geq 5c^{-1}\left(M + |m|\right)$, then $Y_{M,t} = \emptyset$. In fact, if $a \in \omega_{(m_0,k_0)...(m_s,k_s)}$ has an escape situation at $\nu_{s+1} = e_i = \widehat{\mu}_i + t$, then, from Lemma 2.22, it follows that

$$t = e_i - \widehat{\mu}_i = \sum_{i=0}^{s} \left(\nu_{i+1} - \nu_i\right) \leq \frac{5}{c} \sum_{i=0}^{s} |m_i| = \frac{5}{c}(M + |m|).$$

On the other hand, it is clear that

$$\bigcup_{M \geq \frac{1}{5}tc - |m|} Y_{M,t} = \{a \in \langle\omega_{m,k}\rangle : E_i(a) = t\}.$$

Indeed, if $a \in \langle\omega_{m,k}\rangle$ and $E_i(a) = t$, then a belongs to some ω_s. Therefore,

$$m\left(\{a \in \langle\omega_{m,k}\rangle : E_i(a) = t\}\right) \leq \sum_{M \geq \frac{1}{5}tc - |m|} m\left(X_M\right) \leq$$

$$\leq \left|\omega_{m,k}\right| e^{\frac{6\beta}{\beta+c}|m|} \sum_{M \geq \frac{1}{5}tc - |m|} e^{-\frac{3}{4}M} \leq \left|\omega_{m,k}\right| e^{|m| - \frac{3}{20}tc},$$

whenever $|m| \geq \Delta$ and Δ is large enough.

Finally, since $t > 10c^{-1}|m|$, it follows that

$$m\left(\{a \in \langle\omega_{m,k}\rangle : E_i(a) = t\}\right) \leq \left|\omega_{m,k}\right| e^{-\frac{1}{20}ct}.$$

Now, it only remains to prove (2.8) and (2.9):

To bound $\eta(M)$, notice that each component ω_s is determined by the sequence $(m_1, k_1)...(m_s, k_s)$. So, there exist, at the most, $2^s |m_1|^2 ... |m_s|^2$ components of the form $\omega_{(m_0,k_0)...(m_s,k_s)}$. By taking Δ sufficiently large so that $m_i^2 \leq e^{\frac{1}{20}m_i}$, we have $2^s |m_1|^2 ... |m_s|^2 \leq 2^s e^{\frac{1}{20}(|m_1|+...+|m_s|)}$. On the other hand, since $|m_i| \geq \Delta$, it follows that $s\Delta \leq |m_1| + ... + |m_s|$. Hence,

$$\eta_s(M) \leq \rho_s(M) 2^{\Delta^{-1}M} e^{\frac{1}{20}M} \leq \rho_s(M) e^{\frac{1}{16}M},$$

where $\rho_s(M)$ is the number of solutions of the equation $M = |m_1| + ... + |m_s|$. Thus

$$\rho_s(M) \leq \binom{M + s - 1}{s - 1}.$$

If $s \geq 2$, use the Stirling's formula to get

$$\rho_s(M) \leq \frac{e}{\sqrt{2\pi}} \left(\frac{M+s-1}{M(s-1)}\right)^{\frac{1}{2}} \left(\frac{M+s-1}{s-1}\right)^{s-1} \left(\frac{M+s-1}{M}\right)^{M} \leq$$

$$\leq \frac{2e}{\sqrt{2\pi}} \left[\left(\frac{s-1}{M+s-1}\right)^{(1-s)(M+s-1)^{-1}} \left(1 - \frac{s-1}{M+s-1}\right)^{(s-1)(M+s-1)^{-1}-1}\right]^{M+s-1}.$$

Since $(s-1)(M+s-1)^{-1} < sM^{-1} < \Delta^{-1}$ approaches zero as $\Delta \to \infty$, we may take

$$\rho_s(M) \leq 3e^{\frac{1}{32}M},$$

as long as Δ is large enough. This inequality also holds if $s = 1$.

To obtain (2.8), check that

$$\eta(M) \leq \sum_{s=1}^{M\Delta^{-1}} \eta_s(M) \leq \sum_{s=1}^{M\Delta^{-1}} \rho_s(M)e^{\frac{1}{16}M} \leq e^{\frac{1}{8}M}.$$

In order to prove (2.9), we take a neighbourhood $\hat{U}$ of c_λ such that $\hat{U} \subsetneq W$, where W is the neighbourhood of c_λ given by Proposition 2.2. Then, since Proposition 2.20 remains valid under the hypothesis $\xi_n(\omega) \subset \hat{U}$, we obtain that

$$\frac{|\omega_s|}{|\omega_0|} = \prod_{i=1}^{s} \frac{|\omega_i|}{|\omega_{i-1}|} \leq \prod_{i=1}^{s} C \frac{|\xi_{\nu_i}(\omega_i)|}{|\xi_{\nu_i}(\tilde{\omega}_{i-1})|},$$

where $\tilde{\omega}_{i-1} = \omega_{i-1} \cap \xi_{\nu_i}^{-1}(\hat{U})$. On the one hand, we have often seen that

$$|\xi_{\nu_i}(\omega_i)| \leq \left|I_{m_i,k_i}^{+}\right| \leq 5m_i^{-2}e^{-|m_i|}$$

and on the other, we shall check that

$$|\xi_{\nu_i}(\tilde{\omega}_{i-1})| \geq \exp\left\{-\frac{6\beta}{\beta+c}|m_{i-1}|\right\}.$$

Actually, if $\tilde{\omega}_{i-1} \neq \omega_{i-1}$, then

$$|\xi_{\nu_i}(\tilde{\omega}_{i-1})| \geq const - e^{-\Delta} \geq \exp\left\{-\frac{6\beta}{\beta+c}|m_{i-1}|\right\}.$$

If $\tilde{\omega}_{i-1} = \omega_{i-1}$, let us denote by $\nu_{i-1} < \mu_1 < ... < \mu_r < \nu_i$ all the returns of ω between ν_{i-1} and ν_i. Notice that all these returns are inessential and they coincide with the inessential return situations of ω_{i-1} when this is taken as an element of $P_{\nu_{i-1}}$. These return situations are also returns of $\omega_{i-1} \in P_{\nu_i-1}$. From Lemma 2.18(a), we deduce that

$$|\xi_{\nu_i}(\omega_{i-1})| \geq 3^r |\xi_{\mu_1}(\omega_{i-1})| \geq \exp\left\{-\frac{6\beta}{\beta+c}|m_{i-1}|\right\}.$$

Finally,

$$\frac{|\omega_s|}{|\omega_0|} \leq \prod_{i=1}^{s} \exp\left\{\frac{6\beta}{\beta+c}\, |m_{i-1}| - |m_i|\right\} \leq \exp\left\{\frac{6\beta}{\beta+c}\, |m_0| - \left(1 - \frac{6\beta}{\beta+c}\right) \sum_{i=1}^{s} |m_i|\right\}$$

and thus, since $m = m_0$ and $\omega_0 = \omega_{m,k}$, it follows that

$$|\omega_s| \leq |\omega_{m,k}| \exp\left\{\frac{6\beta}{\beta+c}\, |m| - \frac{7}{8}M\right\}.$$

So, Lemma 2.26 holds and Theorem 2.1 is proved.

Chapter 3

CONTRACTIVE DIRECTIONS

Let us take again the family of maps

$$T_{\lambda,a,b} = \left(f_{\lambda,a}(x) + \frac{1}{\lambda} \log \left(1 + \sqrt{b}y\right), \sqrt{b}\left(1 + \sqrt{b}y\right) e^{\lambda x} \sin x \right),$$

defined on the domains $U_m = \left\{ (x,y) \in \mathbf{R}^2 : x \in I, -\sqrt{b}e^{\lambda l}e^{-\frac{1}{2}\pi\lambda} \leq y \leq \sqrt{b}e^{\lambda l}e^{\frac{1}{2}\pi\lambda} \right\}$, where $b = e^{-2\pi\lambda m}$ and $f_{\lambda,a}$ is the family of unimodal maps studied in Chapter 2. The aim of this chapter is to find, for each $T_{\lambda,a,b}$, directions which are maximally contracting for the differential map $DT_{\lambda,a,b}$. This goal is trivially reached in the case of $\Psi_{\lambda,a} = (f_{\lambda,a}, 0)$. So, we begin by recalling the existence of a positive constant $K = K(\lambda)$ such that

$$\|T_{\lambda,b} - \Psi_\lambda\|_{C^3(\widetilde{D})} \leq K\sqrt{b}, \qquad (3.1)$$

as proved in Proposition 1.3. An immediate consequence of (3.1) is

$$\|T_{\lambda,b}\|_{C^3(\widetilde{D})} \leq K. \qquad (3.2)$$

From Proposition 1.3 it also follows that $T_{\lambda,a,b}$ is strongly contractive as $b << 1$. We shall often use that

$$|\det(DT_{\lambda,a,b}(x,y))| \leq Kb, \qquad (3.3)$$

for every $(x,y) \in U_m$. From now on, we shall write $z_j = T^j_{\lambda,a,b}(z)$ for each $z \in U_m$ and $j \geq 0$. Let us set $M^\nu(z) = DT^\nu_{\lambda,a,b}(z) = DT_{\lambda,a,b}(z_{\nu-1})...DT_{\lambda,a,b}(z_1)DT_{\lambda,a,b}(z)$.

Definition 3.1. *A point $z \in U_m$ is said $\xi-$expanding up to time n, if $\|\omega_\nu\| \geq \xi^\nu$ for $1 \leq \nu \leq n$ where $\omega_\nu = \omega_\nu(z) = M^\nu(z)(1,0)$.*

Remark 3.2. *Thought one of the main objectives in this book is the search of expanding points, that is, ξ- expanding for any $n \in \mathbf{N}$ and $\xi > 1$, we shall need results* *about the ξ-expansiveness for $\xi < 1$. So, we shall assume that $\sqrt{b} << \xi < \sqrt{K}$, where K is large enough and b is sufficiently small so that all the forthcoming statemets are true.*

Let z be ξ-expanding up to time n. For $1 \leq \nu \leq n$ we shall construct unit vectors $e^{(\nu)} = e^{(\nu)}(z)$ and $f^{(\nu)} = f^{(\nu)}(z)$ which will be, respectively, maximally contracting and maximally expanding for $M^{\nu}(z)$. For this, we need to solve the equation

$$\frac{d}{d\theta} \left\| M^{\nu}(z)(\cos\theta, \sin\theta) \right\| = 0.$$

If we remove z and write

$$M^{\nu} = \begin{pmatrix} A_{\nu} & B_{\nu} \\ C_{\nu} & D_{\nu} \end{pmatrix}, \tag{3.4}$$

then the solutions are given by

$$\text{tg}\,(2\theta) = \frac{2(A_{\nu}B_{\nu} + C_{\nu}D_{\nu})}{(A_{\nu}^2 + C_{\nu}^2) - (B_{\nu}^2 + D_{\nu}^2)}.$$

Consequently, $e^{(\nu)}$ and $f^{(\nu)}$ are orthogonal and the same holds for their images $M^{\nu}e^{(\nu)}$ and $M^{\nu}f^{(\nu)}$ which are, respectively, maximally expanding and maximally contracting for $(M^{\nu})^{-1} = M^{-\nu}(z_{\nu})$. Then, it follows that $\left\| M^{\nu}e^{(\nu)} \right\| \left\| M^{\nu}f^{(\nu)} \right\| = \left| \det M^{\nu}(z_1) \right| \leq (Kb)^{\nu}$ and, since $\left\| M^{\nu}f^{(\nu)} \right\| \geq \left\| \omega_{\nu} \right\| \geq \xi^{\nu}$, we obtain

$$\left\| M^{\nu}e^{(\nu)} \right\| \leq \left(\frac{Kb}{\xi} \right)^{\nu}. \tag{3.5}$$

This last inequality shows that the direction $e^{(\nu)}$ is exponentially contracted by M^{ν}. Next, we shall prove that it is also exponentially contracted by M^j for $j < \nu$. To this end, denote by $\text{ang}(u, v)$ the angle between the directions of two vectors u and v.

Proposition 3.3. *Let $z \in R_m$ be ξ-expanding up to time n. There exists a constant $K_3 = K_3(K, \xi)$ such that:*
(a) $\left| \text{ang}\left(e^{(\mu)}, e^{(\nu)} \right) \right| \leq (K_3 b)^{\mu}$ for $1 \leq \mu \leq \nu \leq n$.
(b) $\left\| M^{\mu}(z)e^{(\nu)} \right\| \leq (K_3 b)^{\mu}$ for $1 \leq \mu \leq \nu \leq n$.

Proof. The result is evident for $\nu = 1$. Suppose that $\nu \geq 2$ and write $e^{(\nu-1)} = \alpha e^{(\nu)} + \beta f^{(\nu)}$. Then, since $M^{\nu}e^{(\nu)}$ and $M^{\nu}f^{(\nu)}$ are orthogonal, it follows that $\left\| M^{\nu}e^{(\nu-1)} \right\|^2 = \alpha^2 \left\| M^{\nu}e^{(\nu)} \right\|^2 + \beta^2 \left\| M^{\nu}f^{(\nu)} \right\|^2$. From (3.5) and the ξ-expansiveness of z we obtain

$$|\beta|\,\xi^{\nu} \leq |\beta| \left\| M^{\nu}f^{(\nu)} \right\| \leq K \left(\frac{Kb}{\xi} \right)^{\nu-1}$$

and hence,

$$|\beta| \leq \frac{K}{\xi}\left(\frac{Kb}{\xi^2}\right)^{\nu-1}.$$

Therefore, since $\sqrt{b} << \xi$, we have $|\beta| << 1$ and thus $|\alpha| > \frac{1}{2}$. Then

$$\left|\text{ang}\left(e^{(\nu-1)}, e^{(\nu)}\right)\right| < \frac{2K}{\xi}\left(\frac{Kb}{\xi^2}\right)^{\nu-1}.$$

Now, for $1 \leq \mu < \nu \leq n$, we have

$$\left|\text{ang}\left(e^{(\mu)}, e^{(\nu)}\right)\right| \leq \sum_{i=\mu}^{\nu-1} \frac{2K}{\xi}\left(\frac{Kb}{\xi^2}\right)^i \leq \frac{3K}{\xi}\left(\frac{Kb}{\xi^2}\right)^\mu \leq \left(\frac{3K^2 b}{\xi^3}\right)^\mu$$

and statement (a) is proved.

On the other hand, from (3.5) and statement (a), it follows that

$$\left\|M^\mu e^{(\nu)}\right\| \leq K^\mu \left|\text{ang}\left(e^{(\mu)}, e^{(\nu)}\right)\right| + \left(\frac{Kb}{\xi}\right)^\mu < \frac{4K}{\xi}\left(\frac{K^2 b}{\xi^2}\right)^\mu \leq (K_3 b)^\mu$$

and (b) holds. $\square$

Proposition 3.3 states that if z is a ξ-expanding point for every $n \in \mathbf{N}$, then the contracting directions $e^{(n)}(z)$ converge very quickly and the limit direction is exponentially contracted by all the positive iterates of $M(z)$.

Next, we shall prove that maximally expanding directions are nearly horizontal and thus, maximally contracting directions are nearly vertical. Moreover, any vector in a nearly horizontal direction is also expanded.

Proposition 3.4. *Let $z = (x, y) \in U_m$ and $\mathbf{u}_0$ be a unit vector such that $\|M^\nu(z)\mathbf{u}_0\| \geq \xi^\nu$ for $1 \leq \nu \leq n$. Then:*

(a) There exists a constant $K_4 = K_4(K, \xi)$ such that

$$\left|\text{slope } f^{(\nu)}\right| \leq K_4\sqrt{b} \text{ for } 1 \leq \nu \leq n.$$

(b) For any unit vector $\mathbf{u}$ such that $|\text{slope } \mathbf{u}| \leq \frac{1}{10}$ it follows that

$$\|M^\nu(z)\mathbf{u}\| \geq \frac{1}{2}\|M^\nu\| \text{ for } 1 \leq \nu \leq n.$$

Proof. From (3.1) we have $\|M(z)\mathbf{u}_0\| - \|D\Psi_{\lambda,a}(z)\mathbf{u}_0\| \leq K\sqrt{b}$ and consequently $\|D\Psi_{\lambda,a}(z)\mathbf{u}_0\| \geq \xi - K\sqrt{b}$. Therefore, according to the notation used in (3.4),

$$|A_1| = \frac{|\lambda - \text{tg } x|}{\lambda} \geq \xi - K\sqrt{b} \geq \frac{\xi}{5}.$$

Let us write $e^{(1)} = (\alpha_1, \alpha_2)$. From Proposition 1.3 we obtain $\left\| M(z)e^{(1)} \right\| \geq |A_1||\alpha_1| - K\sqrt{b}$ and so, from (3.5), it follows that

$$|\alpha_1| \leq \frac{1}{|A_1|} \left(\frac{Kb}{\xi} + K\sqrt{b} \right) \leq \frac{10K\sqrt{b}}{\xi} << 1.$$

In short, $|\alpha_2| > \frac{1}{2}$, $\left| \text{slope } f^{(1)} \right| \leq 20K\xi^{-1}\sqrt{b}$ and statement (a) is proved for $\nu = 1$.

For $\nu > 1$, let us go back to the proof of Proposition 3.3(a). We had

$$\left| \text{ang} \left(f^{(\nu)}, (1,0) \right) \right| \leq \frac{3K^2 b}{\xi^3} + \frac{20K\sqrt{b}}{\xi} \leq \frac{21K\sqrt{b}}{\xi}.$$

Thus, $\left| \text{slope } f^{(\nu)} \right| \leq 42K\xi^{-1}\sqrt{b} = K_4\sqrt{b}$.

To prove statement (b), let us take an arbitrary unit vector $\mathbf{u} = \alpha e^{(\nu)} + \beta f^{(\nu)}$ such that $|\text{slope } \mathbf{u}| \leq \frac{1}{10}$. For a sufficiently small b we have $|\alpha| \leq \left| \text{ang} \left(\mathbf{u}, f^{(\nu)} \right) \right| \leq \frac{1}{10} + K_4\sqrt{b} \leq \frac{1}{4}$. Then, $|\beta| \geq \frac{1}{2}$ and, since $M^\nu(z)e^{(\nu)}$ and $M^\nu(z)f^{(\nu)}$ are orthogonal, we obtain $\|M^\nu(z)\mathbf{u}\|^2 \geq \beta^2 \left\| M^\nu(z)f^{(\nu)} \right\|^2 = \beta^2 \|M^\nu(z)\|^2$ and consequently $\|M^\nu(z)\mathbf{u}\| \geq \frac{1}{2} \|M^\nu(z)\|$. $\square$

Remark 3.5. *Under the hypotheses of Proposition 3.4, we have*

$$\|M^\nu(z)\| = \left\| M^\nu(z)f^{(\nu)} \right\| \geq \|M^\nu(z)u_0\| \geq \xi^\nu.$$

Therefore, $M^\nu(z)$ expands up to time n every unit vector with slope less than $\frac{1}{10}$.

Proposition 3.6. *Let z_0, $\zeta_0 \in U_m$ and let $\mathbf{u}$ and $\mathbf{v}$ be unit vectors such that $\|z_0 - \zeta_0\| \leq \sigma^n$ and $\|\mathbf{u} - \mathbf{v}\| \leq \sigma^n$ for some $\sigma \leq \left(\frac{\xi}{10K^2} \right)^2$ and for some $n \in \mathbf{N}$. If, in addition, $\|M^\nu(z_1)\mathbf{u}\| \geq \xi^\nu$ for $1 \leq \nu \leq n$, then:*

(a)

$$\frac{1}{2} \leq \frac{\|M^\nu(z_1)\mathbf{u}\|}{\|M^\nu(\zeta_1)\mathbf{v}\|} \leq 2.$$

(b) $|\text{ang} \left(M^\nu(z_1)\mathbf{u}, M^\nu(\zeta_1)\mathbf{v} \right)| \leq (\sqrt{\sigma})^{2n-\nu} \leq (\sqrt{\sigma})^n$.

Proof. Notice that $\|M^\nu(z_1)\mathbf{u} - M^\nu(\zeta_1)\mathbf{v}\| \leq \|M^\nu(z_1) - M^\nu(\zeta_1)\| + \|M^\nu(\zeta_1)(\mathbf{u} - \mathbf{v})\|$ and $\|M^\nu(z_1) - M^\nu(\zeta_1)\| \leq K \|M^{\nu-1}(z_2) - M^{\nu-1}(\zeta_2)\| + K^\nu \|z_1 - \zeta_1\|$. Repeating the estimates for $\nu - 1$ and so on, we obtain

$$\|M^\nu(z_1) - M^\nu(\zeta_1)\| \leq K^\nu \sum_{j=1}^{\nu} \|z_j - \zeta_j\|. \tag{3.6}$$

Since, by hypothesis, $\|M^{\nu}(\varsigma_1)(\mathbf{u} - \mathbf{v})\| \leq K^{\nu}\sigma^n$, it follows that

$$\|M^{\nu}(z_1)\mathbf{u} - M^{\nu}(\varsigma_1)\mathbf{v}\| \leq K^{\nu}\sigma^n \sum_{j=1}^{\nu} K^j + K^{\nu}\sigma^n \leq 3K^{2\nu}\sigma^n.$$

Furthermore, $\sigma \leq \left(\frac{\xi}{10K^2}\right)^2$ leads to $3K^{2\nu}\sigma^n \leq \frac{1}{2}(\sqrt{\sigma})^{2n-\nu}\xi^{\nu}$ and then

$$\|M^{\nu}(z_1)\mathbf{u} - M^{\nu}(\varsigma_1)\mathbf{v}\| \leq \frac{1}{2}(\sqrt{\sigma})^{2n-\nu}\|M^{\nu}(z_1)\mathbf{u}\|. \tag{3.7}$$

From $\|M^{\nu}(\varsigma_1)\mathbf{v}\| \leq \|M^{\nu}(z_1)\mathbf{u}\| + \|M^{\nu}(z_1)\mathbf{u} - M^{\nu}(\varsigma_1)\mathbf{v}\|$ and Remark 3.2 we get

$$\frac{\|M^{\nu}(\varsigma_1)\mathbf{v}\|}{\|M^{\nu}(z_1)\mathbf{u}\|} \leq 1 + \frac{1}{2}(\sqrt{\sigma})^{2n-\nu} \leq 2.$$

On the other hand, since $\|M^{\nu}(\varsigma_1)\mathbf{v}\| \geq \|M^{\nu}(z_1)\mathbf{u}\| - \|M^{\nu}(z_1)\mathbf{u} - M^{\nu}(\varsigma_1)\mathbf{v}\|$, we may use (3.7) to get $\|M^{\nu}(\varsigma_1)\mathbf{v}\| \geq \xi^{\nu}\left(1 - \frac{1}{2}(\sqrt{\sigma})^{2n-\nu}\right) > \frac{1}{2}(\sqrt{\sigma})^{2n-\nu}\xi^{\nu}$ and

$$\frac{\|M^{\nu}(z_1)\mathbf{u}\|}{\|M^{\nu}(\varsigma_1)\mathbf{v}\|} \leq 1 + \frac{\|M^{\nu}(z_1)\mathbf{u} - M^{\nu}(\varsigma_1)\mathbf{v}\|}{\|M^{\nu}(\varsigma_1)\mathbf{v}\|} \leq 2.$$

So, statement (a) holds.

To prove statement (b) notice that

$$|\sin\left(\text{ang}\left(M^{\nu}(z_1)\mathbf{u}, M^{\nu}(\varsigma_1)\mathbf{v}\right)\right)| \leq \frac{\|M^{\nu}(z_1)\mathbf{u} - M^{\nu}(\varsigma_1)\mathbf{v}\|}{\|M^{\nu}(z_1)\mathbf{u}\|} \leq \frac{1}{2}\left(\sqrt{\sigma}\right)^{2n-\nu}. \quad \square$$

Proposition 3.7. *Let $z_1 \in U_m$ be ξ-expanding up to time n. Let $\sqrt{b} < \sigma < \left(\frac{\xi}{10K^2}\right)^4$. Then, for each $\varsigma_1 \in U_m$ such that $\|z_{\nu} - \varsigma_{\nu}\| \leq \sigma^{\nu}$ for $1 \leq \nu \leq n$, and for every pair of unit vectors $\mathbf{u}$ and $\mathbf{v}$ such that $|\text{slope } \mathbf{u}| \leq \frac{1}{10}$ and $|\text{slope } \mathbf{v}| \leq \frac{1}{10}$, it follows that:*

(a)

$$\frac{1}{2} \leq \frac{\|M^{\nu}(z_1)\mathbf{u}\|}{\|M^{\nu}(\varsigma_1)\mathbf{v}\|} \leq 2.$$

(b) There exists a constant $K_7 = K_7(K, \xi)$ such that

$$|\text{ang}\left(M^{\nu}(z_1)\mathbf{u}, M^{\nu}(\varsigma_1)\mathbf{v}\right)| \leq (K_7\sqrt{\sigma})^{\nu+1}.$$

Moreover, $K_7\sqrt{\sigma} \leq \frac{1}{100}K^{-4}\xi < 1$.

Before proving this result we shall state an auxiliary lemma. To this end, we write $\mathbf{u}_{\nu} = M^{\nu}(z_1)\mathbf{u}$, $\mathbf{v}_{\nu} = M^{\nu}(\varsigma_1)\mathbf{v}$. Notice that, from Proposition 3.4, we have

$$\|\mathbf{u}_{\nu}\| \geq \frac{\xi^{\nu}}{2} \text{ for } 1 \leq \nu \leq n. \tag{3.8}$$

Lemma 3.8. *For $1 \leq \nu \leq n$ we may write $\mathbf{v}_\nu = \alpha_\nu \mathbf{u}_\nu + \mathbf{l}_\nu$, with*

$$|\alpha_\nu - 1| \leq \frac{1}{10} + 5K \sum_{i=2}^{\nu} \left(\frac{K^2 \sqrt{\sigma}}{\xi^2}\right)^i \quad and \quad \|\mathbf{l}_\nu\| \leq (K\sqrt{\sigma})^{\nu+1}. \tag{3.9}$$

Proof of Proposition 3.7. For a sufficiently large K we have

$$5K \sum_{i=2}^{\nu} \left(\frac{K^2 \sqrt{\sigma}}{\xi^2}\right)^i \leq 5 \sum_{i=2}^{\nu} \left(\frac{1}{100K}\right)^i < \frac{1}{10}.$$

Then, from (3.9), it follows that $|\alpha_\nu - 1| \leq \frac{1}{5}$.

On the other hand, since $K\sqrt{\sigma} < \frac{1}{100}K^{-3}\xi^2 < \min\left\{1, \frac{1}{10}\xi\right\}$, (3.8) yields

$$\|\mathbf{l}_\nu\| < \left(\frac{\xi}{10}\right)^\nu \leq \frac{1}{5}\|\mathbf{u}_\nu\|.$$

Therefore,

$$\frac{\|\mathbf{v}_\nu\|}{\|\mathbf{u}_\nu\|} \leq |\alpha_\nu| + \frac{\|\mathbf{l}_\nu\|}{\|\mathbf{u}_\nu\|} < 2.$$

Finally, since

$$\frac{\|\mathbf{v}_\nu\|}{\|\mathbf{u}_\nu\|} \geq |\alpha_\nu| - \frac{\|\mathbf{l}_\nu\|}{\|\mathbf{u}_\nu\|} \geq \frac{1}{2},$$

statement (a) holds.

By means of (3.8) and (3.9) we get

$$|\sin(\text{ang}(\mathbf{u}_\nu, \mathbf{v}_\nu))| \leq \frac{\|\mathbf{l}_\nu\|}{\|\mathbf{v}_\nu\|} \leq \frac{4\|\mathbf{l}_\nu\|}{\xi^\nu} \leq \frac{4(K\sqrt{\sigma})^{\nu+1}}{\xi^\nu} \leq \left(\frac{K^2\sqrt{\sigma}}{\xi}\right)^{\nu+1}$$

and statement (b) follows. $\square$

Proof of Lemma 3.8. Let $\mathbf{v} = \alpha_0 \mathbf{u} + \mathbf{l}_0$, where $\mathbf{l}_0$ is a vertical vector. Since $|\text{slope } \mathbf{u}| \leq \frac{1}{10}$ and $|\text{slope } \mathbf{v}| \leq \frac{1}{10}$ it turns out that $|\alpha_0 - 1| \leq \frac{1}{50}$ and $\|\mathbf{l}_0\| \leq \frac{1}{5}$. Let us now split $\mathbf{l}_0 = \overline{\alpha}_0 \mathbf{u} + \overline{\beta}_0 e^{(1)}(z_1)$ and write $\theta_1 = \text{ang}(e^{(1)}(z_1), (1,0))$. From Proposition 3.4(a), we obtain $\left|\theta_1 - \frac{\pi}{2}\right| < K_4\sqrt{b}$. Consequently, $|\overline{\alpha}_0| < K_4\sqrt{b}$ and $|\overline{\beta}_0| < \frac{5}{4}\|\mathbf{l}_0\| \leq \frac{1}{4}$.

On the other hand, setting $e_j^{(i)} = M^j(e^{(i)})$, we may write

$$\mathbf{v}_1 = (\alpha_0 + \overline{\alpha}_0)\mathbf{u}_1 + (M(\zeta_1) - M(z_1))\mathbf{v} + \overline{\beta}_0 e_1^{(1)}(z_1).$$

Then the lemma holds for $\nu = 1$ by defining $\alpha_1 = \overline{\alpha}_0 + \alpha_0$ and

$$\mathbf{l}_1 = (M(\zeta_1) - M(z_1))\mathbf{v} + \overline{\beta}_0 e_1^{(1)}(z_1).$$

Indeed, $|\alpha_1 - 1| \leq |\alpha_0 - 1| + |\overline{\alpha}_0| \leq \frac{1}{10}$ and, from (3.5), it follows that

$$\|l_1\| \leq \|M(\zeta_1) - M(z_1)\| + |\overline{\beta}_0| \left\|e_1^{(1)}(z_1)\right\| < \left(K\sqrt{\sigma}\right)^2.$$

For $\nu \geq 2$, we shall proceed by induction. Let us assume that the statement is true for $\nu - 1$ and take $\mu = \left[\frac{\nu}{2}\right]$. Then, $\mathbf{v}_\mu = \alpha_\mu \mathbf{u}_\mu + \mathbf{l}_\mu$, with

$$|\alpha_\mu - 1| \leq \frac{1}{10} + 5K \sum_{i=2}^{\mu} \left(\frac{K^2\sqrt{\sigma}}{\xi^2}\right)^i \quad \text{and} \quad \|\mathbf{l}_\mu\| \leq (K\sqrt{\sigma})^{\mu+1}. \tag{3.10}$$

Let $e^{(\nu-\mu)} = e^{(\nu-\mu)}(z_{\mu+1})$ and $f^{(\nu-\mu)} = f^{(\nu-\mu)}(z_{\mu+1})$ be unit vectors, maximally contracting and maximally expanding for $M^{\nu-\mu}(z_{\mu+1})$, respectively. From (3.2) and (3.8),

$$\left\|f_{\nu-\mu}^{(\nu-\mu)}\right\| \geq \left\|\frac{\mathbf{u}_\nu}{\|\mathbf{u}_\mu\|}\right\| \geq \frac{\xi^\nu}{2K^\mu},$$

and $\left\|f_{\nu-\mu}^{(\nu-\mu)}\right\| \left\|e_{\nu-\mu}^{(\nu-\mu)}\right\| = |\det(M^{\nu-\mu}(z_{\mu+1}))| \leq (Kb)^{\nu-\mu}$. Hence,

$$\left\|e_{\nu-\mu}^{(\nu-\mu)}\right\| \leq \frac{2K^\mu (Kb)^{\nu-\mu}}{\xi^\nu}. \tag{3.11}$$

Let us now split $\mathbf{l}_\mu = \overline{\alpha}_\mu \mathbf{u}_\mu + \overline{\beta}_\mu e^{(\nu-\mu)}$. Clearly,

$$|\overline{\alpha}_\mu| = \frac{|\det(\mathbf{l}_\mu, e^{(\nu-\mu)})|}{|\det(\mathbf{u}_\mu, e^{(\nu-\mu)})|} \quad \text{and} \quad |\overline{\beta}_\mu| = \frac{|\det(\mathbf{u}_\mu, \mathbf{l}_\mu)|}{|\det(\mathbf{u}_\mu, e^{(\nu-\mu)})|}.$$

We claim that

$$\left|\det\left(\mathbf{u}_\mu, e^{(\nu-\mu)}\right)\right| \geq \frac{\xi^\nu}{3K^{\nu-\mu}}. \tag{3.12}$$

Then, from (3.10), it follows that $|\overline{\alpha}_\mu| \leq 3K^{\nu-\mu}\xi^{-\nu} \|\mathbf{l}_\mu\| \leq 3K^{\nu-\mu}\xi^{-\nu} (K\sqrt{\sigma})^{\mu+1}$ and, since $\nu - \mu \leq \frac{1}{2}\nu + 1 < \mu + 2$, we obtain that

$$|\overline{\alpha}_\mu| \leq 3K \left(\frac{K^2\sqrt{\sigma}}{\xi^2}\right)^{\mu+1}. \tag{3.13}$$

The same arguments lead to

$$|\overline{\beta}_\mu| \leq \frac{3K^{\nu-\mu} \|\mathbf{u}_\mu\| \|\mathbf{l}_\mu\|}{\xi^\nu} \leq \frac{3K^{\nu-\mu}K^\mu (K\sqrt{\sigma})^{\mu+1}}{\xi^\nu} \leq \left(\frac{K^3\sqrt{\sigma}}{\xi^2}\right)^{\mu+1}. \tag{3.14}$$

On the other hand, by applying $M^{\nu-\mu}$ to both sides of the expression $\mathbf{v}_\mu = (\alpha_\mu + \overline{\alpha}_\mu)\,\mathbf{u}_\mu + \overline{\beta}_\mu e^{(\nu-\mu)}$, we get $\mathbf{v}_\nu = (\alpha_\mu + \overline{\alpha}_\mu)\mathbf{u}_\nu + (M^{\nu-\mu}(\zeta_{\mu+1}) - M^{\nu-\mu}(z_{\mu+1}))\,\mathbf{v}_\mu + \overline{\beta}_\mu e_{\nu-\mu}^{(\nu-\mu)}$.

Let us define $\alpha_\nu = \alpha_\mu + \overline{\alpha}_\mu$ and $\mathbf{l}_\nu = (M^{\nu-\mu}(\zeta_{\mu+1}) - M^{\nu-\mu}(z_{\mu+1}))\,\mathbf{v}_\mu + \overline{\beta}_\mu e_{\nu-\mu}^{(\nu-\mu)}$.
Notice that, from (3.10) and (3.13), it follows that

$$|\alpha_\nu - 1| \leq \frac{1}{10} + 5K \sum_{i=2}^{\nu} \left(\frac{K^2 \sqrt{\sigma}}{\xi^2} \right)^i.$$

Furthermore, since $\sigma < \frac{1}{2}$, we obtain

$$\left\| M^{\nu-\mu}(\zeta_{\mu+1}) - M^{\nu-\mu}(z_{\mu+1}) \right\| \leq K^{\nu-\mu} \sum_{j=\mu+1}^{\nu} \sigma^j \leq 2K^{\nu-\mu} \sigma^{\mu+1}$$

in the same way as in (3.6). From this, (3.11) and (3.14) we get

$$\|\mathbf{l}_\nu\| \leq 2K^{\nu-\mu} \sigma^{\mu+1} K^\mu + \left(\frac{K^3 \sqrt{\sigma}}{\xi^2} \right)^{\mu+1} \frac{2K^\mu (Kb)^{\nu-\mu}}{\xi^\nu}.$$

The first term of the rigth side of this inequality is bounded by $\frac{1}{2}(K\sqrt{\sigma})^{\nu+1}$, provided that $K > 4$.

On the other hand, since $\xi^2 < K$, $2\mu \leq \nu \leq 2(\nu - \mu)$ and $\sqrt{b} < \sigma$, it follows that

$$\left(\frac{K^3 \sqrt{\sigma}}{\xi^2} \right)^{\mu+1} \frac{2K^\mu (Kb)^{\nu-\mu}}{\xi^\nu} \leq \frac{1}{2} \left(K\sqrt{\sigma} \right)^{\nu+1}.$$

In short, we conclude that $\|\mathbf{l}_\nu\| \leq (K\sqrt{\sigma})^{\nu+1}$.

Finally, in order to finish the proof we only need to check (3.12). To this end, notice that $\left| \det \left(\mathbf{u}_\mu, e^{(\nu-\mu)} \right) \right| = \left| \left\langle \mathbf{u}_\mu, f^{(\nu-\mu)} \right\rangle \right|$, where $\langle \cdot, \cdot \rangle$ denotes de inner product. Since $\mathbf{u}_\nu = \left\langle \mathbf{u}_\mu, e^{(\nu-\mu)} \right\rangle e_{\nu-\mu}^{(\nu-\mu)} + \left\langle \mathbf{u}_\mu, f^{(\nu-\mu)} \right\rangle f_{\nu-\mu}^{(\nu-\mu)}$ and $e_{\nu-\mu}^{(\nu-\mu)}$ and $f_{\nu-\mu}^{(\nu-\mu)}$ are orthogonal, we may write

$$\left\langle \mathbf{u}_\mu, f^{(\nu-\mu)} \right\rangle^2 = \frac{\|\mathbf{u}_\nu\|^2 - \left\langle \mathbf{u}_\mu, e^{(\nu-\mu)} \right\rangle^2 \left\| e_{\nu-\mu}^{(\nu-\mu)} \right\|^2}{\left\| f_{\nu-\mu}^{(\nu-\mu)} \right\|^2}.$$

Then, from (3.8) and (3.11), since $\left\langle \mathbf{u}_\mu, e^{(\nu-\mu)} \right\rangle^2 \leq K^{2\mu}$ and $2\mu \leq \nu \leq 2(\nu - \mu) \leq 2\nu$, it follows that $\left\{ \det(\mathbf{u}_\mu, e^{(\nu-\mu)}) \right\}^2 \geq \frac{1}{9} K^{-2(\nu-\mu)} \xi^{2\nu}$, provided that $b < \frac{1}{32} K^{-4} \xi^4$. $\quad\square$

Remark 3.9. *All the constants K taking part in the previous estimates do not depend on the parameter a. Thus, in what follows, we may consider a as a new component and write $z = (a, x, y)$ and $z_j = (a, x_j, y_j)$. So, Definition 3.1 and the following statements can be formulated for the new z. Notice that, in our case, $M(z)$ is independent of a. However, $M^{(\nu)}(z)$ depends on a for $\nu > 1$. We shall also write $D_z = D_{(a,x,y)}$.*

The main result of this chapter is the following proposition:

Proposition 3.10. *There exists a constant $K_{10} = K_{10}(K, \xi)$ such that if z is ξ-expanding up to time n, then*

$$\left\| D_z f^{(\nu)}(z) \right\| \le K_{10} \sqrt{b} \ \text{for } 1 \le \nu \le n.$$

Remark 3.11. *Since $f^{(\nu)}(z)$ and $e^{(\nu)}(z)$ are orthogonal, the same statement holds for the vectors $e^{(\nu)}(z)$.*

To prove Proposition 3.10 we introduce new definitions and preliminary results. Let M_* be the adjoint operator of M and define

$$g^{(\nu)} = \frac{M^\nu f^{(\nu)}}{\|M^\nu f^{(\nu)}\|}. \tag{3.15}$$

Since $M_*^{-\nu} f^{(\nu)}$ is colinear to $M^\nu f^{(\nu)}$, we have

$$g^{(\nu)} = \frac{M_*^{-\nu} f^{(\nu)}}{\|M_*^{-\nu} f^{(\nu)}\|}. \tag{3.16}$$

Notice that $1 = \left\langle M_*^{-\nu} f^{(\nu)}, M^\nu f^{(\nu)} \right\rangle = \left\| M_*^{-\nu} f^{(\nu)} \right\| \left\| M^\nu f^{(\nu)} \right\|$. Thus

$$\left\| M_*^{-\nu} f^{(\nu)} \right\| = \left\| M^\nu f^{(\nu)} \right\|^{-1} \le \xi^{-\nu}. \tag{3.17}$$

Lemma 3.12. *There exists a constant $K_{12} = K_{12}(K, \xi) > 0$ such that if z is ξ-expanding up to time n, then, for $1 \le j \le \nu \le n$, it follows that:*

(a) $\left\| D_z f^{(\nu)}(z) \right\| \le K_{12}^\nu,\ \left\| D_z g^{(\nu)}(z) \right\| \le K_{12}^\nu.$

(b) $\left\| D_z^2 f^{(\nu)}(z) \right\| \le K_{12}^\nu,\ \left\| D_z^2 g^{(\nu)}(z) \right\| \le K_{12}^\nu.$

(c) $\left\| D_z \left(M^j f^{(\nu)}(z) \right) \right\| \le K_{12}^\nu,\ \left\| D_z \left(M_*^{-j} f^{(\nu)}(z) \right) \right\| \le K_{12}^\nu.$

(d) $\left\| D_z^2 \left(M^j f^{(\nu)}(z) \right) \right\| \le K_{12}^\nu,\ \left\| D_z^2 \left(M_*^{-j} f^{(\nu)}(z) \right) \right\| \le K_{12}^\nu.$

Proof. Let $\widetilde{M}(z, f) = M_*^\nu(z) M^\nu(z)(f)$ and

$$\widetilde{M}_\sharp(z, f) = \frac{\widetilde{M}(z, f)}{\|\widetilde{M}(z, f)\|}$$

for $z \in \widetilde{D}$ ($\widetilde{D}$ given in Proposition 1.3) and $f \in \mathbf{R}^2$. Since $M^\nu f^{(\nu)}$ is colinear to $M_*^{-\nu} f^{(\nu)}$, we obtain

$$\widetilde{M}_\sharp(z, f^{(\nu)}(z)) = \frac{M_*^\nu M^\nu f^{(\nu)}}{\|M_*^\nu M^\nu f^{(\nu)}\|} = \frac{M_*^\nu M_*^{-\nu} f^{(\nu)}}{\|M_*^\nu M_*^{-\nu} f^{(\nu)}\|} = f^{(\nu)}. \tag{3.18}$$

Let us define the map:

$$(z, f) \in \widetilde{D} \times S^1 \longrightarrow F(z, f) = \widetilde{M}_\sharp(z, f) \in S^1$$

where S^1 is the unit sphere. Then

$$\frac{\partial F}{\partial f}(z, f)(\dot{f}) = \frac{\widetilde{M}(z, \dot{f})}{\left\|\widetilde{M}(z, f)\right\|} - \frac{<\widetilde{M}(z, \dot{f}), \widetilde{M}(z, f)>}{\left\|\widetilde{M}(z, f)\right\|^2} \frac{\widetilde{M}(z, f)}{\left\|\widetilde{M}(z, f)\right\|}, \tag{3.19}$$

where $\dot{f}$ belongs to the tangent space $\{f\}^\perp$. If $f = f^{(\nu)}(z)$ then $\{f\}^\perp \in \left\langle e^{(\nu)}(z) \right\rangle$ and, from (3.18),

$$\left\langle \widetilde{M}\left(z, e^{(\nu)}(z)\right), \widetilde{M}\left(z, f^{(\nu)}(z)\right)\right\rangle = 0. \tag{3.20}$$

Thus

$$\frac{\partial F}{\partial f}\left(z, f^{(\nu)}(z)\right)(\dot{f}) = \frac{\widetilde{M}(z, \dot{f})}{\left\|\widetilde{M}\left(z, f^{(\nu)}(z)\right)\right\|}.$$

Now, since $\left\|M_*^\nu M^\nu f^{(\nu)}\right\| = \left\|M^\nu f^{(\nu)}\right\|^2$ and $\left\|M_*^\nu M^\nu e^{(\nu)}\right\| = \left\|M^\nu e^{(\nu)}\right\|^2$, from (3.5) and (3.17), it follows that

$$\left|\frac{\partial F}{\partial f}\left(z, f^{(\nu)}(z)\right)\right| = \frac{\left\|M^\nu e^{(\nu)}(z)\right\|^2}{\left\|M^\nu f^{(\nu)}(z)\right\|^2} \leq \left(\frac{Kb}{\xi^2}\right)^{2\nu} << 1. \tag{3.21}$$

On the other hand,

$$\frac{\partial F}{\partial z}(z, f)(\dot{z}) = \frac{D_z\widetilde{M}(z, f)(\dot{z})}{\left\|\widetilde{M}(z, f)\right\|} - \frac{\left\langle D_z\widetilde{M}(z, f)(\dot{z}), \widetilde{M}(z, f)\right\rangle}{\left\|\widetilde{M}(z, f)\right\|^2} \frac{\widetilde{M}(z, f)}{\left\|\widetilde{M}(z, f)\right\|}. \tag{3.22}$$

Since $\|F(z, f)\| = 1$ implies $\left\langle \frac{\partial F}{\partial z}(z, f)(\dot{z}), \widetilde{M}(z, f)\right\rangle = 0$, we multiply both sides of (3.22) by $\frac{\partial F}{\partial z}(z, f)(\dot{z})$ to obtain

$$\left\|\frac{\partial F}{\partial z}\left(z, f^{(\nu)}(z)\right)\right\| \leq \frac{\left\|D_z\widetilde{M}\left(z, f^{(\nu)}(z)\right)\right\|}{\left\|\widetilde{M}\left(z, f^{(\nu)}(z)\right)\right\|} \leq \left(\frac{K}{\xi}\right)^{2\nu}. \tag{3.23}$$

Finally, let us express (3.18) as $F\left(z, f^{(\nu)}(z)\right) = f^{(\nu)}(z)$. Then, by applying the implicit function theorem, we deduce that

$$D_z f^{(\nu)} = \left(\frac{\partial F}{\partial f}\left(z, f^{(\nu)}(z)\right) - 1\right)^{-1} \frac{\partial F}{\partial z}\left(z, f^{(\nu)}(z)\right). \tag{3.24}$$

Then, from (3.21) and (3.23), we see that

$$\left\|D_z f^{(\nu)}\right\| \leq 2\left(\frac{K}{\xi}\right)^{2\nu}. \tag{3.25}$$

In the same way, we may bound $\left\|D_z g^{(\nu)}\right\|$ by means of

$$\frac{M^\nu M_*^\nu g^{(\nu)}(z)}{\left\|M^\nu M_*^\nu g^{(\nu)}(z)\right\|} = \frac{M^\nu f^{(\nu)}(z)}{\left\|M^\nu f^{(\nu)}(z)\right\|} = g^{(\nu)}(z).$$

So, statement (a) holds.

To prove statement (b), we derive in (3.24) to obtain

$$D_z^2 f^{(\nu)}(z) = \left(\frac{\partial F}{\partial f}\left(z, f^{(\nu)}(z)\right) - 1\right)^{-1} \frac{\partial^2 F}{\partial z^2}\left(z, f^{(\nu)}(z)\right) -$$

$$- \left(\frac{\partial F}{\partial f}\left(z, f^{(\nu)}(z)\right) - 1\right)^{-3} \left(\frac{\partial F}{\partial z}\left(z, f^{(\nu)}(z)\right)\right)^2 \frac{\partial^2 F}{\partial f^2}\left(z, f^{(\nu)}(z)\right). \tag{3.26}$$

Since (3.21) and (3.23) yield

$$\left|\left(\frac{\partial F}{\partial f}\left(z, f^{(\nu)}(z)\right) - 1\right)^{-1}\right| \le 2 \text{ and } \left\|\left(\frac{\partial F}{\partial z}\left(z, f^{(\nu)}(z)\right)\right)^2\right\| \le \left(\frac{K}{\xi}\right)^{4\nu},$$

it suffices to bound

$$\left\|\frac{\partial^2 F}{\partial z^2}\left(z, f^{(\nu)}(z)\right)\right\| \text{ and } \left\|\frac{\partial^2 F}{\partial f^2}\left(z, f^{(\nu)}(z)\right)\right\|.$$

By deriving (3.22) and taking into account (3.17) and the fact that $\left\|\widetilde{M}\left(z, f^{(\nu)}(z)\right)\right\| = \left\|M^\nu f^{(\nu)}\right\|^2$, we get

$$\left\|\frac{\partial^2 F}{\partial z^2}\left(z, f^{(\nu)}(z)\right)\right\| \le \left(\frac{2K^4}{\xi^4}\right)^\nu. \tag{3.27}$$

To bound

$$\left\|\frac{\partial^2 F}{\partial f^2}\left(z, f^{(\nu)}(z)\right)\right\|$$

let us write (3.19) in the following way:

$$\frac{\partial F}{\partial f}(z, f)\,(\dot{f}) = \frac{1}{\left\|\widetilde{M}(z, f)\right\|}\left\{\widetilde{M}(z, \dot{f}) - < \widetilde{M}(z, \dot{f}), F(z, f) > F(z, f)\right\}.$$

Now, a few more estimates are necessary. First, derive the above equation and use (3.20). Then, since $\left\|F(z, f^{(\nu)}(z))\right\| = 1$ for every $z \in \widetilde{D}$, it follows from (3.5) and (3.17) that

$$\left\|\frac{\partial^2 F}{\partial f^2}\left(z, f^{(\nu)}(z)\right)\right\| \le 2\left(\frac{K^2 b}{\xi^2}\right)^{4\nu}. \tag{3.28}$$

Finally, putting (3.27) and (3.28) in (3.26), we obtain

$$\left\|D_z^2 f^{(\nu)}(z)\right\| \le 2\left(\frac{2K^4}{\xi^4}\right)^\nu + 16\left(\frac{K}{\xi}\right)^{2\nu}\left(\frac{K^2 b}{\xi^2}\right)^{4\nu} \le \left(\frac{5K^4}{\xi^4}\right)^\nu. \tag{3.29}$$

In a similar way we get

$$\left\|D_z^2 g^{(\nu)}(z)\right\| \leq \left(\frac{5K^4}{\xi^4}\right)^\nu$$

and statement (b) holds.

To prove statement (c) let us write

$$D_z\left(M^j f^{(\nu)}(z)\right) = D_z\left(M(z_{j-1})M^{j-1}(z)f^{(\nu)}(z)\right) =$$

$$= D_z M(z_{j-1})D_z(z_{j-1})M^{j-1}(z)f^{(\nu)}(z) + M(z_{j-1})D_z\left(M^{j-1}(z)f^{(\nu)}(z)\right). \qquad (3.30)$$

From this expression we obtain

$$\left\|D_z\left(M^j f^{(\nu)}(z)\right)\right\| \leq K^{2j} + K\left\|D_z\left(M^{j-1}f^{(\nu)}(z)\right)\right\|. \qquad (3.31)$$

Then, from (3.25), it follows that

$$\left\|D_z\left(M^j f^{(\nu)}(z)\right)\right\| \leq K^{2j} + K^{2j-1} + \ldots + K^j\left\|D_z f^{(\nu)}(z)\right\| \leq \left(\frac{2K^2}{\xi}\right)^{2\nu}. \qquad (3.32)$$

To get the second inequality of (c) notice that (3.16) and (3.17) lead to

$$M_*^{-j}f^{(\nu)} = \left\|M_*^{-\nu}f^{(\nu)}\right\|M_*^{\nu-j}g^{(\nu)} = \frac{M_*^{\nu-j}g^{(\nu)}}{\|M^\nu f^{(\nu)}\|}.$$

Then,

$$D_z(M_*^{-j}f^{(\nu)}) = \frac{D_z\left(M_*^{\nu-j}g^{(\nu)}\right)}{\|M^\nu f^{(\nu)}\|} - \frac{M_*^{\nu-j}g^{(\nu)}D_z\left(\|M^\nu f^{(\nu)}\|\right)}{\|M^\nu f^{(\nu)}\|^2}. \qquad (3.33)$$

Now, just as above, we get

$$\left\|D_z\left(M_*^{\nu-j}g^{(\nu)}\right)\right\| \leq \left(\frac{2K^2}{\xi}\right)^{2\nu}. \qquad (3.34)$$

On the other hand, from (3.32) it follows that

$$\left\|D_z\left(\|M^\nu f^{(\nu)}\|\right)\right\| = \frac{\left|\left\langle D_z\left(M^\nu f^{(\nu)}\right), M^\nu f^{(\nu)}\right\rangle\right|}{\|M^\nu f^{(\nu)}\|} \leq \left(\frac{2K^2}{\xi}\right)^{2\nu}. \qquad (3.35)$$

So,

$$\left\|D_z\left(M_*^{-j}f^{(\nu)}\right)\right\| \leq \left(\frac{5K^5}{\xi^4}\right)^\nu.$$

The proof of statement (d) requires also too long estimates to be done in detail. Therefore, we shall only summarize them. Deriving (3.30) and using (3.2) we get

$$\left\|D_z^2\left(M^j f^{(\nu)}(z)\right)\right\| \leq 4K^{4j} + 2K^j\left\|D_z\left(M^{j-1}f^{(\nu)}(z)\right)\right\| + K\left\|D_z^2\left(M^{j-1}f^{(\nu)}(z)\right)\right\|.$$

Then, once again as in (3.31), we deduce that

$$\left\| D_z^2 \left(M^j f^{(\nu)}(z) \right) \right\| \le 4K^{4j} + 4K^{4j-1} + \ldots + 4K^{3j} + 2^j K^{2j} \left\| D_z f^{(\nu)}(z) \right\| + K^j \left\| D_z^2 f^{(\nu)}(z) \right\|.$$

Finally, from (3.25) and (3.29), it follows that

$$\left\| D_z^2 \left(M^j f^{(\nu)}(z) \right) \right\| \le \left(\frac{10K^8}{\xi^4} \right)^\nu. \tag{3.36}$$

To prove the second inequality of statement (d), notice that

$$\left\| D_z^2 \left(M_*^{\nu-j} g^{(\nu)} \right) \right\| \le \left(\frac{10K^8}{\xi^4} \right)^\nu$$

follows as (3.36). Furthermore, (3.32) and (3.36) lead to

$$\left\| D_z^2 \left(\left\| M^\nu f^{(\nu)} \right\| \right) \right\| \le \left(\frac{12K^9}{\xi^5} \right)^\nu.$$

Then, deriving (3.33), using these two last estimates, (3.17), (3.34) and (3.35) we get

$$\left\| D_z^2 \left(M_*^{-j} f^{(\nu)} \right) \right\| \le \left(\frac{20K^{10}}{\xi^7} \right)^\nu$$

and the proof of the lemma is complete. $\square$

Proof of Proposition 3.10. Let us first check the result for $\nu = 1$. For this, we copy (3.22) with $\widetilde{M} = M_* M$. Then, from (3.18), we have

$$\frac{\partial F}{\partial z} \left(z, f^{(1)} \right) (\overset{\bullet}{z}) = \frac{D_z \widetilde{M} \left(z, f^{(1)} \right) (\overset{\bullet}{z})}{\| M f^{(1)} \|^2} - \left\langle \frac{D_z \widetilde{M} \left(z, f^{(1)} \right) (\overset{\bullet}{z})}{\| M f^{(1)} \|^2}, f^{(1)} \right\rangle f^{(1)}.$$

Let us now write $\widetilde{M} = \widehat{M} + \Theta$, where, according to (3.1),

$$\widehat{M} = \begin{pmatrix} p(x) & 0 \\ 0 & 0 \end{pmatrix} \quad \text{and} \quad \| \Theta \|_{C^2(\widetilde{D})} \le K\sqrt{b}. \tag{3.37}$$

Proposition 3.4 also allows us to write $f^{(1)} = (1,0) + \left(const\sqrt{b}, const\sqrt{b} \right)$, where, from now on, *const* means any constant independent of a and b. Therefore,

$$\left(\frac{\partial F}{\partial z} \left(z, f^{(1)} \right) (\overset{\bullet}{z}) \right) = \frac{1}{\| M f^{(1)} \|^2} \left(const\sqrt{b}, const\sqrt{b} \right)$$

and from (3.17), it follows that

$$\left\| \frac{\partial F}{\partial z} \left(z, f^{(1)} \right) (\overset{\bullet}{z}) \right\| \le const\sqrt{b}. \tag{3.38}$$

In summary, from (3.21) and (3.24) we obtain

$$\left\| D_z f^{(1)}(z) \right\| = \left| \frac{\partial F}{\partial f}\left(z, f^{(1)}\right) - 1 \right|^{-1} \left\| \frac{\partial F}{\partial z}\left(z, f^{(1)}\right)(\dot{z}) \right\| \leq const\sqrt{b}.$$

Next, we shall complete the proof by checking that

$$\left\| D_z f^{(\nu+1)}(z) \right\| \leq \left\| D_z f^{(\nu)}(z) \right\| + (const\ b)^\nu, \tag{3.39}$$

for $1 \leq \nu \leq n-1$. To this end, let us write $f^{(\nu+1)} = \left\langle f^{(\nu+1)}, f^{(\nu)} \right\rangle f^{(\nu)} \pm \mathbf{h}^{(\nu)}$, where $\mathbf{h}^{(\nu)}$ is colinear to $e^{(\nu)}$. Then,

$$D_z f^{(\nu+1)} = \left\langle f^{(\nu+1)}, f^{(\nu)} \right\rangle D_z f^{(\nu)} + D_z \left(\left\langle f^{(\nu+1)}, f^{(\nu)} \right\rangle \right) f^{(\nu)} \pm D_z \mathbf{h}^{(\nu)}. \tag{3.40}$$

Since $\left\langle f^{(\nu)}, D_z f^{(\nu)} \right\rangle = 0$, we have $\left| \left\langle f^{(\nu+1)}, D_z f^{(\nu)} \right\rangle \right| \leq \left| ang\left(f^{(\nu+1)}, f^{(\nu)} \right) \right| \left\| D_z f^{(\nu)} \right\|$. From Proposition 3.3 and Lemma 3.12, we get that $\left| \left\langle f^{(\nu+1)}, D_z f^{(\nu)} \right\rangle \right| \leq (const\ b)^\nu$, $\left| \left\langle D_z f^{(\nu+1)}, f^{(\nu)} \right\rangle \right| \leq (const\ b)^\nu$ and consequently,

$$\left| D_z \left(\left\langle f^{(\nu+1)}, f^{(\nu)} \right\rangle \right) \right| \leq (const\ b)^\nu. \tag{3.41}$$

Let us now bound $\left\| D_z \mathbf{h}^{(\nu)} \right\|$. Let A be the matrix associated to M_*^ν with respect to the basis

$$\left\{ g^{(\nu)}, \frac{M^\nu e^{(\nu)}}{\| M^\nu e^{(\nu)} \|} \right\} \quad \text{and} \quad \left\{ f^{(\nu)}, e^{(\nu)} \right\}.$$

Since from (3.16) and (3.17) it follows that

$$M_*^\nu g^{(\nu)} = M_*^\nu \frac{M_*^{-\nu} f^{(\nu)}}{\| M_*^{-\nu} f^{(\nu)} \|} = \left\| M^\nu f^{(\nu)} \right\| f^{(\nu)}$$

and since $\left\| M_*^\nu M^\nu e^{(\nu)} \right\| = \left\| M^\nu e^{(\nu)} \right\|^2$ implies

$$M_*^\nu \frac{M^\nu e^{(\nu)}}{\| M^\nu e^{(\nu)} \|} = \left\| M^\nu e^{(\nu)} \right\| e^{(\nu)},$$

it is clear that

$$A = \begin{pmatrix} \left\| M^\nu f^{(\nu)} \right\| & 0 \\ 0 & \left\| M^\nu e^{(\nu)} \right\| \end{pmatrix}.$$

Now, let us consider the following vector

$$\overline{\mathbf{h}}^{(\nu)} = M_*^{-\nu} \mathbf{h}^{(\nu)} = M_*^{-\nu} f^{(\nu+1)} - \left\langle f^{(\nu+1)}, f^{(\nu)} \right\rangle M_*^{-\nu} f^{(\nu)}. \tag{3.42}$$

Since $\mathbf{h}^{(\nu)}$ is colinear to $e^{(\nu)}$, $\overline{\mathbf{h}}^{(\nu)}$ is colinear to $M_*^{-\nu} e^{(\nu)}$ and, consequently, to $M^\nu e^{(\nu)}$. Therefore,

$$\mathbf{h}^{(\nu)} = M_*^\nu \overline{\mathbf{h}}^{(\nu)} = \left\| M^\nu e^{(\nu)} \right\| \overline{\mathbf{h}}^{(\nu)} = (det\ M^\nu) \left\| M^\nu f^{(\nu)} \right\|^{-1} \overline{\mathbf{h}}^{(\nu)} \tag{3.43}$$

and hence,

$$D_z \mathbf{h}^{(\nu)} = D_z \left(\det M^\nu\right) \left\| M^\nu f^{(\nu)} \right\|^{-1} \overline{\mathbf{h}}^{(\nu)} + \left(\det M^\nu\right) D_z \left(\left\| M^\nu f^{(\nu)} \right\|^{-1}\right) \overline{\mathbf{h}}^{(\nu)} +$$

$$+ \left(\det M^\nu\right) \left\| M^\nu f^{(\nu)} \right\|^{-1} D_z \overline{\mathbf{h}}^{(\nu)}. \tag{3.44}$$

To complete the proof we shall bound the terms which appear in this expression.

From (3.17) and (3.42) it follows that

$$\left\| \overline{\mathbf{h}}^{(\nu)} \right\| \leq K \left\| M^{(\nu+1)} f^{(\nu+1)} \right\|^{-1} + \left\| M^\nu f^{(\nu)} \right\|^{-1} \leq (const)^\nu. \tag{3.45}$$

Deriving (3.42), we obtain

$$D_z \overline{\mathbf{h}}^{(\nu)} = D_z M_*^{-\nu} f^{(\nu+1)} -$$

$$- \left\langle f^{(\nu+1)}, f^{(\nu)} \right\rangle D_z M_*^{-\nu} f^{(\nu)} - D_z \left(\left\langle f^{(\nu+1)}, f^{(\nu)} \right\rangle\right) M_*^{-\nu} f^{(\nu)} \tag{3.46}$$

and hence, from Lemma 3.12 and (3.41),

$$\left\| D_z \overline{\mathbf{h}}^{(\nu)} \right\| \leq (const)^\nu. \tag{3.47}$$

Lemma 3.12, togheter with (3.17), also enables us to get

$$\left\| D_z \left(\left\| M^\nu f^{(\nu)} \right\|^{-1}\right) \right\| = \left\| D_z \left(\left\| M_*^{-\nu} f^{(\nu)} \right\|\right) \right\| \leq \left\| D_z \left(M_*^{-\nu} f^{(\nu)}\right) \right\| \leq (const)^\nu. \tag{3.48}$$

Recall that Proposition 1.3 states that $|(\det M^\nu)| \leq (const\, b)^\nu$. Thus, after checking that $|D_z (\det M)| \leq const\, b$, we have

$$|D_z (\det M^\nu)| \leq \nu \left\| D_z (\det M) \right\| \left| \det M \right|^{\nu-1} \leq (const\, b)^\nu. \tag{3.49}$$

Finally, putting these last estimates in (3.44) we get

$$\left\| D_z \mathbf{h}^{(\nu)} \right\| \leq (const\, b)^\nu. \tag{3.50}$$

This completes the proof. $\square$

Statements (b) and (d) in Lemma 3.12 have not been used in the proof of Proposition 3.10, but they will be used to prove the following result:

Proposition 3.13. *There exists a constant* $K_{13} = K_{13}(K, \xi)$ *such that if z is ξ-expanding up to time n, then*

$$\left\| D_z^2 f^{(\nu)}(z) \right\| \leq K_{13} \sqrt{b} \text{ for every } 1 \leq \nu \leq n.$$

Proof. The proof of this statement involves second order derivatives as in Lemma 3.12(d). Therefore, some estimates become too long to be developed in detail. So, once again, we shall just outline them.

For $\nu = 1$, we derive (3.22) and take $\widetilde{M} = M_* M$. Then, by applying (3.37) and in the same way as (3.38) was obtained, we prove that

$$\left\| \frac{\partial^2 F}{\partial z^2} \left(z, f^{(1)} \right) \right\| \leq const\sqrt{b}.$$

Finally, using (3.21), (3.28) and (3.38) in (3.26) we get

$$\left\| D_z^2 f^{(1)} \right\| \leq \left(const\sqrt{b} \right).$$

To complete the proof it suffices to show that $\left\| D_z^2 f^{(\nu+1)}(z) \right\| \leq \left\| D_z^2 f^{(\nu)}(z) \right\| + (const\ b)^\nu$ for $1 \leq \nu \leq n-1$. For this, we shall proceed as in the proof of Proposition 3.10. Let $f^{(\nu+1)} = \left\langle f^{(\nu+1)}, f^{(\nu)} \right\rangle f^{(\nu)} \pm \mathbf{h}^{(\nu)}$, where $\mathbf{h}^{(\nu)}$ is colinear to $e^{(\nu)}$. Deriving (3.40) we have

$$
\begin{aligned}
D_z^2 f^{(\nu+1)} = \ & 2D_z \left(\left\langle f^{(\nu+1)}, f^{(\nu)} \right\rangle \right) D_z f^{(\nu)} + \\
& + \left\langle f^{(\nu+1)}, f^{(\nu)} \right\rangle D_z^2 f^{(\nu)} + D_z^2 \left(\left\langle f^{(\nu+1)}, f^{(\nu)} \right\rangle \right) f^{(\nu)} \pm D_z^2 \mathbf{h}^{(\nu)}.
\end{aligned}
$$

Now, according to (3.41) and Lemma 3.12, it suffices to verify that

$$\left| D_z^2 \left(\left\langle f^{(\nu+1)}, f^{(\nu)} \right\rangle \right) \right| \leq (const\ b)^\nu \ \text{ and } \ \left\| D_z^2 \mathbf{h}^{(\nu)} \right\| \leq (const\ b)^\nu.$$

In order to check the first inequality it is helpful to see that

$$D_z^2 \left(\left\langle f^{(\nu+1)}, f^{(\nu)} \right\rangle \right) =$$

$$= \left\langle D_z^2 f^{(\nu+1)} - D_z^2 f^{(\nu)}, f^{(\nu)} - f^{(\nu+1)} \right\rangle + \left\langle D_z f^{(\nu+1)} - D_z f^{(\nu)}, D_z f^{(\nu+1)} - D_z f^{(\nu)} \right\rangle.$$

The bound

$$\left| \left\langle D_z^2 f^{(\nu+1)} - D_z^2 f^{(\nu)}, f^{(\nu)} - f^{(\nu+1)} \right\rangle \right| \leq (const\ b)^\nu$$

for the first term follows from Lemma 3.12(b) and Proposition 3.3. With respect to the second term, (3.40) yields

$$D_z f^{(\nu+1)} - D_z f^{(\nu)} = \left(\left\langle f^{(\nu)}, f^{(\nu+1)} \right\rangle - 1 \right) D_z f^{(\nu)} + D_z \left(\left\langle f^{(\nu)}, f^{(\nu+1)} \right\rangle \right) f^{(\nu)} + D_z \mathbf{h}^{(\nu)}.$$

Now,

$$\left|\left\langle f^{(\nu)}, f^{(\nu+1)}\right\rangle - 1\right| \leq (const\ b)^{\nu}$$

and, in (3.41) and (3.50), we proved that

$$\left|D_z\left(\left\langle f^{(\nu)}, f^{(\nu+1)}\right\rangle\right)\right| \leq (const\ b)^{\nu} \quad \text{and} \quad \left\|D_z\mathbf{h}^{(\nu)}\right\| \leq (const\ b)^{\nu},$$

respectively. Therefore, from Lemma 3.12, we conclude that

$$\left\|D_z f^{(\nu+1)} - D_z f^{(\nu)}\right\| \leq (const\ b)^{\nu}$$

and consequently,

$$\left|D_z^2\left(\left\langle f^{(\nu+1)}, f^{(\nu)}\right\rangle\right)\right| \leq (const\ b)^{\nu}. \tag{3.51}$$

Finally, to prove that

$$\left\|D_z^2\mathbf{h}^{(\nu)}\right\| \leq (const\ b)^{\nu},$$

we derive (3.44) to obtain

$$D_z^2\mathbf{h}^{(\nu)} = D_z^2\left(\det M^{\nu}\right)\left\|M^{\nu}f^{(\nu)}\right\|^{-1}\overline{\mathbf{h}}^{(\nu)} + 2D_z\left(\det M^{\nu}\right)D_z\left(\left\|M^{\nu}f^{(\nu)}\right\|^{-1}\right)\overline{\mathbf{h}}^{(\nu)}+$$

$$+\left(\det M^{\nu}\right)D_z^2\left(\left\|M^{\nu}f^{(\nu)}\right\|^{-1}\right)\overline{\mathbf{h}}^{(\nu)} + 2D_z\left(\det M^{\nu}\right)\left\|M^{\nu}f^{(\nu)}\right\|^{-1}D_z\overline{\mathbf{h}}^{(\nu)}+$$

$$+2\left(\det M^{\nu}\right)D_z\left(\left\|M^{\nu}f^{(\nu)}\right\|^{-1}\right)D_z\overline{\mathbf{h}}^{(\nu)} + \left(\det M^{\nu}\right)\left\|M^{\nu}f^{(\nu)}\right\|^{-1}D_z^2\overline{\mathbf{h}}^{(\nu)}.$$

Notice that $|(\det M^{\nu})| \leq (const\ b)^{\nu}$ follows from (3.3) and recall that some other terms have already been bounded. Namely, see (3.17), (3.45), (3.47), (3.48) and (3.49). Therefore, the proof ends if we show that $\|D_z^2(\det M^{\nu})\| \leq (const\ b)^{\nu}$, $\left\|D_z^2\overline{\mathbf{h}}^{(\nu)}\right\| \leq (const)^{\nu}$ and $\left\|D_z^2\left(\left\|M^{\nu}f^{(\nu)}\right\|^{-1}\right)\right\| \leq (const)^{\nu}$.

Since

$$\left\|D_z^2\left(\det M\right)\right\| = \left\|D_z^2\left(-\frac{be^{\lambda x}}{\lambda\cos x}\right)\right\| \leq const\ b,$$

going a step beyond (3.49) we obtain

$$\left\|D_z^2\left(\det M^{\nu}\right)\right\| \leq \nu\left\|D_z^2\left(\det M\right)\right\||\det M|^{\nu-1}+$$

$$+\nu\left(\nu-1\right)\left\|D_z\left(\det M\right)\right\|^2|\det M|^{\nu-2} \leq (const\ b)^{\nu}.$$

On the other hand, by means of (3.48) we obtain

$$D_z^2\left(\left\|M^{\nu}f^{(\nu)}\right\|^{-1}\right) =$$

$$= \frac{\left\langle D_z^2 M_*^{-\nu} f^{(\nu)}, M_*^{-\nu} f^{(\nu)} \right\rangle + \left\| D_z M_*^{-\nu} f^{(\nu)} \right\|^2}{\left\| M_*^{-\nu} f^{(\nu)} \right\|} - \frac{\left\langle D_z M_*^{-\nu} f^{(\nu)}, M_*^{-\nu} f^{(\nu)} \right\rangle^2}{\left\| M_*^{-\nu} f^{(\nu)} \right\|^3}.$$

Using $\left\| M_*^{-\nu} f^{(\nu)} \right\|^{-1} \leq K^\nu$ and Lemma 3.12, we get $\left\| D_z^2 \left(\left\| M^\nu f^{(\nu)} \right\|^{-1} \right) \right\| \leq (const)^\nu$.

Finally, deriving (3.46) and using (3.17), (3.41), (3.51) and Lemma 3.12 we get $\left\| D_z^2 \overline{h}^{(\nu)} \right\| \leq (const)^\nu$. So, the proposition holds. $\square$

Remark 3.14. *The fact that $f^{(\nu)}$ and $e^{(\nu)}$ are orthogonal, together with Proposition 3.10, allows us to claim that Proposition 3.13 is also true for the maximally contracting directions.*

We end this chapter with a result which will only be used in Chapter 7:

Proposition 3.15. *There exists a constant $K_{15} = K_{15}(K, \xi)$ such that if z is ξ-expanding up to time n, then*

$$\left\| D_z M^i e^{(\nu)}(z) \right\| \leq (K_{15} b)^i \text{ for every } 1 \leq i \leq \nu \leq n.$$

Proof. Let us write $e^{(\nu)} = \left\langle e^{(\nu)}, f^{(i)} \right\rangle f^{(i)} \pm \mathbf{k}_i$, with $\mathbf{k}_i$ colinear to $e^{(i)}$. Then,

$$D_z M^i e^{(\nu)} = \left\langle e^{(\nu)}, f^{(i)} \right\rangle D_z M^i f^{(i)} + D_z \left(\left\langle e^{(\nu)}, f^{(i)} \right\rangle \right) M^i f^{(i)} \pm D_z M^i \mathbf{k}_i,$$

with $\left\| M^i f^{(i)} \right\| \leq (const)^i$. From Lemma 3.12 it follows that $\left\| D_z M^i f^{(i)} \right\| \leq (const)^i$ and, from Proposition 3.3, $\left| \left\langle e^{(\nu)}, f^{(i)} \right\rangle \right| \leq (const \, b)^i$. So, it suffices to verify that

$$\left\| D_z \left(\left\langle e^{(\nu)}, f^{(i)} \right\rangle \right) \right\| \leq (const \, b)^i \text{ and } \left\| D_z M^i \mathbf{k}_i \right\| \leq (const \, b)^i. \tag{3.52}$$

To prove the first inequality, let us write $f^{(\nu)} = \left\langle f^{(\nu)}, f^{(\nu-1)} \right\rangle f^{(\nu-1)} \pm \mathbf{h}^{(\nu-1)}$, where $\mathbf{h}^{(\nu-1)}$ is colinear to $e^{(\nu-1)}$.

From Proposition 3.3 it follows that $\left| \left\langle f^{(\nu)}, f^{(\nu-1)} \right\rangle \right| \geq 1 - (const \, b)^{\nu-1}$. Now, repeating the above splitting and setting $c_j = \left\langle f^{(j+1)}, f^{(j)} \right\rangle$, we have

$$f^{(\nu)} = C_i f^{(i)} + \mathbf{l}, \tag{3.53}$$

where $C_i = c_{\nu-1} c_{\nu-2} \dots c_i$ and

$$\mathbf{l} = \pm \mathbf{h}^{(\nu-1)} \pm C_{\nu-1} \mathbf{h}^{(\nu-2)} \pm C_{\nu-2} \mathbf{h}^{(\nu-3)} \pm \dots \pm C_{i+1} \mathbf{h}^{(i)}.$$

Since $1 - (const\ b)^j \le |c_j| \le 1$, it follows that $C_i \ge \prod_{j=i}^{\nu-1} \left(1 - (const\ b)^j\right) > \frac{1}{2}$. Furthermore, the vectors $\mathbf{h}^{(j)}$ were already given by (3.42) and (3.43). So, from (3.45), we deduce $\left\|\mathbf{h}^{(j)}\right\| \le (const\ b)^j$. Consequently,

$$\|\mathbf{1}\| \le \sum_{j=i}^{\nu-1} \left\|\mathbf{h}^{(j)}\right\| \le \sum_{j=i}^{\infty} (const\ b)^j \le (const\ b)^i. \tag{3.54}$$

From (3.41) and (3.50) it follows that

$$\|D_z(C_j)\| \le (const\ b)^j \text{ and } \left\|D_z\mathbf{h}^{(j)}\right\| \le (const\ b)^j,$$

respectively. Thus

$$\|D_z\mathbf{1}\| \le \sum_{j=i}^{\nu-1} (const\ b)^j + \sum_{j=i}^{\nu-2} (const\ b)^{2j+1} \le (const\ b)^i. \tag{3.55}$$

Since $C_i > \frac{1}{2}$, from (3.53) we obtain that $f^{(i)} = C_i^{-1}f^{(\nu)} \pm C_i^{-1}\mathbf{1}$. Then, $\left\langle e^{(\nu)}, f^{(i)}\right\rangle = \pm C_i^{-1}\left\langle \mathbf{1}, e^{(\nu)}\right\rangle$ and

$$D_z\left(\left\langle e^{(\nu)}, f^{(i)}\right\rangle\right) = \mp\frac{1}{C_i^2}D_z(C_i)\left\langle \mathbf{1}, e^{(\nu)}\right\rangle \pm \frac{1}{C_i}\left(\left\langle D_z\mathbf{1}, e^{(\nu)}\right\rangle + \left\langle \mathbf{1}, D_z e^{(\nu)}\right\rangle\right).$$

Finally, from (3.54), (3.55) and Remark 3.11, we get $\left\|D_z\left(\left\langle e^{(\nu)}, f^{(i)}\right\rangle\right)\right\| \le (const\ b)^i$.

Now, to obtain the second inequality in (3.52) let us consider the matrix associated to M^i with respect to the basis

$$\left\{f^{(i)}, e^{(i)}\right\} \text{ and } \left\{g^{(i)}, \frac{M^i e^{(i)}}{\|M^i e^{(i)}\|}\right\}.$$

Clearly, from (3.15), this matrix is given by

$$A = \begin{pmatrix} \|M^i f^{(i)}\| & 0 \\ 0 & \|M^i e^{(i)}\| \end{pmatrix},$$

and, since $\mathbf{k}_i$ is colinear to $e^{(i)}$, $M^i\mathbf{k}_i = \|M^i e^{(i)}\|\,\mathbf{k}_i = \|M^i f^{(i)}\|^{-1}(\det M^i)\,\mathbf{k}_i$. Notice that among the six terms that appear in $D_z\left(M^i\mathbf{k}_i\right)$ we know that $|\det M^i| \le (const\ b)^i$, $\|M^i f^{(i)}\|^{-1} \le (const)^i$, $\|\mathbf{k}_i\| \le 1$. From (3.48) and (3.49) it follows that

$$\left\|D_z\left(\|M^i f^{(i)}\|^{-1}\right)\right\| \le (const)^i \text{ and } \left\|D_z\left(\det M^i\right)\right\| \le (const\ b)^i.$$

Therefore, to bound $\|D_z\left(M^i\mathbf{k}_i\right)\|$ it is enough to prove that $\|D_z\mathbf{k}_i\| \le (const)^i$. To this end, since

$$\|D_z\mathbf{k}_i\| \le \left\|D_z\left(\left\langle e^{(\nu)}, e^{(i)}\right\rangle\right)\right\| + \left\|D_z e^{(i)}\right\|$$

and
$$D_z\left(\left\langle e^{(\nu)}, e^{(i)}\right\rangle\right) = \left\langle D_z e^{(\nu)}, e^{(i)}\right\rangle + \left\langle e^{(\nu)}, D_z e^{(i)}\right\rangle,$$

from Proposition 3.3 and Remark 3.11, we conclude that

$$\left|\left\langle D_z e^{(\nu)}, e^{(i)}\right\rangle\right| \le \left\|D_z e^{(\nu)}\right\| \left|\sin\left(\text{ang}(e^{(\nu)}, e^{(i)})\right)\right| \le (const\ b)^i$$

and $\left\langle e^{(\nu)}, D_z e^{(i)}\right\rangle \le const\sqrt{b}$. Hence, $\|D_z \mathbf{k}_i\| \le (const)^i$ and $\|D_z M^i \mathbf{k}_i\| \le (const\ b)^i$. This completes the proof of the proposition. □

Chapter 4

CRITICAL POINTS OF THE BIDIMENSIONAL MAP

The properties of the contractive directions stated in Chapter 3 enable us to extend the notion of critical points to the bidimensional case. This extension will be the starting point to prove the main theorem following the techniques used in Chapter 2 for proving Theorem 2.1. Let us first study the behaviour of the orbits of $T_{\lambda,a,b}$ when they stay away from the line $x = \text{arctg } \lambda$.

4.1 Preliminary results

Once again, we take λ sufficiently small and write

$$M(z) = DT_{\lambda,a,b}(z) = \begin{pmatrix} A(z) & B(z) \\ C(z) & D(z) \end{pmatrix}$$

for each $z \in U_m$. We also take $\delta > 0$ small enough and define

$$B_\delta = \left\{ (x,y) \in \mathbf{R}^2 : |\, x - \text{arctg } \lambda\,| < \delta \right\}$$

and $V_\delta = B_\delta \cap \{y = 0\}$.

Lemma 4.1. *Let $z_i = (x_i, y_i)$ be a piece of any orbit of $T_{\lambda,a,b}$ such that $z_i \notin B_\delta$ for $i = 1, ..., k$. If $\mathbf{v}$ is a unit vector with $|slope\ \mathbf{v}| \leq \frac{1}{10}$, then it follows that*

$$\left| slope\ M^i(z_1)\mathbf{v} \right| \leq \sqrt[4]{b}.$$

for $i = 1, ..., k$. Moreover, for $i = 2, ..., k$, we have

$$\left\| M^i(z_1)\mathbf{v} \right\| \geq (1 - \sqrt[5]{b}) \left| f'_{\lambda,a}(x_i) \right| \left\| M^{i-1}(z_1)\mathbf{v} \right\|.$$

Proof. Let $\mathbf{v} = (v_1, v_2)$ be a unit vector such that $|v_2| \leq \frac{1}{10}|v_1|$. Since $|A(z_1)| > 3\delta$, from Proposition 1.3 we obtain that $|\text{slope } M(z_1)\mathbf{v}| < \sqrt[4]{b}$, provided that b is small enough. Using this argument along the whole orbit, we conclude the first part of the statement.

To prove the second part, let us write $\|M^i(z_1)\mathbf{v}\| = \|M(z_i)M^{i-1}(z_1)\mathbf{v}\|$. Because of the C^3-closeness between $T_{\lambda,a,b}$ and $\Psi_{\lambda,a}$, we may also write

$$\left\|M(z_i)M^{i-1}(z_1)\mathbf{v} - D\Psi_{\lambda,a}(z_i)M^{i-1}(z_1)\mathbf{v}\right\| \leq K\sqrt{b}\left\|M^{i-1}(z_1)\mathbf{v}\right\|$$

and so,

$$\left\|M^i(z_1)\mathbf{v}\right\| \geq \left\|D\Psi_{\lambda,a}(z_i)M^{i-1}(z_1)\mathbf{v}\right\| - K\sqrt{b}\left\|M^{i-1}(z_1)\mathbf{v}\right\|. \tag{4.1}$$

Let us check that

$$\left\|D\Psi_{\lambda,a}(z_i)M^{i-1}(z_1)\mathbf{v}\right\| \geq \left(1 - \left|\text{slope } M^{i-1}(z_1)\mathbf{v}\right|\right)\left|f'_{\lambda,a}(x_i)\right|\left\|M^{i-1}(z_1)\mathbf{v}\right\|.$$

Indeed, let

$$\mathbf{w} = (w_1, w_2) = \frac{M^{i-1}(z_1)\mathbf{v}}{\|M^{i-1}(z_1)\mathbf{v}\|}.$$

Then,

$$\|D\Psi_{\lambda,a}(z_i)\mathbf{w}\| = \left(\sqrt{1 + (\text{slope } \mathbf{w})^2}\right)^{-1}\left\|D\Psi_{\lambda,a}(z_i)(1, \frac{w_2}{w_1})\right\| =$$

$$= \left(\sqrt{1 + (\text{slope } \mathbf{w})^2}\right)^{-1}\left|f'_{\lambda,a}(x_i)\right| > (1 - |\text{slope } \mathbf{w}|)\left|f'_{\lambda,a}(x_i)\right|.$$

Now, since $\left|f'_{\lambda,a}(x_i)\right| > |x_i - c_\lambda| > \delta$, if we take $b << 1$ such that $K\sqrt[4]{b} < \delta$, then (4.1) leads to

$$\left\|M^i(z_1)\mathbf{v}\right\| \geq \left(1 - \left|\text{slope } M^{i-1}(z_1)\mathbf{v}\right| - \sqrt[4]{b}\right)\left|f'_{\lambda,a}(x_i)\right|\left\|M^{i-1}(z_1)\mathbf{v}\right\| \geq$$

$$\geq \left(1 - 2\sqrt[4]{b}\right)\left|f'_{\lambda,a}(x_i)\right|\left\|M^{i-1}(z_1)\mathbf{v}\right\| \geq \left(1 - \sqrt[5]{b}\right)\left|f'_{\lambda,a}(x_i)\right|\left\|M^{i-1}(z_1)\mathbf{v}\right\|,$$

and the lemma is proved. $\square$

Recall that in Chapter 2 we denoted by $\{\xi_i(a)\}_{i=1}^{\infty}$ the $f_{\lambda,a}$-orbit of $c_\lambda = \text{arctg } \lambda$. In Chapter 3 we defined $\omega_i(z) = M^i(z)(1, 0)$ for each point $z \in U_m$.

Lemma 4.2. *Let $\{z_i = (x_i, y_i)\}_{i=0}^{k}$ be a piece of any orbit of $T_{\lambda,a,b}$ such that $z_i \notin B_\delta$ for $1 \leq i \leq k$. There exists $b_0 = b_0(k)$ sufficiently small such that if $b \leq b_0$ and $\mathbf{v}$ is a unit vector with $|\text{slope } \mathbf{v}| \leq \sqrt[4]{b}$, then*

$$\left\|M^i(z_1)\dot{\mathbf{v}}\right\| \geq \left(1 - \sqrt[9]{b}\right)\left|f'_{\lambda,a}(f^i_{\lambda,a}(x_0))\right|\left\|M^{i-1}(z_1)\mathbf{v}\right\| \text{ for } 1 \leq i \leq k.$$

Moreover, if $z_0 \in R = \left\{ (x, y) \in U_m : |x - \operatorname{arctg} \lambda| < K b^{\frac{1}{20}} \right\}$, then

$$\frac{\|\omega_i(z_1)\|}{\|\omega_{i-1}(z_1)\|} \geq \left(1 - \sqrt[6]{b}\right) \left|f'_{\lambda,a}(\xi_i(a))\right| \text{ for } 1 \leq i \leq k.$$

Proof. From Lemma 4.1 we have that

$$\frac{\|M^i(z_1)\mathbf{v}\|}{\|M^{i-1}(z_1)\mathbf{v}\|} \geq \left(1 - \sqrt[5]{b}\right) \left|f'_{\lambda,a}(x_i)\right|$$

for $1 \leq i \leq k$ and b small enough. Of course, the result follows if we prove that

$$\left|f'_{\lambda,a}(x_i)\right| > \left(1 - \frac{\sqrt[4]{b}}{2}\right) \left|f'_{\lambda,a}(f^i_{\lambda,a}(x_0))\right|.$$

For this, notice that

$$\left|x_j - f^j_{\lambda,a}(x_0)\right| \leq$$

$$\leq K^j \sqrt{b} + \left\|T^{j-1}_{\lambda,a,b}(\Psi_{\lambda,a}(z_0)) - \Psi^{j-1}_{\lambda,a}(\Psi_{\lambda,a}(z_0))\right\| \leq \dots \leq \sum_{i=1}^{j} K^i \sqrt{b} < K^{2j}\sqrt{b}. \quad (4.2)$$

Hence, since $\left|f'_{\lambda,a}(x_i) - f'_{\lambda,a}(f^i_{\lambda,a}(x_0))\right| \leq K \left|x_i - f^i_{\lambda,a}(x_0)\right| \leq K^{2i+1}\sqrt{b}$, it follows that

$$\left|f'_{\lambda,a}(x_i)\right| \geq \left|f'_{\lambda,a}(f^i_{\lambda,a}(x_0))\right| - K^{2i+1}\sqrt{b} > \left|f'_{\lambda,a}(f^i_{\lambda,a}(x_0))\right| - K^{2k+1}\sqrt{b} \quad (4.3)$$

for $i \in \{1, ..., k\}$. Let us now take $b < b_0(k) = \delta^4 \left(4\lambda K^{2k+1}\right)^{-4}$. Then (4.2) leads to $\left|f'_{\lambda,a}(f^i_{\lambda,a}(x_0))\right| \geq \lambda^{-1} \left|c_\lambda - f^i_{\lambda,a}(x_0)\right| \geq \lambda^{-1} \left(\delta - K^{2k+1}\sqrt{b}\right) > \frac{1}{2}\delta\lambda^{-1}$ and thus,

$$\frac{\sqrt[4]{b}}{2} \left|f'_{\lambda,a}(f^i_{\lambda,a}(x_0))\right| > \frac{\delta\sqrt[4]{b}}{4\lambda} > K^{2k+1}\sqrt{b}.$$

Finally, by using the above inequality in (4.3), we conclude that

$$\left|f'_{\lambda,a}(x_i)\right| \geq \left(1 - \frac{\sqrt[4]{b}}{2}\right) \left|f'_{\lambda,a}\left(f^i_{\lambda,a}(x_0)\right)\right|,$$

and the first part of the statement holds. To complete the proof note that if $z_0 \in R$, then (4.2) in the above argument can be replaced by

$$\left|\xi_j(a) - x_j\right| \leq \left|\xi_j(a) - f^j_{\lambda,a}(x_0)\right| + \left|f^j_{\lambda,a}(x_0) - x_j\right| \leq K^{2j}b^{\frac{1}{20}}. \quad \square \quad (4.4)$$

Remark 4.3. According to (4.2), it follows that if $\{z_i = (x_i, y_i)\}_{i=1}^{k}$ is a piece of any orbit outside B_δ, then $x_0, f_{\lambda,a}(x_0), ..., f^k_{\lambda,a}(x_0)$ is a piece of an unidimensional orbit outside $V_{\frac{\delta}{2}}$. In the same way, if for some $i \in \{1, ..., k\}$ $z_i \in B_\delta$, then $f_{\lambda,a}^i(x_0) \in V_{2\delta}$ whenever $b < b_0(k)$ is small enough.

Proposition 4.4. *For each λ sufficiently small there exist positive constants C_0 and c_0 such that for every sufficiently small $\delta > 0$, there exist $a_1 = a_1(c_0, \delta) < a(\lambda)$ close to $a(\lambda)$ and $b_0 = b_0(c_0, \delta)$ small enough such that for every $(a, b) \in [a_1, a(\lambda)] \times [0, b_0)$ the following statement holds:*

If $\{z_i = (x_i, y_i)\}_{i=1}^k$ is a piece of any orbit of $T_{\lambda,a,b}$ outside B_δ and $\mathbf{v}$ is a unit vector with $|\text{slope } \mathbf{v}| \leq \sqrt[4]{b}$, then:

(a) $\|M^i(z_1)\mathbf{v}\| \geq C_0 e^{c_0 i} \min\left\{\left|f_{\lambda,a}'(\eta)\right| : \eta \notin V_{\frac{\delta}{2}}\right\}$ *for $1 \leq i \leq k$.*

(b) If $z_{k+1} \in B_\delta$, then $\left\|M^k(z_1)\mathbf{v}\right\| \geq C_0 e^{c_0 k}$.

(c) If $z_0 \in B_\delta$, then $\|M^i(z_1)\mathbf{v}\| \geq C_0 e^{c_0 i}$ for $1 \leq i \leq k$.

Proof. Let C_0, $\tilde{c}_0$ and W be as in Proposition 2.2. Let $\delta > 0$ be small enough so that $V_{2\delta} \subset W$. Let us take $\Delta \in \mathbf{N}$ sufficiently large to get $U_\Delta = \left(c_\lambda - e^{-\Delta}, c_\lambda + e^{-\Delta}\right) \subset V_{\frac{\delta}{2}}$ and let $a_1 = a_1(\lambda, \tilde{c}_0, \Delta)$ be the value of the parameter obtained in Proposition 2.2. Let us choose $c_0' \in (0, \tilde{c}_0)$ arbitrarily close to $\tilde{c}_0$ and take $k_0 = k_0(c_0', \delta) \in \mathbf{N}$ large enough so that

$$C_0 e^{\tilde{c}_0 k_0} \min\left\{\left|f_{\lambda,a}'(\eta)\right| : \eta \notin V_{\frac{\delta}{2}}\right\} \geq e^{c_0' k_0}. \tag{4.5}$$

Finally, let $b_0 = b_0(k_0) << 1$ such that, for every $b \leq b_0$, Lemmas 4.1 and 4.2 hold for pieces of orbits of length less than or equal to k_0.

For $1 \leq i \leq k$, let us write $i = m k_0 + p$ with $p < k_0$. Then,

$$\left\|M^i(z_1)\mathbf{v}\right\| = \prod_{j=1}^{k_0} \frac{\|M^j(z_1)\mathbf{v}\|}{\|M^{j-1}(z_1)\mathbf{v}\|} \cdots \prod_{j=(m-1)k_0+1}^{mk_0} \frac{\|M^j(z_1)\mathbf{v}\|}{\|M^{j-1}(z_1)\mathbf{v}\|} \prod_{j=mk_0+1}^{i} \frac{\|M^j(z_1)\mathbf{v}\|}{\|M^{j-1}(z_1)\mathbf{v}\|}$$

and from Lemma 4.2, it follows that

$$\left\|M^i(z_1)\mathbf{v}\right\| \geq \left(1 - \sqrt[6]{b}\right)^i \prod_{l=0}^{m-1}\prod_{j=1}^{k_0} \left|f_{\lambda,a}'(f_{\lambda,a}^{lk_0+j}(x_0))\right| \prod_{j=1}^{p} \left|f_{\lambda,a}'(f_{\lambda,a}^{mk_0+j}(x_0))\right|.$$

Now, from Proposition 2.2(a) and Remark 4.3, we obtain

$$\left\|M^i(z_1)\mathbf{v}\right\| \geq \left(1 - \sqrt[6]{b}\right)^i e^{\tilde{c}_0(mk_0+p)} \left(C_0 \min\left\{\left|f_{\lambda,a}'(\eta)\right| : \eta \notin V_{\frac{\delta}{2}}\right\}\right)^{m+1},$$

and, from (4.5), we conclude that

$$\left\|M^i(z_1)\mathbf{v}\right\| \geq \left(1 - \sqrt[6]{b}\right)^i e^{c_0' mk_0} C_0 e^{\tilde{c}_0 p} \min\left\{\left|f_{\lambda,a}'(\eta)\right| : \eta \notin V_{\frac{\delta}{2}}\right\} \geq$$

$$\geq \left(1 - \sqrt[6]{b}\right)^i e^{c_0' i} C_0 \min\left\{\left|f_{\lambda,a}'(\eta)\right| : \eta \notin V_{\frac{\delta}{2}}\right\}.$$

Then, for any $c_0 \in (0, c_0')$, it follows that $\left\|M^i(z_1)\mathbf{v}\right\| \geq C_0 e^{c_0 i} \min\left\{\left|f_{\lambda,a}'(\eta)\right| : \eta \notin V_{\frac{\delta}{2}}\right\}$, provided that $\left(1 - \sqrt[6]{b}\right) e^{c_0'} > e^{c_0}$. Statement (a) is proved.

Notice that statement (b) follows in a similar way because if $z_{k+1} \in B_\delta$, then, from Remark 4.3, $f_{\lambda,a}^k(x_0) \in V_{2\delta} \subset W$ and Proposition 2.2 yields

$$\prod_{j=1}^{p} \left|f_{\lambda,a}'(f_{\lambda,a}^{mk_0+j}(x_0))\right| \geq C_0 e^{\tilde{c}_0 p}.$$

Finally, to prove statement (c) we go back to the proof of Proposition 2.2(c). For each $\delta > 0$ and $\hat{c}_0 > 0$, there exist $n_1 = n_1(\hat{c}_0, \delta) \in \mathbf{N}$ and $a_1(\lambda)$ close to $a(\lambda)$, such that if $k \leq n_1$ and $a \in [a_1(\lambda), a(\lambda)]$, then $\left|(f_a')^k(f_a(x_0))\right| \geq e^{\hat{c}_0 k}$ holds for every $x_0 \in V_\delta$. Taking $b \leq b(n_1)$ sufficiently small, Lemma 4.2 allows us to state

$$\left\|M^k(z_1)\mathbf{v}\right\| = \prod_{j=1}^{k} \frac{\|M^j(z_1)\mathbf{v}\|}{\|M^{j-1}(z_1)\mathbf{v}\|} \geq e^{\frac{3}{4}\hat{c}_0 k},$$

for $1 \leq k \leq n_1$. So, statement (c) is proved for $k \leq n_1$.

If $k > n_1$, then

$$\left\|M^i(z_1)\mathbf{v}\right\| = \prod_{j=1}^{n_1} \frac{\|M^j(z_1)\mathbf{v}\|}{\|M^{j-1}(z_1)\mathbf{v}\|} \prod_{j=n_1+1}^{i} \frac{\|M^j(z_1)\mathbf{v}\|}{\|M^{j-1}(z_1)\mathbf{v}\|}$$

and, proceeding as in the proof of statement (a), we conclude that

$$\left\|M^i(z_1)\mathbf{v}\right\| \geq e^{\frac{3}{4}\hat{c}_0 n_1} C_0 e^{c_0(i-n_1)} \min\left\{\left|f_{\lambda,a}'(\eta)\right| : \eta \notin V_{\frac{\delta}{2}}\right\} \geq \frac{C_0}{2} e^{\frac{3}{4}\hat{c}_0 n_1} \delta e^{c_0(i-n_1)}.$$

As in the proof of Proposition 2.2, $\delta e^{\frac{1}{2}G_0 n_1} > const$ for a constant G_0 close to $\hat{c}_0$. Therefore, $\left\|M^i(z_1)\mathbf{v}\right\| \geq \hat{C}_0 e^{\frac{1}{2}\hat{c}_0 n_1} e^{c_0(i-n_1)} > \hat{C}_0 e^{c_0' i}$ for $c_0' = \min\left\{\frac{1}{5}\hat{c}_0, c_0\right\}$. Redefining the constants the result follows. $\square$

Proposition 4.5. Let $R = \left\{(x,y) \in U_m : |x - c_\lambda| < Kb^{\frac{1}{20}}\right\}$ and $\tilde{B}_\lambda = \left(\frac{3}{4}c_\lambda, \frac{5}{4}c_\lambda\right) \times \mathbf{R}$. For any sufficiently large natural number N, there exist a set $\Omega_N \subset [a_1(\lambda), a(\lambda)]$ and $b_0 = b_0(\lambda, N)$ small enough such that if $(a, b) \in \Omega_N \times [0, b_0)$, then for each $z_0 \in R$ the following statements hold:

(a) $z_j(a) = T_{\lambda,a,b}^j(z_0) \notin \tilde{B}_\lambda$ for $1 \leq j \leq N - 1$.

(b) $\|\omega_j(a)\| = \|M^j(z_1; a)(1, 0)\| \geq (1 + \lambda^2)^j$ for $1 \leq j \leq N - 1$.

Proof. Let N_0 be the natural number given by Proposition 2.3. Let $\Omega_N \subset [a_1(\lambda), a(\lambda)]$

be the interval given by Proposition 2.5 for $N \geq N_0$. If $a \in \Omega_N$ and $1 \leq j \leq N - 1$, then $\xi_j(a) \leq -\text{arctg } \lambda^2$. Let us now take $b_0 = b_0(N)$ sufficiently small so that Lemma 4.2 holds for every $a \in \Omega_N$ and write $z_j(a) = (x_j(a), y_j(a))$. From (4.4) it follows that $|\xi_j(a) - x_j(a)| \leq K^{2j} b^{\frac{1}{20}}$ and this allows us to set

$$x_j(a) < -\text{arctg } \lambda^2 + K^{2N} b^{\frac{1}{20}} < 0$$

for $1 \leq j \leq N - 1$, provided that $b_0 = b_0(\lambda, N)$ is small enough. So, statement (a) is proved. Statement (b) is immediate from Lemma 4.2. In fact, for $1 \leq j \leq N - 1$, we have

$$\|\omega_j(a)\| \geq \prod_{i=1}^{j} \left(1 - \sqrt[8]{b}\right) \left|f'_{\lambda,a}(\xi_i(a))\right| \geq \prod_{i=1}^{j} \left(1 - \sqrt[8]{b}\right)(1 + \lambda) \geq \left(1 + \lambda^2\right)^j,$$

whenever b_0 is small enough. $\square$

Remark 4.6. *Let $P_{a,m}$ and $Q_{a,m}$ be the analytical continuations of the fixed points of $\Psi_{\lambda,a}$ introduced in Chapter 1. Proposition 1.4 yields a sequence $a_m(\lambda)$ of values of the parameter a such that, for any sufficiently large m, $W^u\left(P_{a_m(\lambda),m}\right)$ and $W^s\left(Q_{a_m(\lambda),m}\right)$ have a homoclinic tangency. Furthermore, $\lim_{m \to \infty} a_m(\lambda) = a(\lambda)$ and there is not other homoclinic tangency for $1 \leq a < a_m(\lambda)$. Henceforth, we shall consider values of the parameter a in a subset Ω_0 which, for the moment, we only need to take close to $a(\lambda)$ and such that $\sup \Omega_0 < a(\lambda)$. Moreover, we shall always take Ω_0 so that $a_1 < \inf \Omega_0$, where a_1 is the value of the parameter given by Proposition 4.4. As soon as Ω_0 has been fixed, we shall take $b = e^{-2\pi\lambda m}$ sufficiently small so that $\sup \Omega_0 < a_m(\lambda)$.*

The next result, whose proof can be seen in [8], will allow us to state properties of the invariant manifolds of $P_{a,m}$ and $Q_{a,m}$ when m is large enough.

Proposition 4.7. *(Dependence of invariant manifolds). Let $U \subset \mathbf{R}^2$ be an open set and $g = (g_t : U \to \mathbf{R}^2)_t$ be a C^k one-parameter family of (not necessarily invertible) maps. Assume that $\theta \in U$ is a hyperbolic fixed point for g_0 and split $\mathbf{R}^2 = E^u \oplus E^s$. Then, for $h = (h_t)_t$ close to g and t close to 0, h_t has a unique hyperbolic fixed point near θ whose unstable manifold may be written as the graph of a C^k map*

$$\Phi_h(t, \cdot) : x \in B_\epsilon(\theta) \subset E^u \to \Phi_h(t, x) \in E^s,$$

where $B_\epsilon(\theta)$ is a neighbourhood of θ in E^u such that $graph(\Phi_g(0, \cdot)) = W^u_{loc}(\theta)$.

Moreover, denoting by $C^{k-1}(V, E^s)$ the space of C^{k-1} maps defined from a neighbourhood V of $(0, \theta) \in \mathbf{R}^3$ to E^s, endowed with the C^{k-1} topology, then Φ_h may be defined in such a way that $h \longrightarrow \Phi_h(\cdot, \cdot) \in C^{k-1}(V, E^s)$ is a class one map.

4.2 Construction of critical points

In this section we shall describe two algorithms for the construction of critical approximations on $C^2(b)$-curves of $W_{a,m}^u$.

Definition 4.8. *We say that γ is a $C^2(b)$-curve if it is the graph of a function $y = y(x)$ such that $|y'(x)| \leq \sqrt[4]{b}$ and $|y''(x)| \leq \sqrt[4]{b}$.*

Definition 4.9. *A point $z_0 \in W^u(P_{a,m}) = W_{a,m}^u$ is said to be a critical approximation of order n if $e^{(n)}(z_1)$ is tangent to $W_{a,m}^u$ at $z_1 = T_{\lambda,a,b}(z_0)$.*

Let $z_0^{(0)}$ be the point of $W_{a,m}^u \cap \{x = c_\lambda\}$ closest to $P_{a,m}$ in $W_{a,m}^u$, and set $z_i^{(0)} = (x_i^{(0)}, y_i^{(0)}) = T_{\lambda,a,b}^i(z_0^{(0)})$. Let $G_0 = \left[z_2^{(0)}, z_1^{(0)}\right]$ be the arc in $W_{a,m}^u$ joining $z_2^{(0)}$ and $z_1^{(0)}$ and define $G_g = T_{\lambda,a,b}^g(G_0) \setminus T_{\lambda,a,b}^{g-1}(G_0)$ for $g \geq 1$.

Definition 4.10. *A critical approximation is said to be of generation g if it belongs to G_g.*

We begin by constructing critical approximations of generation 0 and 1 close to $x = \text{arctg } \lambda$. For this, we need to prove that G_0 and G_1 are $C^2(b)$-curves defined on a suitable domain. This fact will be stated in Proposition 4.13. Now, we set two auxiliary results:

Proposition 4.11. *For each $a_M(\lambda)$ close enough to $a(\lambda)$ there exist $a_0(\lambda) < a_M(\lambda)$, $\delta_0 = \delta_0(a_0) > 0$ and $\delta_1 = \delta_1(a_0) > \delta_0$ such that, for every $a \in (a_0(\lambda), a_M(\lambda))$, the following statements hold:*

(a) $x_i^{(0)} \notin I_0 = [Q_{a_0} + \delta_0, \xi_1(a_0) - \delta_0]$ for $i = 1, 2, 3$.

(b) $x_4^{(0)} \in I_1 = [Q_{a_0} + \delta_1, \xi_1(a_0) - \delta_1]$.

Proof. First, we claim that the result is true in a unidimensional framework: Once $a_M(\lambda)$ is fixed to be close enough to $a(\lambda)$, there exist $a_0 = a_0(\lambda)$, $\delta_0 = \delta_0(a_0)$ and $\delta_1 = \delta_1(a_0) > \delta_0$ such that if $a \in (a_0(\lambda), a_M(\lambda))$, then $\xi_i(a) = f_{\lambda,a}^i(c_\lambda) \notin I_0$ for $i = 1, 2, 3$ and $\xi_4(a) \in I_1$. Indeed, let $F(\lambda, a)$ be the function defined in the proof of Proposition 1.4. Since $F_\lambda : a \to F(\lambda, a)$ is strictly decreasing in a neighbourhood of $a(\lambda)$ and $F(\lambda, a(\lambda)) = 0$, we may take a_0 sufficiently close to $a_M(\lambda)$ so that

$$\frac{40}{\lambda^4}(a_M - a_0) < \frac{1}{4}F(\lambda, a_M). \tag{4.6}$$

Let $\delta_0 = \left(1 + \lambda^{-1}\sqrt{a^2(\lambda) - 1}\right) F(\lambda, a_0)$. Notice that if a_M is sufficiently close to $a(\lambda)$, then $Q_{a_0} + \delta_0 < 0 < \arccos a(\lambda)^{-1} < \xi_1(a_0) - \delta_0$ and hence, we may define $I_0 = [Q_{a_0} + \delta_0, \xi_1(a_0) - \delta_0]$. Now, if $a \in (a_0(\lambda), a_M(\lambda))$, then $\xi_1(a) > \xi_1(a_0) > \xi_1(a_0) - \delta_0$ and $\xi_2(a) - Q_a = F(\lambda, a) < F(\lambda, a_0) < \delta_0$. Therefore, $\xi_1(a) \notin I_0$ and $\xi_2(a) \notin I_0$.

On the other hand, for each $a \in (a_0(\lambda), a_M(\lambda))$, there exists $y \in (Q_a, \xi_2(a))$ such that $\xi_3(a) - Q_a = f'_{\lambda,a}(y)(\xi_2(a) - Q_a) < \left(1 + \lambda^{-1}\sqrt{a^2(\lambda) - 1}\right) F(\lambda, a_0) = \delta_0$ and hence $\xi_3(a) \notin I_0$.

Now, let us see that $\xi_4(a_0) \in I_1$. From the mean value theorem, there exists $y \in (Q_{a_0}, \xi_2(a_0))$ such that $\xi_4(a_0) - Q_{a_0} = \left(f^2_{\lambda,a_0}\right)'(y)F(\lambda, a_0) > \left\{f'_{\lambda,a_0}(f_{\lambda,a_0}(y))\right\}^2 F(\lambda, a_0)$. By applying again the mean value theorem to $f'_{\lambda,a}$ and taking a_M sufficiently close to $a(\lambda)$, it follows that $\left|f'_{\lambda,a_0}(f_{\lambda,a_0}(y))\right| \geq \frac{3}{4}\left|f'_{\lambda,a(\lambda)}(Q_{a(\lambda)})\right|$. Then, since $a(\lambda) > \sqrt{1 + 4\lambda^2}$, we get $\xi_4(a_0) - Q_{a_0} > \left(\frac{5}{2} + \lambda^{-1}\sqrt{a^2(\lambda) - 1}\right) F(\lambda, a_0) > \delta_0$.

To prove that $\xi_4(a) \in I_1$ for every $a \in (a_0, a_M)$, apply the mean value theorem to both terms in the right side of the following inequality

$$|\xi_4(a) - \xi_4(a_0)| \leq |f_{\lambda,a_M}(\xi_3(a_M)) - f_{\lambda,a_0}(\xi_3(a_M))| + |f_{\lambda,a_0}(\xi_3(a_M)) - f_{\lambda,a_0}(\xi_3(a_0))|.$$

So, $|\xi_4(a) - \xi_4(a_0)| < \lambda^{-1}(a_M - a_0) + 3\lambda^{-1}|\xi_3(a_M) - \xi_3(a_0)| \leq \ldots \leq 40\lambda^{-4}(a_M - a_0)$ and, from (4.6), $|\xi_4(a) - \xi_4(a_0)| > \left(\frac{9}{4} + \lambda^{-1}\sqrt{a^2(\lambda) - 1}\right) F(\lambda, a_0) = \delta_1 > \delta_0$.

Finally, according to (4.2), the bidimensional statement holds for

$$\delta_0 = \left(\frac{5}{4} + \frac{\sqrt{a^2(\lambda) - 1}}{\lambda}\right) F(\lambda, a_0) \text{ and } \delta_1 = \left(2 + \frac{\sqrt{a^2(\lambda) - 1}}{\lambda}\right) F(\lambda, a_0)$$

whenever $b << F(\lambda, a_0)$. □

Lemma 4.12. *Let $J \subset \mathbb{R}$ be an interval such that $J \cap (c_\lambda - \kappa, c_\lambda + \kappa) = \emptyset$ for some $\kappa > 0$. Let U be a neighbourhood of a_0 and $\{C_a : a \in U\}$ be a family of curves defined as the graphs of functions $x \in J \rightarrow y(a, x)$, where $y : U \times J \rightarrow \mathbb{R}$ is C^2. If $\|y\|_{C^2(a,x)} \leq const\sqrt{b}$ and b is sufficiently small $(0 < b << \kappa^2)$, then each $T_{\lambda,a,b}(C_a)$ is given by the graph of a map $x \in I_a \rightarrow \tilde{y}(a, x)$ such that:*

(a) Each interval $I_a = [\alpha(a), \beta(a)]$ verifies $\lim_{a \to a_0} \alpha(a) = \alpha(a_0)$ and $\lim_{a \to a_0} \beta(a) = \beta(a_0)$.

(b) If I is an interval such that $I \subset I_a$, for each $a \in U$, then $\tilde{y} : U \times I \rightarrow \mathbb{R}$ is a C^2 map such that $\|\tilde{y}\|_{C^2(a,x)} \leq const\sqrt{b}$.

Proof. Let $(1, y'_a(x))$ be a tangent vector to the curve C_a at $z = (x, y(a, x))$. Then,

$$|\text{slope } D_z T_{\lambda,a,b}(z)(1, y'_a(x))| = \frac{|C + Dy'_a(x)|}{|A + By'_a(x)|} < \frac{K\sqrt{b} + constKb}{3\kappa - constKb} < const\sqrt{b}.$$

Therefore, $T_{\lambda,a,b}(C_a)$ is the graph of a function $x \in I_a \to \tilde{y}(a,x)$ defined on an interval $I_a = [\alpha(a), \beta(a)]$ and such that $\left|\frac{\partial}{\partial x}\tilde{y}(a,x)\right| < const\sqrt{b}$. Statement (a) follows from the continuity of $y(a,x)$ and $T_{\lambda,a,b}$. Since $T_{\lambda,a,b}$ is a Lipschitz function, it also follows that $\|\tilde{y}\|_{C^0(a,x)} \le const\sqrt{b}$.

To prove statement (b) write

$$T_{\lambda,a,b}^{(2)}(x, y(a,x)) - \tilde{y}\left(a, T_{\lambda,a,b}^{(1)}(x, y(a,x))\right) = T_a^{(2)}(x,y) - \tilde{y}(a, T_a^{(1)}(x,y)) = 0,$$

where $T_a^{(i)}$ denotes the components of T_a ($i = 1, 2$). It will be also helpful to write $\overline{x} = T_a^{(1)}(x,y)$. Deriving with respect to x, we obtain

$$F_x^{(2)} - F_x^{(1)}\frac{\partial \tilde{y}}{\partial \overline{x}}(a, \overline{x}) = 0, \tag{4.7}$$

where $F_x^{(i)} = D_x T_a^{(i)}(x,y) + D_y T_a^{(i)}(x,y)\frac{\partial y}{\partial x}$.

Since $\left|F_x^{(2)}\right| \le const\sqrt{b}$ and $const\,\kappa < \left|F_x^{(1)}\right| < const$, we get

$$\left|\frac{\partial \tilde{y}}{\partial \overline{x}}\right| \le const\sqrt{b}.$$

On the other hand, deriving with respect to a we have

$$F_a^{(2)} - \frac{\partial \tilde{y}}{\partial a}(a, \overline{x}) - F_a^{(1)}\frac{\partial \tilde{y}}{\partial \overline{x}}(a, \overline{x}) = 0, \tag{4.8}$$

where $F_a^{(i)} = D_a T_a^{(i)}(x,y) + D_y T_a^{(i)}(x,y)\frac{\partial y}{\partial a}$. Since $\left|F_a^{(1)}\right| \le const$, $\left|F_a^{(2)}\right| \le const\sqrt{b}$ we also obtain

$$\left|\frac{\partial \tilde{y}}{\partial a}\right| \le const\,b.$$

Now, derive (4.7) and (4.8) with respect to both x and a to get

$$F_{xx}^{(2)} - \left[F_x^{(1)}\right]^2 \frac{\partial^2 \tilde{y}}{\partial \overline{x}^2}(a, \overline{x}) - F_{xx}^{(1)}\frac{\partial \tilde{y}}{\partial \overline{x}}(a, \overline{x}) = 0$$

$$F_{xa}^{(2)} - F_x^{(1)}\frac{\partial^2 \tilde{y}}{\partial \overline{x}\partial a}(a, \overline{x}) - \left[F_x^{(1)}\right]^2 \frac{\partial^2 \tilde{y}}{\partial \overline{x}^2}(a, \overline{x}) - F_{xa}^{(1)}\frac{\partial \tilde{y}}{\partial \overline{x}}(a, \overline{x}) = 0$$

$$F_{aa}^{(2)} - \frac{\partial^2 \tilde{y}}{\partial a^2}(a, \overline{x}) - 2F_a^{(1)}\frac{\partial^2 \tilde{y}}{\partial \overline{x}\partial a}(a, \overline{x}) - \left[F_a^{(1)}\right]^2 \frac{\partial^2 \tilde{y}}{\partial \overline{x}^2}(a, \overline{x}) - F_{aa}^{(1)}\frac{\partial \tilde{y}}{\partial \overline{x}}(a, \overline{x}) = 0.$$

Since $\left|F_{xx}^{(1)}\right| \le const$, $\left|F_{xx}^{(2)}\right| \le const\sqrt{b}$, $\left|F_{xa}^{(i)}\right| \le const\sqrt{b}$ and $const\,\kappa < \left|F_x^{(1)}\right|$, it follows that

$$\left|\frac{\partial^2 \tilde{y}}{\partial \overline{x}^2}\right| \le const\sqrt{b}, \quad \left|\frac{\partial^2 \tilde{y}}{\partial \overline{x}\partial a}\right| \le const\sqrt{b} \text{ and } \left|\frac{\partial^2 \tilde{y}}{\partial a^2}\right| \le const\sqrt{b}.$$

The proof is now complete. $\square$

Proposition 4.13. *Let $a_M(\lambda)$, $a_0(\lambda)$ and I_0 be as in Proposition 4.11. Then, for each $a \in (a_0(\lambda), a_M(\lambda))$ and $i = 0, 1$, the curves $G_i(a) \cap (I_0 \times \mathbf{R})$ are the graphs of maps*

$$(a, x) \in (a_0(\lambda), a_M(\lambda)) \times I_0 \rightarrow y(a, x) \in \mathbf{R}$$

satisfying $\|y\|_{C^2(a,x)} \leq const\sqrt{b}$.

Proof. According to Proposition 4.7, the statement holds for G_0 in a sufficiently small neighbourhood of $P_{a,m}$. Indeed, there exists C^2 functions $\Phi_{T_\lambda} = y_{T_\lambda}$ such that $graph(y_{T_\lambda}) = W^u_{loc}(P_{a,m})$ and, moreover, since $T_{\lambda,a,b}$ and $\Psi_{\lambda,a}$ are C^3 close, we obtain

$$\|\Phi_{T_\lambda} - \Phi_{\Psi_\lambda}\|_{C^2(a,x)} \leq const\sqrt{b}.$$

Hence, since $\Phi_{\Psi_\lambda} \equiv 0$, it follows that $\|y_{T_\lambda}\|_{C^2(a,x)} \leq const\sqrt{b}$.

On the other hand, from Proposition 4.11(a), there exists a constant $\kappa = \kappa(\delta_0)$ such that if $x \in [c_\lambda - \kappa, c_\lambda + \kappa]$, then $f^i_{\lambda,a}(x) \notin I_0$ for $i = 1, 2, 3$. Thus, there exists a constant, which we also denote by $\kappa = \kappa(\delta_0)$, such that if $z_0 \in B_\kappa$, then $z_i = T^i_{\lambda,a,b}(z_0) \notin I_0 \times \mathbf{R}$ for $i = 1, 2, 3$.

Let $z^+ = G_0(a) \cap \{x = c_\lambda + \kappa\}$ and $z^- = G_0(a) \cap \{x = c_\lambda - \kappa\}$. For $i > 0$, let $z^+_i = T^i_{\lambda,a,b}(z^+)$ and $z^-_i = T^i_{\lambda,a,b}(z^-)$. Let us consider $\left[z^+, z^+_1\right] \subset G_0(a)$. From Proposition 4.7, there exists a natural number N_1 such that $T^{-N_1}_{\lambda,a,b}\left(\left[z^+, z^+_1\right]\right)$ is the graph of a function $(a, x) \rightarrow y(a, x)$ satisfying $\|y(a, x)\|_{C^2(a,x)} \leq const\sqrt{b}$. Then, for b small enough, Lemma 4.12 implies that the same holds if we replace $T^{-N_1}_{\lambda,a,b}\left(\left[z^+, z^+_1\right]\right)$ by $\left[z^+, z^+_1\right]$. Furthermore,

$$G_0(a) \cap (I_0 \times \mathbf{R}) \subset \left[z^+_2, z^+_1\right] = T_{\lambda,a,b}\left(\left[z^+, z^+_1\right]\right)$$

and $\left[z^+, z^+_1\right] \cap B_\kappa = \emptyset$. Therefore, once again from Lemma 4.12, we get the result for $G_0(a) \cap (I_0 \times \mathbf{R})$. Since

$$G_1(a) \cap (I_0 \times \mathbf{R}) \subset \left[z^+_3, z^-_1\right] = T_{\lambda,a,b}\left(\left[z^+_2, z^-\right]\right)$$

the above argument can be applied to $G_1(a) \cap (I_0 \times \mathbf{R})$ and the proof ends. $\square$

Remark 4.14. *From Proposition 4.13 it follows that the curves $G_i(a) \cap (I_0 \times \mathbf{R})$, for $i = 0, 1$, are contained in a horizontal strip with width $const\sqrt{b}$ centered at $y = 0$. Furthermore, it is easy to check that the first three folds of $W^u_{a,m}$ occur as in Figure 4.1. Therefore, the whole $W^u_{a,m}$ is contained in the afore-said strip.*

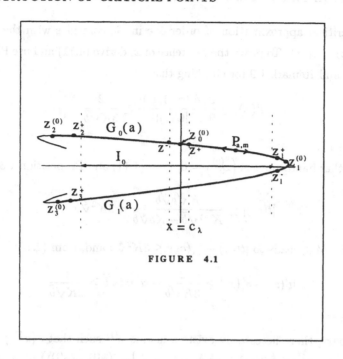

FIGURE 4.1

4.2.1 Critical approximations of generation zero: The algorithm A

We choose $a \in \Omega_0 = \Omega_0(\lambda)$ to be an interval of parameter values close enough to $a(\lambda)$ so that the previous results hold. Let $z(x) = (x, y(x))$ be a parametrization of G_0 defined on I_0 such that $|y'(x)| < \sqrt[4]{b}$ and $|y''(x)| < \sqrt[4]{b}$. Let $L = \lambda b^{\frac{1}{20}}$. From Proposition 4.5, it follows that $z_1(x) = T_{\lambda,a,b}(z(x))$ is e^c-expanding up to time $N - 1$ for each $x \in V_L = B_L \cap \{y = 0\}$ and any $c \in (0, \log(1 + \lambda^2))$. Then, whenever $1 \le n \le N - 1$ and $x \in (c_\lambda - L, c_\lambda + L)$, we may define $q_n(x)$ so that the vector $(q_n(x), 1)$ is colinear to $e^{(n)}(z_1(x))$. Thus, according to Proposition 3.4, we get

$$|q_n(x)| = \left| \text{slope } f^{(n)}(z_1(x)) \right| \le K_4 \sqrt{b} \tag{4.9}$$

and taking into account Remark 3.11, we obtain

$$|q_n'(x)| \le const\sqrt{b}. \tag{4.10}$$

Now, since $(1, y'(x))$ is tangent to G_0 at $z(x)$, the tangent space to $W_{a,m}^u$ at $z_1(x)$ is generated by $(t(x), 1)$, with

$$t(x) = \frac{A(z(x)) + B(z(x))y'(x)}{C(z(x)) + D(z(x))y'(x)}. \tag{4.11}$$

Hence, the critical approximation of order one in G_0 coincides with the point $z(x)$ such that $t(x) = q_1(x)$. To prove the existence of x, derive (4.11) and use Propositions 1.3 and 4.13 and Remark 1.9 for checking that

$$|t'(x)| \geq \frac{2}{9} \frac{|A'(z)(1, y')|}{|C(z)|} \geq \frac{2}{3K\sqrt{b}}$$

for every $x \in V_L$.

On the other hand, since $A\left(z_0^{(0)}\right) = 0$, from (4.11) and Proposition 1.3, we obtain

$$|t(c_\lambda)| \leq \frac{K\sqrt{b}\sqrt[4]{b}}{K^{-1}\sqrt{b} - Kb\sqrt[4]{b}} < 2K^2\sqrt[4]{b}.$$

Consequently, (4.9) leads to $|t(c_\lambda) - q_1(c_\lambda)| \leq 3K^2\sqrt[4]{b}$ and, from (4.10),

$$|t'(x) - q_1'(x)| \geq \frac{2}{3K\sqrt{b}} - const\sqrt{b} \geq \frac{1}{2K\sqrt{b}}$$

for each $x \in V_L$.

In summary, there necessarily exists a unique $x^{(1)}$ such that $t\left(x^{(1)}\right) = q_1\left(x^{(1)}\right)$, which defines a unique critical approximation $z_0^{(1)} = \left(x^{(1)}, y(x^{(1)})\right)$ of order one. This approximation verifies

$$\left|x^{(1)} - c_\lambda\right| < 3K^2\sqrt[4]{b}\left(\frac{1}{2K\sqrt{b}}\right)^{-1} = 6K^3\sqrt[4]{b}\sqrt{b}$$

Next, we shall construct, from $z_0^{(1)}$, a critical approximation of order two in G_0. For this, notice that $\left|q_1(x^{(1)}) - q_2(x^{(1)})\right| < 2\left|ang\left(e^{(2)}(z_1^{(1)}), e^{(1)}(z_1^{(1)})\right)\right|$. So, according to the proof of Proposition 3.3(a), we have $\left|q_2(x^{(1)}) - t(x^{(1)})\right| = \left|q_2(x^{(1)}) - q_1(x^{(1)})\right| \leq 6K(Kb)$. Therefore, we may argue as above to obtain

$$|t'(x) - q_2'(x)| \geq \frac{1}{2K\sqrt{b}}$$

for every $x \in V_L$. Then, there exists a unique $x^{(2)}$ such that $t\left(x^{(2)}\right) = q_2\left(x^{(2)}\right)$. So, we get a unique critical approximation of order two, $z_0^{(2)} = \left(x^{(2)}, y(x^{(2)})\right)$, which satisfies $\left|x^{(1)} - x^{(2)}\right| \leq 12K^3b\sqrt{b}$.

In fact, the previous process can be repeated whenever e^c-expansiveness holds. Notice that, for $1 \leq n \leq N - 1$, Proposition 3.3 implies

$$\left|ang\left(e^{(n)}(z_1^{(n-1)}), e^{(n-1)}(z_1^{(n-1)})\right)\right| \leq 3K(Kb)^{n-1},$$

which, together with the inequality $|t'(x) - q'_n(x)| \geq \frac{1}{2} \left(K\sqrt{b} \right)^{-1}$, enables us to obtain

$$\left| x^{(n)} - x^{(n-1)} \right| \leq 12K^2 \left(Kb \right)^{n-1} \sqrt{b}$$

and so,

$$\left| x^{(n)} - c_\lambda \right| \leq 12K^3 \sqrt[4]{b}\sqrt{b} + 12K^2\sqrt{b} \sum_{i=1}^{n-1} (Kb)^i \; (<< \lambda b^{\frac{1}{20}}).$$

Therefore, up to time $N-1$, we shall find points $z_0^{(n)} = \left(x^{(n)}, y(x^{(n)}) \right)$, with $e^{(n)} \left(z_1^{(n)} \right)$ tangent to $W_{a,m}^u$ at $z_1^{(n)}$, which are n-th critical approximations. Furthermore, we may bound the distance in $W_{a,m}^u$ between critical approximations of consecutive orders according to the following inequality:

$$\left| z_0^{(n)} - z_0^{(n-1)} \right| \leq \left(1 + \sqrt[4]{b} \right) \left| x^{(n)} - x^{(n-1)} \right| \leq 15K^2 \left(Kb \right)^{n-1} \sqrt{b} < (Kb)^{n-1}.$$

In general, for $1 \leq i < j \leq N-1$, we have

$$\left| z_0^{(i)} - z_0^{(j)} \right| \leq (Kb)^i. \tag{4.12}$$

This process which from now on will be called algorithm A, does not depend on G_0. It can be used on any long enough $C^2(b)$-curve contained in B_L, provided that there exists a initial point $(x, y(x))$, far enough from the boundary of B_L, such that $|t(x) - q_1(x)|$ is small. For instance, this algorithm can be iterated in order to obtain a sequence of critical approximations $\left\{ \omega_0^{(n)} \right\}_{n=1}^{N-1}$ in G_1. Henceforth, for $n \leq N-1$, $C_n = \left\{ z_0^{(n-1)}, \omega_0^{(n-1)} \right\}$ will be called the n-th critical set.

Remark 4.15. *The images of the elements of C_n are expansive up to time n for $n \leq N-1$. The expansiveness of $z_1^{(N-2)}$ and $\omega_1^{(N-2)}$ up to time $N-1$ enables us to construct the points $z_0^{(N-1)}$ and $\omega_0^{(N-1)}$ for every $a \in \Omega$. Nevertheless, the N-th iterate of these points may be too close to the line $x = c_\lambda$ for some values of the parameter a and so, we cannot guarantee expansiveness up to time N. Thus, we do not define the set C_N yet.*

4.2.2 Critical approximations of higher generations: The algorithm B

In this section, we describe an algorithm to construct, from known approximations, critical approximations of the same order but higher generation.

Let x_0 be such that $|x_0 - c_\lambda| << \lambda b^{\frac{1}{20}}$ and let $l > 0$. Assume that there exist two $C^2(b)$-curves contained in $W^u_{a,m}$, defined by

$$\gamma : x \in (x_0 - l, x_0 + l) \to z(x) = (x, y(x)),$$

$$\widetilde{\gamma} : x \in (x_0 - l, x_0 + l) \to \widetilde{z}(x) = (x, \widetilde{y}(x)).$$

Let us set $z_0 = z(x_0)$ and $\zeta_0 = \widetilde{z}(x_0)$ and assume that $\zeta_1 = T_{\lambda,a,b}(\zeta_0)$ is e^c-expanding up to time $\mu \geq 1$. Moreover, we suppose that ζ_0 is a critical approximation of order μ, that is, $\widetilde{t}(x_0) = \widetilde{q}_\mu(x_0)$, where $(\widetilde{q}_\mu(x_0), 1)$ is colinear to $e^{(\mu)}(\zeta_1)$.

Let $d = \|z_0 - \zeta_0\|$. Remark 4.14 implies $d < const\sqrt{b} << \lambda b^{\frac{1}{20}}$ and thus,

$$\left(x_0 - \sqrt{d}, x_0 + \sqrt{d}\right) \subset \left(c_\lambda - \lambda b^{\frac{1}{20}}, c_\lambda + \lambda b^{\frac{1}{20}}\right).$$

Now, and this is crucial, we assume that

$$d = \|z_0 - \zeta_0\| < \min\left\{\frac{\sigma_0^{2\mu}}{4}, l^2\right\}, \tag{4.13}$$

where $\sigma_0 = \left(\frac{1}{10K^2}\right)^2$. Then, for every $x \in \left(x_0 - \sqrt{d}, x_0 + \sqrt{d}\right)$, we get $\|z(x) - \zeta_0\| \leq \left(1 + \sqrt[4]{b}\right)\sqrt{d} + d < 2\sqrt{d} < \sigma_0^\mu$ and, from Proposition 3.6, it follows that $z_1(x)$ is e^c-expanding up to time μ.

Since γ and $\widetilde{\gamma}$ are disjoint, then $|y'(x_0) - \widetilde{y}'(x_0)| \leq 2\sqrt{d}$. Furthermore, since $z_0 \in B_L$, we have $\|DT_{\lambda,a,b}(z_0)\| \leq const\, b^{\frac{1}{20}}$. Therefore we obtain

$$\left|t(x_0) - \widetilde{t}(x_0)\right| \leq \frac{2K}{\sqrt{b}}\left\{const\, b^{\frac{1}{20}}\sqrt{d} + Kd\sqrt{1 + \sqrt{b}}\right\} \leq constK\frac{b^{\frac{1}{20}}}{\sqrt{b}}\sqrt{d} + 4K^2\frac{d}{\sqrt{b}} \tag{4.14}$$

from

$$|C(z) + D(z)y'| \geq \frac{\sqrt{b}}{2K}.$$

On the other hand, since $e^{(\mu)}$ is nearly vertical, from Proposition 3.10 it follows that $|q_\mu(x_0) - \widetilde{q}_\mu(x_0)| \leq 4\left\|e^{(\mu)}(z_1) - e^{(\mu)}(\zeta_1)\right\| \leq 4K_{10}Kd\sqrt{b}$. Using (4.14) and the fact that ζ_0 is a μ-th critical approximation, we get

$$|t(x_0) - q_\mu(x_0)| \leq constK\frac{b^{\frac{1}{20}}}{\sqrt{b}}\sqrt{d} + 4K^2\frac{d}{\sqrt{b}} + 4K_{10}Kd\sqrt{b}.$$

Finally, recall that in the construction of algorithm A we proved that $\left|t'(x) - q'_\mu(x)\right| \geq \frac{1}{2}\left(K\sqrt{b}\right)^{-1}$ for every $x \in V_L$. Therefore, we have precisely those conditions which allow the use of algorithm A. So, we may claim that there exists a μ-th critical approximation $z_0^{(\mu)} = \left(x^{(\mu)}, y(x^{(\mu)}) \right) \in \gamma$, which satisfies

$$\left| x^{(\mu)} - x_0 \right| \leq const K b^{\frac{1}{20}} \sqrt{d} + 8K^2 d + 8K_{10} K db < \frac{\sqrt{d}}{4}. \tag{4.15}$$

We shall say that $z_0^{(\mu)}$ has been constructed by means of algorithm B.

4.3 The contractive fields

We finish this chapter by introducing a concept which will be needed to define the binding period in dimension two.

Let ξ be such that $\sqrt{b} << \xi < \sqrt{K}$ as in Chapter 3. Let us consider the interval $I_1 = (Q_{a_0} + \delta_1, \xi_1(a_0) - \delta_1)$ given in Proposition 4.11 and let $z = (x, y)$ be a ξ-expanding point up to time τ, with $x \in I_1$. As in the proof of Proposition 3.4, it follows that $z \notin B_{2\delta}$ for some $\delta = const \, \xi$. Furthermore, if $z \notin B_\delta$, then z is $C(\delta)$-expanding up to time 1. Therefore, z belongs to the region where the C^2 vector field $z \rightarrow e^{(1)}(z)$ is defined on. So, there exists a unique integral curve Γ^1 passing through z. According to Proposition 3.4(a), we know that Γ^1 is a nearly vertical curve. Thus, it can be given by $(x_1(t), t)$, where

$$x_1 : t \in U_y \subset \left(-const\sqrt{b}, const\sqrt{b} \right) \rightarrow x_1(t) \in \mathbf{R}$$

is a C^2 function satisfying $|x_1'(t)| = \left| slope \, f^{(1)}(x_1(t), t) \right| \leq K_4 \sqrt{b}$. In this way, for every point $z_1 \in \Gamma^1$, it follows that

$$|z_1 - z| \leq \int_{U_y} \|(x_1'(t), 1)\| \, dt \leq const\sqrt{b},$$

where $|z_1 - z|$ denotes the distance in Γ^1. Therefore, Γ^1 does not intercept B_δ and consequently, it can be extended to G_0 and G_1, which are $C^2(b)$-curves defined on the interval $I_0 \supset I_1$ given in Proposition 4.11.

From Proposition 3.3, we have $|T_{\lambda,a,b}(z) - T_{\lambda,a,b}(z_1)| < const\sqrt{b}K_3 b < \left(const\sqrt{b} \right)^2$ for every $z \in \Gamma^1$. Then, Proposition 3.7 implies that z_1 is also expanding up to time two and, from Proposition 3.6, the same holds for every $\overline{z}_1 \in U_1 = \bigcup_{z_1 \in \Gamma^1}$ $[z_1 - \sigma, z_1 + \sigma]$, where $\sigma < \left(\frac{\xi}{10K^2} \right)^2$ is a fixed positive real number and $[z_1 - \sigma, z_1 + \sigma]$ is a horizontal segment of length 2σ centered at z. Notice that σ can be chosen sufficiently small so that the boundary of U_1 cuts G_0 and G_1 inside $(I_0 \times \mathbf{R}) \setminus B_\delta$.

Now, we may consider the vector field $z \to e^{(2)}(z)$ defined on U^1 and the respective integral curve passing through z, which we denote by Γ^2. Using once again Proposition 3.4, Γ^2 is a nearly vertical curve which can be parametrized by $(x_2(t), t)$. Repeating the previous arguments, we check that each $z_2 \in \Gamma^2$ satisfies $|z_2 - z| \le const\sqrt{b}$ and so, in the joint domain of definition of Γ^1 and Γ^2 we have $|x_1(t) - x_2(t)| \le const\sqrt{b}$. Then, from Propositions 3.3 and 3.10, we obtain $|x_1'(t) - x_2'(t)| \le const\sqrt{b} < \sigma$. From the mean value theorem it follows that $|x_1(t) - x_2(t)| \le const\sqrt{b}\sigma << \sigma$ and hence Γ^2 can be extended to G_0 and G_1 without leaving U^1.

Let $U^2 = \underset{z_2 \in \Gamma^2}{\cup} [z_2 - \sigma^2, z_2 + \sigma^2]$. As mentioned above and since $\sigma << 1$, we have $U^2 \subset U^1$. The previous arguments allow us to claim that every point of U^2 is ξ-expanding up to time 3. Indeed, if $z_2 \in \Gamma^2$, since Proposition 3.3 implies

$$\left\| T_{\lambda,a,b}(z_2) - T_{\lambda,a,b}(z) \right\| \le const\sqrt{b}K_3b < \left(const\sqrt{b} \right)^2$$

and

$$\left\| T^2_{\lambda,a,b}(z_2) - T^2_{\lambda,a,b}(z) \right\| \le const\sqrt{b}\,(K_3b)^2 < \left(const\sqrt{b} \right)^3,$$

the claim follows from Proposition 3.7. If $z_2 \notin \Gamma^2$, the claim follows from Proposition 3.6.

In summary, we may find, by recurrence, a nearly vertical curve Γ^τ, passing through z, which intercepts G_0 and G_1 inside $(I_0 \times \mathbf{R}) \setminus B_\delta$. Moreover, for each $z_\tau \in \Gamma^\tau$, we obtain

$$\left\| T^\nu_{\lambda,a,b}(z_\tau) - T^\nu_{\lambda,a,b}(z) \right\| \le const\sqrt{b}\,(K_3b)^\nu < \left(const\sqrt{b} \right)^{\nu+1} \tag{4.16}$$

for $0 \le \nu \le \tau$. If $\mathbf{u}$ and $\mathbf{v}$ are unit vectors such that $|\text{slope } \mathbf{u}| < \frac{1}{10}$ and $|\text{slope } \mathbf{v}| < \frac{1}{10}$, then Proposition 3.7 yields

$$\frac{1}{2}\left\| DT^\nu_{\lambda,a,b}(z_\tau)\mathbf{v} \right\| \le \left\| DT^\nu_{\lambda,a,b}(z)\mathbf{u} \right\| \le 2\left\| DT^\nu_{\lambda,a,b}(z_\tau)\mathbf{v} \right\| \tag{4.17}$$

and

$$\left| \text{ang}\left(DT^\nu_{\lambda,a,b}(z_\tau)\mathbf{v}, DT^\nu_{\lambda,a,b}(z)\mathbf{u} \right) \right| \le \left(K_7\sqrt[4]{b} \right)^{\nu+1}, \tag{4.18}$$

for $1 \le \nu \le \tau$. A particular case, which will be useful later, is given by $z_\tau = \Gamma^\tau \cap G_1$, $\mathbf{u} = (1, 0)$ and $\mathbf{v}$ the unit vector tangent to G_1 at z_τ. See Proposition 4.13 in order to check that the assumptions in Proposition 3.7 are fulfilled.

Chapter 5

THE INDUCTIVE PROCESS

Let us recall that in Chapter 4 we constructed, for each $n \leq N - 1$, the n-th critical set $C_n = \left\{ z_0^{(n-1)}, \omega_0^{(n-1)} \right\}$, where $z_0^{(n-1)}$ and $\omega_0^{(n-1)}$ are the critical approximations of generations zero and one, respectively.

The main goal of this chapter is to inductively define the sets C_n for every $n \in \mathbf{N}$. So, for $n \geq N$, we assume that C_{n-1} satisfies the hypotheses I.H.1, I.H.2,...below, which allow us to derive C_n from C_{n-1}.

Let us assume that, for every $k \leq n - 1$, the k-th critical set C_k has been constructed and its elements are critical approximations of order $k - 1$ and generation lower than or equal to θk, where $\theta = \theta(b) << 1$ is given by

$$\theta = \frac{20 \log \left(2\sigma_0^{-1} \right)}{\log b^{-1}}, \text{ with } \sigma_0 = \left(\frac{1}{10K^2} \right)^2.$$

Moreover $C_k \subset B_L$, with L as in 4.2.1.

From the upper bound for the generation of the elements of C_k we get an upper bound for the cardinal of C_k. This is crucial to show that the set of parameters excluded during the proof of the main theorem has small Lebesgue measure.

Next, we introduce the first three induction hypotheses:

I.H.1. The image $z_1^{(k-1)}$ of each $z_0^{(k-1)} \in C_k$ is e^c-expanding up to time k, where c is a constant such that $0 < c < min\{\frac{1}{2}c_0, log(1 + \lambda^2)\}$.

I.H.2. If $\gamma(\zeta, r)$ denotes the segment of $W_{a,m}^u$ of radius r centered at ζ, we assume that $\gamma(z_0^{(k-1)}, \rho_0^{\theta k})$ is a $C^2(b)$-curve for each $z_0^{(k-1)} \in C_k$, where $\rho_0 = \rho_0(\lambda, K, \delta)$ is a small positive constant which will be determined later. If the generation of $z_0^{(k-1)}$ is zero or one, then $\rho_0^{\theta k}$ is replaced by $\frac{1}{4}\Lambda$, where $\Lambda = f_{\lambda, a(\lambda)}(c_\lambda) + arccos\, a(\lambda)^{-1}$ is the length of the unstable manifold of the fixed point $P_{a(\lambda)}$ of $f_{\lambda, a(\lambda)}$.

I.H.3. *If $g \geq 1$ is the generation of $z_0^{(k-1)}$, then $T_{\lambda,a,b}^{1-g}\left(\gamma(z_0^{(k-1)}, \rho_0^{\theta k})\right)$ is contained in $G_1 \cap (I_0 \times \mathbf{R})$, where I_0 is the interval in Proposition 4.11. Furthermore, if C_0 is the constant in Proposition 4.4, then the tangent vectors to $T_{\lambda,a,b}^{1-g}\left(\gamma(z_0^{(k-1)}, \rho_0^{\theta k})\right)$ are $\frac{1}{4}C_0$ expanded by $DT_{\lambda,a,b}^{g-1}$. For $g = 1$, $\rho_0^{\theta k}$ is replaced again by $\frac{1}{4}\Lambda$.*

Remark 5.1. *With regard to I.H.3, notice that if g is the generation of $z_0^{(k-1)}$, then $T_{\lambda,a,b}^{1-g}\left(z_0^{(k-1)}\right) \in G_1$. Since this preimage could be near the folds of $W_{a,m}^u$, avoiding this fact will be very important. Therefore, we need $T_{\lambda,a,b}^{1-g}\left(z_0^{(k-1)}\right)$ to be contained in the nearly horizontal part of G_1.*

Before going on, notice that for $n \leq N-1$ the elements of C_n satisfy I.H.1 to I.H.3. We already know that $z_1^{(n-1)}$ and $\omega_1^{(n-1)}$ are e^c-expanding up to time n. The second hypothesis is immediately fulfilled, because both critical approximations belong to $G_i \cap B_L$ for $i = 0,1$ and these curves are $C^2(b)$ on $I_0 \times \mathbf{R}$. The third hypothesis is also trivially true.

Lemma 5.2. *$z_0^{(k-1)}$ is the unique point of C_k in $\gamma\left(z_0^{(k-1)}, \frac{1}{2}\rho_0^{\theta k}\right)$.*

Proof. The proof is based on the following fact: If $z_0^{(l)}$ and $\zeta_0^{(l)}$ are critical approximations of the same order, contained in the same $C^2(b)$-curve of $W_{a,m}^u$, and their images are e^c-expanding up to time l, then $z_0^{(l)} = \zeta_0^{(l)}$ whenever $\left|z_0^{(l)} - \zeta_0^{(l)}\right| \leq \sigma_0^l$. In fact, let us consider the parametrization of $\left[z_0^{(l)}, \zeta_0^{(l)}\right]$ defined by $x \in (x_1, x_2) \to z(x) = (x, y(x))$, with $z(x_1) = z_0^{(l)}$, $z(x_2) = \zeta_0^{(l)}$ and $|y'(x)| \leq \sqrt[4]{b}$, $|y''(x)| \leq \sqrt[4]{b}$ for every $x \in (x_1, x_2)$. Then

$$\left|\xi - z_0^{(l)}\right| \leq \sigma_0^l \leq \left(\frac{e^c}{10K^2}\right)^{2l}$$

for every $\xi \in \left[z_0^{(l)}, \zeta_0^{(l)}\right]$. So, from Proposition 3.6, $\xi_1 = T_{\lambda,a,b}(\xi)$ is also e^c-expanding up to time l. Thus, we may define the map $x \in (x_1, x_2) \to q_l(x)$, where $(q_l(x), 1)$ is colinear to the l-th contractive direction at $T_{\lambda,a,b}(z(x)) = z_1(x)$. If $(t(x), 1)$ is the tangent vector to $W_{a,m}^u$ at $z_1(x)$, then $t(x_1) - q_l(x_1) = t(x_2) - q_l(x_2) = 0$. Moreover, as in Chapter 4,

$$\left|t'(x) - q_l'(x)\right| \geq \frac{1}{2K\sqrt{b}}$$

for every $x \in (x_1, x_2)$. Therefore, $x_1 = x_2$ and consequently, $z_0^{(l)} = \zeta_0^{(l)}$.

Now, we shall prove the lemma. Assume that $\zeta_0^{(k-1)} \in \gamma\left(z_0^{(k-1)}, \frac{1}{2}\rho_0^{\theta k}\right) \cap C_k$. Take a small enough b so that $\sigma_0^k < \rho_0^{\theta k}$. There are two cases:

1. There exists $l \leq k-1$ such that $\sigma_0^{l+1} < \rho_0^{\theta k} \leq \sigma_0^l$.

2. $\sigma_0 < \rho_0^{\theta k}$.

To demonstrate the lemma in the first case notice that, from Proposition 3.6, the image of every point in $\gamma\left(z_0^{(k-1)}, \sigma_0^{k-1}\right)$ is e^c-expanding up to time $k-1$. Reversing the arguments in algorithm A, we get a unique critical approximation $z_0^{(k-2)}$. Now (4.12) leads to $\left|z_0^{(k-1)} - z_0^{(k-2)}\right| \le (Kb)^{k-2} << \frac{1}{4}\sigma_0^{k-1}$. In the same way we find a critical approximation $\zeta_0^{(k-2)}$ such that $\left|\zeta_0^{(k-1)} - \zeta_0^{(k-2)}\right| \le (Kb)^{k-2} << \frac{1}{4}\sigma_0^{k-1}$. Repeating the arguments up to l, we obtain unique critical approximations $z_0^{(l)}$ and $\zeta_0^{(l)}$ satisfying

$$\left|z_0^{(k-1)} - z_0^{(l)}\right| \le (Kb)^l << \frac{1}{4}\sigma_0^{l+1} < \frac{1}{4}\rho_0^{\theta k} \text{ and } \left|\zeta_0^{(k-1)} - \zeta_0^{(l)}\right| << \frac{1}{4}\rho_0^{\theta k}.$$

Thus, $\left|\zeta_0^{(l)} - z_0^{(l)}\right| \le \rho_0^{\theta k} < \sigma_0^l$. Furthermore, since $z_1^{(k-1)}$ and $\zeta_1^{(k-1)}$ are e^c-expanding up to time l, the same follows for $z_1^{(l)}$ and $\zeta_1^{(l)}$ from Proposition 3.6. On the other hand, $\left[z_0^{(l)}, \zeta_0^{(l)}\right] \subset \gamma\left(z_0^{(k-1)}, \rho_0^{\theta k}\right)$ is a $C^2(b)$-curve. Then, $z_0^{(l)} = \zeta_0^{(l)}$ and thus $z_0^{(k-1)} = \zeta_0^{(k-1)}$.

In the second case, the lemma follows in a similar way taking into account the fact that the image of every point of $\left[z_0^{(1)}, \zeta_0^{(1)}\right]$ is e^c-expanding up to time one,

$$\left|z_0^{(k-1)} - z_0^{(1)}\right| \le Kb << \frac{1}{4}\sigma_0 < \frac{1}{4}\rho_0^{\theta k}$$

and the same holds for $\zeta_0^{(1)}$. $\quad\square$

From $C_k \subset \bigcup_{g \le \theta k} G_g$,

$$length\left(\bigcup_{g \le \theta k} G_g\right) \le K^{\theta k} length\,(G_0) \le 2\Lambda K^{\theta k}$$

and Lemma 5.2 we get

$$card\,(C_k) \le 4\Lambda \left(\frac{K}{\rho_0}\right)^{\theta k}. \tag{5.1}$$

5.1 The construction of C_n

We construct the set C_n as the union of two sets C_n' and C_n'', which are obtained in the following way:

(a) The elements of C_n' are obtained by applying algorithm A to replace each $z_0^{(n-2)} \in C_{n-1}$ by the corresponding critical approximation of order $n-1$ and the same generation as $z_0^{(n-2)}$. Notice that algorithm A can be applied because $z_1^{(n-2)} = T_{\lambda,a,b}(z_0^{(n-2)})$ is e^c-expanding up to time $n-1$ for every $z_0^{(n-2)} \in C_{n-1}$. Furthermore, from Proposition 3.6, the same can be said for the image of every point in

$\gamma\left(z_0^{(n-2)}, \rho_0^{\theta(n-1)}\right)$, provided that $b << 1$. Therefore, there exists a unique critical approximation of order $n-1$, $z_0^{(n-1)} \in \gamma\left(z_0^{(n-2)}, \sigma_0^{n-1}\right)$, such that

$$\left|z_0^{(n-2)} - z_0^{(n-1)}\right| \le (Kb)^{n-2} \le \frac{1}{2}\left(\frac{1}{4K}\right)^{n-1}. \tag{5.2}$$

Finally, notice that $\gamma\left(z_0^{(n-1)}, \rho_0^{\theta n}\right) \subset \gamma\left(z_0^{(n-2)}, \rho_0^{\theta(n-1)}\right)$ and this allows us to claim that $\gamma\left(z_0^{(n-1)}, \rho_0^{\theta n}\right)$ satisfies I.H.2 and I.H.3, whenever algorithm A is applied.

(b) C_n'' consists of critical approximations $z_0^{(n-1)}$ of generation $g \in (\theta(n-1), \theta n]$, such that $\gamma\left(z_0^{(n-1)}, \rho_0^{\theta n}\right)$ satisfies I.H.2 and I.H.3. Moreover, we assume that $z_0^{(n-1)}$ can be obtained by applying algorithm B to a point $\zeta_0^{(n-1)} \in C_n'$ with

$$\left\|\zeta_0^{(n-1)} - z_0^{(n-1)}\right\| \le b^{\frac{g}{10}}. \tag{5.3}$$

Then, from the definition of θ we deduce that

$$\left\|\zeta_0^{(n-1)} - z_0^{(n-1)}\right\| \le b^{\frac{g}{10}} \le b^{\frac{\theta}{10}(n-1)} < \frac{1}{2}\left(\frac{1}{4K}\right)^{n-1}. \tag{5.4}$$

Remark 5.3. *When we apply algorithm A to construct C_n', each critical approximation of order $n-2$ is replaced by a unique critical approximation of order $n-1$ in the same branch of $W_{a,m}^u$. However, when we apply algorithm B to get an element of C_n'' on a different branch of $W_{a,m}^u$, we first use algorithm A and thus, we obtain two new critical approximations of order $n-1$ from one of order $n-2$. In either case, every new critical approximation comes from one critical approximation of an inferior order.*

Remark 5.4. *The condition given by (5.3) is stronger than the one needed to apply algorithm B, that is, taking into account that $\rho_0^{\theta n}$ now plays the role that l does in the display of such algorithm, we may easily check that*

$$b^{\frac{g}{10}} \le min\left\{\frac{\sigma_0^{2(n-1)}}{4}, \rho_0^{2\theta n}\right\}.$$

In order to prove the main theorem, we also need to complete the condition given in (5.3): More precisely, it will be necessary for this condition to remain true when the parameter a varies slightly. We shall go back to this point in Chapter 8, where the definitive condition is given.

Next, we shall check that $C_n \subset B_L$. Let $z_0^{(n-1)} = (x_0, y_0)$ be a critical approxima-tion of generation g. Then, its construction involves at the most $n-1$ times algorithm A and $g-1$ times algorithm B. Thus, according to (4.12) and (5.4),

$$|x_0 - c_\lambda| \leq 12\sqrt{b}K^2 \left(K\sqrt[4]{b} + \sum_{i=1}^{n-2} (Kb)^i \right) + \sum_{j=2}^{g} b^{\frac{1}{20}} << \lambda b^{\frac{1}{20}}.$$

To formulate the fourth induction hypothesis we need some new concepts.

Definition 5.5. *Let* $\beta = \beta(\lambda, K) > 0$ *be a small constant and let* $\tilde{\Lambda} = \tilde{\Lambda}(\lambda) = f_{\lambda,a(\lambda)}(c_\lambda) - c_\lambda$. *We say that a point* ξ_0 *is bound to* C_k *up to time* p *if there exists* $z_0 = z_0^{(k-1)} \in C_k$ *such that*

$$\left\| T_{\lambda,a,b}^j(\xi_0) - T_{\lambda,a,b}^j(z_0) \right\| = \|\xi_j - z_j\| \leq h_k e^{-\beta j} \text{ for } 1 \leq j \leq p,$$

where

$$h_k = \tilde{\Lambda} \left(1 - \sum_{i=1}^{k-1} \left(\frac{e^\beta}{4} \right)^i \right) \in \left(\frac{\tilde{\Lambda}}{2}, \frac{3\tilde{\Lambda}}{4} \right).$$

If $p \geq k$, *we simply say that* ξ_0 *is bound to* C_k.

I.H.4. *For* $1 \leq k \leq n - 1$ *and for any* ξ_0 *bound to* C_k, ξ_1 *is* e^c-*expanding up to time* k.

Remark 5.6. *Clearly I.H.4 is stronger than I.H.1. Nevertheless, I.H.4 does not follow from I.H.1 and Proposition 3.7. In our case, said proposition guarantees that if* ξ_0 *satisfies*

$$\|\xi_j - z_j\| \leq \left(\frac{e^c}{10K^2} \right)^{4j} \text{ for } 1 \leq j \leq k,$$

then ξ_1 *is* e^c-*expanding up to time* k. *However, I.H.4 extends this result, because*

$$h_k e^{-\beta j} \geq \frac{\tilde{\Lambda}}{2} e^{-\beta j} \geq \left(\frac{e^c}{10K^2} \right)^{4j}$$

provided that K *is large enough.*

To begin the inductive process we need C_n to satisfy I.H.4 for $n \leq N - 1$. To this end, notice that the orbits of $z_0^{(n-1)}$ and $w_0^{(n-1)}$ are close to $x = -\arccos a\,(\lambda)^{-1}$ during a large number of iterates N_1, whenever a is near $a(\lambda)$ and b is small enough. From Definition 5.5, $\|\xi_j - z_j\| < \frac{3}{4}\tilde{\Lambda} < \frac{3}{8}\Lambda$ for $1 \leq j \leq N_1$ and thus, ξ_j stays away from $x = f_{\lambda,a(\lambda)}(c_\lambda)$. Therefore, if $\delta << 1$, then ξ_j cannot belong to B_δ for $j \leq N_1 - 1$.

On the other hand, if $N_1 \leq j \leq n$, then $\|\xi_j - z_j\| \leq h_k e^{-\beta N_1} << 1$ and so, from Proposition 4.5(a), $\xi_j \notin B_\delta$ for $N_1 < j \leq n$. Now, the claim follows from the proof of Proposition 4.4(c). Notice that N_1 is chosen to be large enough after taking β and δ.

Proposition 5.7. *If ξ_0 is bound to C_{k+1} up to time p, then ξ_0 is bound to C_k up to time p. Consequently, every point bound to C_{k+1} is also bound to C_k.*

Proof. Let $z_0 = z_0^{(k)} \in C_{k+1}$ be such that $\|\xi_j - z_j\| \leq h_{k+1} e^{-\beta j}$ for $1 \leq j \leq p$. From (5.2) or (5.4), there exists $\tilde{z}_0 \in C_k$ such that $\|\xi_j - \tilde{z}_j\| \leq h_{k+1} e^{-\beta j} + 4^{-k} K^{j-k}$ for $1 \leq j \leq p$. So the statement holds by taking $K^{-1} < \tilde{\Lambda}$. $\square$

From the critical set C_{n-1} satisfying I.H.2, I.H.3 and I.H.4, we construct, using algorithms A and B, critical approximations of order $n - 1$ which define C_n. We already know that C_n satisfies I.H.2 and I.H.3. From Proposition 5.7, if ξ_0 is bound to C_n, then it is also bound to C_{n-1} and, since C_{n-1} satisfies I.H.4, ξ_1 is e^c-expanding up to time $n-1$. To prove that ξ_1 is e^c-expanding up to time n, we must remove some parameters. The main goal of the remainder of this book is to check this exclusion of parameters in such a way that a positive Lebesgue measure set survives regardless of $b << 1$.

5.2   Returns, binding points and binding periods

In the unidimensional case, it was necessary to pay special attention to the iterates μ such that $|\xi_\mu(a) - c_\lambda| < e^{-\Delta} \cong \delta$, for which an important loss of the exponential growth occurs. We show that $\xi_{\mu+j}(a)$ stays near $\xi_j(a)$ during a certain period of time $1 \leq j \leq p < \mu$. This fact allows us to inductively prove that the loss of exponential growth is eventually compensated. In the bidimensional case, the iterates μ of a critical approximation z_0 such that z_μ is near the line $x = c_\lambda$ are also bad iterates, but now, because of new facts. In addition to the loss of exponential growth of $\|\omega_i(z_1)\| = \left\| DT_{\lambda,a,b}^i (z_1)(1,0) \right\|$, the slope of $\omega_\mu(z_1)$ can be high and we cannot apply the previous results which state exponential growth for vectors with a small slope.

Now, it is also necessary to find a critical approximation ζ_0 to bind the "return" z_μ and to compare the behaviour of $\omega_{\mu+j}(z_1)$ and $\omega_j(\zeta_1)$. The existence of this binding point ζ_0 will be studied in Chapter 6, but this is not an obstacle to introducing the concept of binding period $[\mu + 1, \mu + p]$ in this chapter. To compare the behaviour of $\omega_{\mu+j}(z_1)$ and $\omega_j(\zeta_1)$ along the binding period, it seems necessary to state a bound distortion result prevailing during this period, that is, to find a constant $B > 0$ such that

$$\frac{1}{B} \leq \frac{\|\omega_j(\zeta_1)\|}{\|\omega_{\mu+j}(z_1)\|} \leq B \text{ for } 1 \leq j \leq p.$$

Nevertheless, this parallelism with Chapter 2 is not possible mainly because the slopes of $\omega_\mu(z_1)$ and $\omega_0(\zeta_1) = (1,0)$ can be very different. This problem will be solved in the following way. We split $\omega_\mu(z_1) = h_\mu(z_1) + \sigma_\mu(z_1)$, where $\sigma_\mu(z_1)$ is a vector on a contractive direction and $h_\mu(z_1)$ is a horizontal vector, which will be accurately defined in the next section. According to Proposition 3.3, after a number l of iterates we have

$$\|\sigma_{\mu+l}(z_1)\| = \left\| DT^l_{\lambda,a,b}(z_{\mu+1})\sigma_\mu(z_1) \right\| \cong 0$$

and

$$h_{\mu+l}(z_1) = DT^l_{\lambda,a,b}(z_{\mu+1})h_\mu(z_1) \cong \omega_{\mu+l}(z_1).$$

Therefore, we hope that, in order to prove the main theorem, it suffices to obtain exponential growth on the vectors $h_\mu(z_1)$. The period of time $[\mu+1, \mu+l]$ is called the folding period associated to the "return" μ. We shall also prove that $l << p$.

Unfortunately, we need not only a bound distortion result during binding periods but also an expansive behaviour of $\omega_j(\zeta_1)$ to compensate for the loss of growth on returns. To this end, since ζ_0 is a critical approximation of an order lower than or equal to the order of z_0, we can use the inductive hypotheses. At this stage, we deem it wise to think retrospectively about what we have to do to obtain this expansiveness: On each return μ_* of ζ_0 we split $\omega_{\mu_*}(\zeta_1)$ into a horizontal vector and a vector on a contractive direction. If $\mu_* \in [\mu+1, \mu+p]$ and if we still require the bound distortion to prevail, it is also necessary to split $\omega_{\mu+\mu_*}(z_1)$. Therefore, we have to identify returns with the iterates on which we need to split $\omega_\mu(z_1)$. Then, unlike the unidimensional case, returns μ can belong to binding periods associated to previous returns. Such returns μ are called bound returns and they are not defined by the orbit of z_1 but by the orbits of binding points associated to previous returns. Next, we state recurrently the definition of returns:

Let us assume that, for every $1 \leq k \leq n-1$, we have constructed the sets C_k and that, for every $z_0 \in C_k$, we have defined the returns of z_0 in $[1, k]$. Moreover, suppose that for every return $\mu \in [1, k]$ of z_0 we have found a binding point $\zeta_{0,\mu} \in \overset{\mu}{\underset{j=1}{\cup}} C_j$ and defined its binding period $[\mu+1, \mu+p]$.

Now, let $z_0 = z_0^{(n-1)} \in C_n$. In the construction of C_n, according to Remark 5.3, $z_0^{(n-1)}$ is generated by a unique $z_0^{(n-2)} \in C_{n-1}$. We define the returns of $z_0^{(n-1)}$ in $[1, n-1]$ as the returns of $z_0^{(n-2)}$ with the same binding points and binding periods. To establish when n is a return of $z_0^{(n-1)}$ we distinguish between two cases:

A. Suppose that n belongs to the binding period associated to a previous return ν of z_0. Let ν be the maximum with this property and let $\tilde{\zeta}_{0,\nu} \in \bigcup_{j=1}^{\nu} C_j$ be the binding point associated to the return ν of z_0. Then, n is a return of z_0 if and only if $n - \nu$ is a return of $\tilde{\zeta}_{0,\nu}$. Furthermore, if $\zeta_{0,n-\nu} \in \bigcup_{j=1}^{n-\nu} C_j$ is the binding point associated to the return $n-\nu$ of $\tilde{\zeta}_{0,\nu}$ and $[n - \nu + 1, n - \nu + p]$ is its binding period, then $\zeta_{0,n-\nu} \in \bigcup_{j=1}^{n} C_j$ is the binding point associated to the return n of z_0 and $[n + 1, n + p]$ is its binding period. We say that n is a bound return of z_0.

Remark 5.8. *We may say that $n - \nu$ is a return of $\tilde{\zeta}_{0,\nu} \in \bigcup_{j=1}^{\nu} C_j$ because, if $\tilde{\zeta}_{0,\nu} \in C_d$, then $n - \nu < d$. This claim will be checked later. In fact, we shall prove that if $\tilde{\zeta}_{0,\nu} \in C_d$, then the length of the binding period associated to the return ν of z_0 is less than d.*

B. Suppose that no binding period contains n. Then, n is a return of z_0 if and only if $z_n = T_{\lambda,a,b}^n(z_0) \in B_\delta$. Such iterates are called free returns. In Chapter 6 we shall describe a method for choosing the respective binding point $\zeta_0 \in C_n$.

If n is a free return of $z_0 \in C_n$, we only consider the parameter values a for which

$$d_n(z_0) = \|z_n - \zeta_0\| \geq e^{-\alpha n} \qquad \text{(BA)}$$

holds, where $\alpha = \alpha(\lambda, K, c, \beta)$ is a small positive constant. This condition allows us to obtain properties on the relative position between z_n and ζ_0, which will be useful to estimate the loss of expansiveness of $\|\omega_n(z_1)\|$.

The binding period associated to a free return n of z_0 is defined in two stages. The primary binding period is $[n + 1, n + p_0]$, where p_0 is the largest natural number such that

$$\|z_{n+j} - \zeta_j\| \leq h e^{-\beta j} \text{ for } 1 \leq j \leq p_0, \qquad (5.5)$$

$h = h(\lambda, K, \alpha) \leq \frac{1}{8}\tilde{\Lambda}$ is a constant to be fixed in Chapter 7 and $\beta = \beta(\lambda, K)$ is the constant given in Definition 5.5. In Chapter 7 we shall also prove that $p_0 \leq 2c^{-1} \log d_n(z_0)^{-1}$ and therefore, from (BA), it follows that $p_0 \leq 2c^{-1}\alpha n < n$, provided that $2\alpha < c$.

Therefore, the basic assumption (BA) and the uniqueness of the binding point associated to a free return of a point z_0 of C_n, allow us to define the returns, binding points and binding periods of z_n in $[1, p_0]$, which coincide with the returns, binding points and binding periods of ζ_0. Then, the binding period $[n+1, n+p]$ associated to the free return n of z_0 is given by the following condition:

- p *is the largest natural number belonging to* $[1, p_0]$ *such that* $p+1$ *is an iterate of* z_n *which does not belong to any binding period associated to returns of* z_n.

This definition guarantees that, at the end of a binding period associated to a free return, z_0 is again in a free iterate. Hence, if μ and ν are two returns of z_0 in $[1, n]$, with binding periods $[\mu+1, \mu+p_\mu]$ and $[\nu+1, \nu+p_\nu]$, respectively, then $\nu + p_\nu \leq \mu + p_\mu$ whenever $\nu \in [\mu+1, \mu+p_\mu]$. We say that the set of binding periods associated to $z_0 \in C_n$ has a *stair structure*.

In order to get expansiveness for every point bound to C_n, we extend the aforementioned concepts to such points. To this end, for every ξ_0 bound to C_n, we take one of the points $z_0 \in C_n$ given in Definition 5.5. Then we let returns, binding points and binding periods of ξ_0 coincide with those of z_0. Notice that the choice of z_0 will not be essential to obtain expansiveness for ξ_0. To make our exposition more readable, we shall develop part of the following results for critical approximations only. The extensions to points bound to critical sets are immediate.

5.2.1 Comments and properties

1. Firstly, let us remark that, in the definition of bound return, $n - \nu$ is a free return of $\tilde{\zeta}_{0,\nu}$. Indeed, assume that $n - \nu$ is a bound return and let $\mu \in (0, n - \nu)$ be the largest return of $\tilde{\zeta}_{0,\nu}$ such that $n - \nu \in [\mu+1, \mu+p_\mu]$. From the stair structure of binding periods, it follows that ν is the largest return of z_0 whose binding period contains $\nu + \mu$. Then, by definition, $\nu + \mu$ is a (bound) return of z_0 whose binding period contains n and this contradicts the maximality of ν.

2. Let n be a bound return of $z_0 \in C_n$. We claim that, for every return ν of z_0 in $[1, n)$ such that $n \in [\nu+1, \nu+p_\nu]$, $n - \nu$ is a return of $\zeta_{0,\nu}$, where $\zeta_{0,\nu}$ is the binding point associated to z_ν. To prove this, let us denote by $\nu_1 > \nu_2 > \ldots > \nu_s$ all the returns of z_0 for which $n \in [\nu_i + 1, \nu_i + p_i]$, where $p_i = p_{\nu_i}$. Let $\zeta_{0,i} = \zeta_{0,\nu_i}$

be the respective binding points. From the definition of bound return, $n - \nu_1$ is a return of $\zeta_{0,1}$. Now, if $n \in [\nu_2 + 1, \nu_2 + p_2]$, then ν_1 is a bound return of z_0. Let $\mu \in [0, \nu_1)$ be the largest return of z_0 whose binding period contains ν_1. Notice that $\mu \geq \nu_2$. If $\mu > \nu_2$, then $n \notin [\mu + 1, \mu + p_\mu]$ and, by the definition of bound return, if $\zeta_{0,\mu}$ is the binding point associated to the return μ of z_0, then $\nu_1 - \mu$ is a return of $\zeta_{0,\mu}$ with binding point $\zeta_{0,1}$ and binding period $[\nu_1 + 1, \nu_1 + p_1]$. This contradicts the stair structure of binding periods. Thus, $\mu = \nu_2$. Then, $\nu_1 - \nu_2$ is a return of $\zeta_{0,2}$ and, by definition, $\zeta_{0,1}$ and $[\nu_1 + 1, \nu_1 + p_1]$ are its binding point and binding period, respectively. Finally, from the stair structure of binding periods we get that $n - \nu_2$ is a (bound) return of $\zeta_{0,2}$. The same arguments extend the claim to all the returns ν_i.

3. Let us now demonstrate the claim in Remark 5.8. Let ν be a return of z_0 and $\tilde{\zeta}_{0,\nu}$ its binding point. Let $d \in \{1, ..., \nu\}$ be such that $\tilde{\zeta}_{0,\nu} \in C_d$. We shall check that $d > p_\nu$, where p_ν is the length of the binding period associated to ν. We distinguish between two cases:

(a) If ν is a free return of z_0, then $d = \nu$ and (BA) leads to $p_\nu < d$.

(b) If ν is a bound return of z_0, let μ be the largest return of z_0 whose binding period contains ν and let $\zeta_{0,\mu}$ be the binding point associated to z_μ. Hence, $\nu - \mu$ is a free return of $\zeta_{0,\mu}$ with binding point $\tilde{\zeta}_{0,\nu}$ and binding period $[\nu + 1, \nu + p_\nu]$. Then, $p_\nu < \nu - \mu = d$.

4. Finally, in the definition of binding period associated to a free return n of $z_0 \in C_n$, we have $p \approx p_0$. Indeed, suppose that $p < p_0$, then $1 + p = \nu_1$ has to be a return of z_n such that $p_0 + 1 \leq \nu_1 + p_1$, where p_1 is the length of the binding period associated to ν_1. Hence, $p_0 - p \leq p_1$. Notice that ν_1 has to be a free return of z_n. Thus, from (BA), $p_1 \leq 2c^{-1}\alpha\nu_1 \leq 2c^{-1}\alpha p_0$ and consequently,

$$p_0 - p \leq 2c^{-1}\alpha p_0. \tag{5.6}$$

In the one-dimensional case, the length of the binding period associated to a return n depends on the proximity between ξ_n and the critical point c_λ. The following result states that, in the two-dimensional case, the length of the binding period associated to a free return n of one point $z_0 \in C_n$ depends on $d_n(z_0)$.

Proposition 5.9. *For every $z_0 \in C_n$ the following statements hold:*

(a) If n is a free return of z_0 and p is the length of its binding period, then

$$p \geq \frac{\log d_n(z_0)^{-1}}{2\,(\log K + 2\beta)}.$$

(b) If $\nu < n$ is a return of z_0 and p_ν is the length of its binding period, then

$$p_\nu \geq \frac{\log d_\nu(z_0)^{-1}}{4\,(\log K + 2\beta)}.$$

Proof. If ζ_0 denotes the binding point associated to z_n, then, from (5.6),

$$\|z_{n+p+1} - \zeta_{p+1}\| \geq \|z_{n+p_0+1} - \zeta_{p_0+1}\| K^{-2c^{-1}\alpha p_0},$$

From the definition of p_0, $\|z_{n+p_0+1} - \zeta_{p_0+1}\| \geq he^{-\beta(p_0+1)}$. Therefore, since (5.6) leads to

$$\frac{p}{p_0} \geq 1 - \frac{2\alpha}{c} > \frac{2}{3} \quad \text{and} \quad \frac{p+1}{p_0} \geq \frac{1}{2},$$

by taking $\alpha = \alpha\,(K, c, \beta)$ small enough so that $8\alpha \log K \leq \beta c$, we obtain

$$\|z_{n+p+1} - \zeta_{p+1}\| \geq he^{-\beta(p_0+1)} K^{-2\alpha c^{-1} p_0} \geq he^{-2\beta(p+1)}. \tag{5.7}$$

Furthermore, the inequality $\|z_{n+p+1} - \zeta_{p+1}\| \leq K^{p+1} \|z_n - \zeta_0\| = K^{p+1} d_n(z_0)$, together with (5.7), implies that $(p+1)\,(\log K + 2\beta) \geq \log h - \log d_n(z_0)$. Taking δ and b small enough and bearing in mind that $h = h(\lambda, K, \alpha) < 1$, we may assume that

$$\log d_n(z_0) < 4 \log h < \frac{2p}{p-1} \log h.$$

So, statement (a) follows from $(p+1)\,(\log K + 2\beta) \geq \frac{1}{2}p^{-1}\,(p+1) \log d_n(z_0)^{-1}$.

To prove statement (b), notice that, from the definition of returns, it follows that there exist $\mu \leq \nu$ and $\tilde{z}_0 \in C_\mu$ such that μ is a free return of $\tilde{z}_0$ with binding period $[\mu+1, \mu+p_\nu]$. Then, from (a),

$$p_\nu \geq \frac{\log d_\mu(\tilde{z}_0)^{-1}}{2\,(\log K + 2\beta)}.$$

On the other hand, whether ν is a free return or whether ν is a bound return, Proposition 5.21 and Remark 5.22, which will be stated in the appendix of this chapter, allow us to claim that $\log d_\mu(\tilde{z}_0)^{-1} \geq \log\left(\frac{6}{7} d_\nu(z_0)^{-1}\right)$. Finally, since ν is a return of z_0, $d_\nu(z_0)^{\frac{1}{2}} < \frac{6}{7}$ and then $\log d_\mu(\tilde{z}_0)^{-1} \geq \log d_\nu(z_0)^{-\frac{1}{2}}$. Thus, statement (b) is proved. $\square$

5.3   The folding period

As we said at the beginning of Section 5.2, given $z_0 = z_0^{(n-1)} \in C_n$, on each return μ
of z_0 we split the vector $\omega_\mu(z_1) = DT_{\lambda,a,b}^\mu(z_1)(1,0)$ into a horizontal vector $h_\mu(z_1)$ and
a vector $\sigma_\mu(z_1)$ on the l-th contractive direction at $z_{\mu+1}$. According to Proposition
3.3(b), we may take l large enough so that the vector $\omega_{\mu+l}$ has a slope of order $\sqrt{b}$.
Let p be the length of the binding period associated to μ. Since z_μ is bound to C_s up
to time p with $p < s \leq \mu$, if $l < p$, then I.H.4 yields the expansiveness of $z_{\mu+1}$ up to
time p, and consequently the existence of the l-th contractive direction. The aim of
this section is to accurately define the folding period $[\mu + 1, \mu + l]$ associated to the
return μ of z_0 and the splitting algorithm outlined above.

Let us define $h_{\mu+j} = DT_{\lambda,a,b}^j(z_{\mu+1})h_\mu$ and $\sigma_{\mu+j} = DT_{\lambda,a,b}^j(z_{\mu+1})\sigma_\mu$. According to
Lemma 4.1, if $z_{\mu+j} \notin B_\delta$, then $h_{\mu+j}$ is a nearly horizontal vector. Furthermore, from
Proposition 3.3, $\sigma_{\mu+l} \approx \theta$ and then $h_{\mu+l} \approx \omega_{\mu+l}$. So, no meaning error is introduced
in considering $h_{\mu+j}$ instead of $\omega_{\mu+j}$. Nevertheless, new returns ν of z_0 can take place
in $[\mu + 1, \mu + l]$ creating higher order folds. Then, we once more take $\omega_\nu(z_1)$ as a
horizontal vector by neglecting $\sigma_\nu(z_1)$. This fact can happen reiteratively, giving rise
to folds inside folds, each one of them furnishing a correction term to σ. The whole
process defines what will be called the splitting algorithm. To formulate it, we first
need to define the folding period $[\mu + 1, \mu + l]$ associated to a return μ. This period
must satisfy the following properties:

(a) If ν is a return of z_0, with $\nu \in [\mu + 1, \mu + l]$, then $\nu + l_\nu \leq \mu + l$, where l_ν is
the length of the folding period associated to ν. That is, folding periods display a
stair structure.

(b) $z_{\mu+l+1} \in \tilde{V} = \{(x,y) \in \mathbf{R}^2 : x \in I_2\}$, where $I_2 = [Q_{a_0} + \delta_2, \xi_1(a_0) - \delta_2]$ and
$\delta_2 = R\left(\frac{3}{2} + \lambda^{-1}\sqrt{a^2(\lambda) - 1}\right) F(\lambda, a_0)$.

The constant $R > 1$, which plays a technical role, is chosen close enough to 1 so
that $\delta_2 < \delta_1$, where δ_1 is the constant in Proposition 4.11. Notice that Proposition
4.11(b) implies that $T_{\lambda,a,b}^i(z) \in \tilde{V}$ for some $i \in \{0, ..., 4\}$.

Now, we define the folding period associated to a free return $\mu = n$ of $z_0 \in C_n$.
This definition is extended to all the returns of points ξ_0 bound to C_n as in the case
of binding periods. It is also necessary to set the concept of primary folding period.

Let n be a free return of a point $z_0 \in C_n$ and let us write $\omega_n = h_n + \sigma_n$ and

$\theta_n = \text{ang}(h_n, \omega_n)$. Since σ_n is a nearly vertical vector,

$$\text{tg } \theta_n \approx \frac{\|\sigma_n\|}{\|h_n\|}.$$

In Chapter 6 we shall prove that $\text{tg } \theta_n \approx d_n(z_0)^{-1}$. Furthermore, Propositions 3.3 and 4.4 imply that $\|\sigma_{n+j}\| \approx \|\sigma_n\| (K_3 b)^j$ and $\|h_{n+j}\| \geq \|h_n\|$, respectively. Thus, if we want to find a natural number l for which

$$\frac{\|\sigma_{n+l}\|}{\|h_{n+l}\|} < \frac{\|\sigma_n\| (K_3 b)^l}{\|h_n\|} \ll 1,$$

we must guarantee that $d_n(z_0)^{-1} (K_3 b)^l \ll 1$. Hence, l depends mainly on b and $d_n(z_0)$. The primary folding period associated to the free return n is $[n+1, n+l_0]$, where

$$l_0 = \max \left\{ \frac{10 \log K}{c \log b^{-1}} \log d_n(z_0)^{-1}, 4 \right\}.$$

First, let us check that $l_0 \ll p$. From Proposition 5.9, it suffices to show that

$$\frac{10 \log K \log d_n(z_0)^{-1}}{c \log b^{-1}} \ll \frac{\log d_n(z_0)^{-1}}{2(\log K + 2\beta)} \quad \text{and} \quad 4 \ll \frac{\log d_n(z_0)^{-1}}{2(\log K + 2\beta)}.$$

The first inequality holds by taking $b \ll 1$ and the second one follows from Proposition 5.23, by taking δ small enough.

Finally, the folding period associated to a return n of z_0 is $[n+1, n+l]$, where $l \leq l_0$ is the largest natural number for which $z_{n+l+1} \in \tilde{V}$ and $l+1$ is neither a return nor belongs to any of the folding periods associated to returns of z_n.

Proposition 5.10. *The length of a folding period satisfies*

$$l \geq \frac{1}{2} l_0.$$

Proof. Write $l_0^{(n)} = l_0$ and $l^{(n)} = l$, where the superscript denotes the return iterate. First, assume that $l_0^{(n)} < 20$. Then, taking Ω_0 close enough to $a(\lambda)$, the number of iterates between two consecutives returns of z_0 is greater than 20 and all these iterates, except maybe the first three, belong to $\tilde{V}$. Therefore, $l^{(n)} = l_0^{(n)}$.

Now assume that $l_0^{(n)} \geq 20$. Since $z_{n+l+1} \in \tilde{V}$, either $l^{(n)} \geq l_0^{(n)} - 3$ or there exists a return ν with folding period $\left[\nu + 1, \nu + l^{(\nu)} \right]$ such that $l^{(n)} + 1 = \nu - 1$ and $\nu + l^{(\nu)} \geq l_0^{(n)} \geq \nu$. In the first case, the proposition follows from $l_0^{(n)} \geq 20$. In the second one notice that

$$l_0^{(\nu)} = \max \left\{ \frac{10 \log K}{c \log b^{-1}} \log d_\mu(\zeta_0)^{-1}, 4 \right\},$$

where $\mu \le \nu$ is a free return of $\zeta_0 \in C_\mu$. From (BA), $\log d_\mu(\zeta_0)^{-1} \le \alpha\mu \le \alpha\nu$. So,

$$l_0^{(\nu)} \le \max\left\{\frac{10\alpha\nu\log K}{c\log b^{-1}}, 4\right\}$$

and consequently,

$$l_0^{(n)} - l^{(n)} \le l_0^{(\nu)} + 2 \le \frac{10\alpha l_0^{(n)}\log K}{c\log b^{-1}} + 6 < \frac{l_0^{(n)}}{2},$$

provided that b is small enough. $\square$

In the next result, we get a relationship between the whole contraction of $e^{(l)}(z_{n+1})$ during the folding period and $d_n(z_0)$. This statement is crucial to prove that, at the end of the folding period, the slope of $\omega_{n+l}(z_1)$ is again of order $\sqrt{b}$.

Proposition 5.11. *Let $\mu \in [1, n]$ be a free return of $z_0 \in C_n$ and l be the length of its folding period. Then*

$$\left\|DT_{\lambda,a,b}^l(z_{\mu+1})e^{(l)}(z_{\mu+1})\right\| \le \left(\sqrt{b}\right)^l d_\mu(z_0)^2.$$

Proof. From (3.5) we have $\left\|DT_{\lambda,a,b}^l(z_{\mu+1})e^{(l)}(z_{\mu+1})\right\| \le (Kb)^l$. So, the proposition follows if $\left(K\sqrt{b}\right)^l \le d_\mu(z_0)^2$. Let $\tilde{z}_0 \in C_\mu$ be the critical approximation which has a free return at time μ. From Remark 5.22 in appendix,

$$\frac{d_\mu(\tilde{z}_0)}{d_\mu(z_0)} \le \frac{7}{6}.$$

Therefore it suffices to check that $\left(K\sqrt{b}\right)^l \le \left(\frac{3}{4}d_\mu(\tilde{z}_0)\right)^2$ or, taking into account that μ is a return, $\left(K\sqrt{b}\right)^l \le d_\mu(\tilde{z}_0)^4$. From Proposition 5.10, the last inequality holds if we get

$$\frac{5\log K}{c\log b^{-1}}\log d_\mu(\tilde{z}_0)^{-1}\log\left(K\sqrt{b}\right) < 4\log d_\mu(\tilde{z}_0).$$

So the statement holds by taking b small enough. $\square$

5.3.1   The splitting algorithm

We now give the definition of the splitting algorithm outlined previously.

For $0 \le \mu \le k \le n-1$ and for every ξ_0 bound to C_k, write $\omega_\mu = \omega_\mu(\xi_1) = h_\mu + \sigma_\mu$, where h_μ and σ_μ are inductively defined in the following way:

S.A.1. $h_0 = \omega_0 = (1, 0)$, $\sigma_0 = \theta$.

S.A.2. Let $\tilde{h}_\mu = DT_{\lambda,a,b}(\xi_\mu)h_{\mu-1}$ and $\tilde{\sigma}_\mu = DT_{\lambda,a,b}(\xi_\mu)\sigma_{\mu-1}$.

S.A.3. If μ is neither a return of ξ_0 nor the end of a folding period, then $h_\mu = \tilde{h}_\mu$ and $\sigma_\mu = \tilde{\sigma}_\mu$.

S.A.4. If μ is a return of ξ_0, then $\tilde{h}_\mu = \alpha_\mu(\xi_1)e^{(l)}(\xi_{\mu+1}) + \beta_\mu(\xi_1)(1,0)$, where $e^{(l)} = (q_l, 1)$ is a vector on the l-th contractive direction at $\xi_{\mu+1}$ and l is the length of the folding period associated to μ. In this case, we take

$$h_\mu = \tilde{h}_\mu - \alpha_\mu(\xi_1)e^{(l)}(\xi_{\mu+1}) = \beta_\mu(\xi_1)(1,0)$$

$$\sigma_\mu = \tilde{\sigma}_\mu + \alpha_\mu(\xi_1)e^{(l)}(\xi_{\mu+1}).$$

S.A.5. If $\mu = \mu_1 + l_1$ is the end of a folding period, then we take

$$h_\mu = \tilde{h}_\mu + \alpha_{\mu_1}(\xi_1)DT^{l_1}_{\lambda,a,b}(\xi_{\mu_1+1})e^{(l_1)}(\xi_{\mu_1+1}),$$

$$\sigma_\mu = \tilde{\sigma}_\mu - \alpha_{\mu_1}(\xi_1)DT^{l_1}_{\lambda,a,b}(\xi_{\mu_1+1})e^{(l_1)}(\xi_{\mu_1+1}).$$

In general, if s folding periods $[\mu_1 + 1, \mu_1 + l_1], ..., [\mu_s + 1, \mu_s + l_s]$ end at time μ, then

$$h_\mu = \tilde{h}_\mu + \sum_{i=1}^{s} \alpha_{\mu_i}(\xi_1)DT^{l_i}_{\lambda,a,b}(\xi_{\mu_i+1})e^{(l_i)}(\xi_{\mu_i+1}),$$

$$\sigma_\mu = \tilde{\sigma}_\mu - \sum_{i=1}^{s} \alpha_{\mu_i}(\xi_1)DT^{l_i}_{\lambda,a,b}(\xi_{\mu_i+1})e^{(l_i)}(\xi_{\mu_i+1}).$$

Notice that, from the definition of folding periods, S.A.4 and S.A.5 never apply simultaneously. Now, let us distinguish between two kinds of iterates.

A. For iterates μ which do not belong to any folding period we have $\omega_{\mu-1} = h_{\mu-1}$ and $\sigma_{\mu-1} = \theta$. We shall check that $\omega_{\mu-1}$ is a nearly horizontal vector and hence, from Proposition 4.4, it will be exponentially expanded up to the next return situation. We say that μ is a fold-free iterate.

B. We introduce a new classification in the set L of the iterates which belong to a folding period. Write $L = \bigcup_j L_j$, where $L_j = [\mu_j + 1, \mu_j + l_j]$ and $\{\mu_j\}$ is the set of returns of ξ_0 such that μ_j and $\mu_j + l_j + 1$ are fold-free iterates. If $\mu \in L$, take the unique j with $\mu \in L_j$ and consider the following situations:

B1. If $\mu = \mu_j$, then $h_{\mu_j-1} = \omega_{\mu_j-1}$ and $\sigma_{\mu_j-1} = \theta$. Therefore, by applying S.A.4, $\sigma_{\mu_j} = \alpha_{\mu_j}(\xi_1)e^{(l_j)}(\xi_{\mu_j+1})$ and $h_{\mu_j} = \beta_{\mu_j}(\xi_1)(1,0)$.

B2. If μ is neither a return nor the end of a folding period we apply S.A.3 to get $\sigma_\mu = DT_{\lambda,a,b}(\xi_\mu)\sigma_{\mu-1}$ and $h_\mu = DT_{\lambda,a,b}(\xi_\mu)h_{\mu-1}$.

B3. Now, let us assume that $\mu = \nu$ is a return with $\nu \neq \mu_j$ and let l_ν be the length of the folding period associated to ν. We apply S.A.4 but, since $\sigma_{\nu-1} \neq \theta$ we have $\tilde{\sigma}_\nu \neq \theta$ and σ_ν is not a vector on the direction of $e^{(l_\nu)}(\xi_{\nu+1})$, that is, σ_ν is the sum of the correction terms introduced at previous returns to ν. However, by definition, h_ν is a horizontal vector.

B4. Let us assume that μ is the end of a folding period $\nu + l_1$, with $\nu \neq \mu_j$. We apply S.A.5. So, we no longer consider the correction term introduced at the return ν. Nevertheless, since there exist previous returns to ν whose folding periods contain μ, $\sigma_\mu \neq \theta$.

B5. Finally, if $\mu = \mu_j + l_j$ then we apply S.A.5 again. In this case, $\sigma_\mu = \theta$ and thus, at the end of L_j we have $\omega_\mu = h_\mu$.

Next, we shall prove the result which states the frequency of fold-free iterates.

Proposition 5.12. *Let ξ_0 be bound to C_k and let μ be an iterate of ξ_0 with $0 \leq \mu \leq k \leq n-1$. There exist fold-free iterates μ_1 and μ_2 of ξ_0 with $\mu_1 \leq \mu \leq \mu_2$ such that*

(a)
$$\mu_2 - \mu_1 \leq max\left\{\frac{10\alpha logK}{c\,logb^{-1}}\mu, 4\right\} + 1.$$

(b) For any free iterate $\nu > \mu$, it follows that
$$\mu_2 - \mu_1 \leq max\left\{\frac{50log^2K}{c\,logb^{-1}}(\nu - \mu), 4\right\} + 1.$$

Proof. It suffices to prove the statement for critical approximations $z_0 \in C_k$. Assume that μ is not a fold-free iterate, that is, $\mu \in L_j = [\mu_j + 1, \mu_j + l_j]$ for some L_j as above. From the definition of returns, there exist $\tilde{\mu} \leq \mu_j$ and $\overline{z}_0 \in C_{\tilde{\mu}}$ such that $\tilde{\mu}$ is a free return for $\overline{z}_0$, whose folding period is $[\tilde{\mu} + 1, \tilde{\mu} + l_j]$. To prove statement (a), choose $\mu_1 = \mu_j$ and $\mu_2 = \mu_j + l_j + 1$. Let $l_{0,j}$ be the length of the primary folding period of μ_j. Since $l_j \leq l_{0,j}$, (BA) leads to

$$\mu_2 - \mu_1 \leq max\left\{\frac{10\alpha \log K}{c\log b^{-1}}\mu, 4\right\} + 1$$

and hence statement (a) follows.

To prove statement (b), denote by p_j the length of the binding period of μ_j. From Proposition 5.9,

$$\nu - \mu_j \geq p_j \geq \frac{\log d_{\tilde{\mu}}(\overline{z}_0)^{-1}}{2(\log K + 2\beta)} \geq \frac{\log d_{\tilde{\mu}}(\overline{z}_0)^{-1}}{3\log K}.$$

Take again $\mu_1 = \mu_j$ and $\mu_2 = \mu_j + l_j + 1$ to obtain

$$\mu_2 - \mu_1 \leq \max\left\{\frac{30\log^2 K}{c\log b^{-1}}(\nu - \mu_j), 4\right\} + 1.$$

Since we have already proved that $l_j << p_j$ we may assume $5l_j < 2p_j$. Therefore, $3(\nu - \mu_j) < 5(\nu - \mu_j - l_j) < 5(\nu - \mu)$ and the proposition is proved. □

5.4   The remainder of the induction hypotheses

We introduce the two remaining induction hypotheses on the set C_k for $k \leq n - 1$.

I.H.5. *If k is a free return of a point $z_0 \in C_k$, then there exists a binding point $\zeta_0 \in C_k$ associated to z_k. We remove the parameter values a in such a way that the following condition holds*

$$d_k(z_0) = \|z_k - \zeta_0\| \geq e^{-\alpha k}. \quad \text{(BA)}$$

Moreover, for every ξ_0 bound to C_k and any return $\nu \leq k$ of ξ_0

(a)

$$|a_\nu(\xi_1)| \leq 3K\sqrt{b}\,\|h_{\nu-1}(\xi_1)\|,$$

(b)

$$\frac{d_\nu(\xi_0)}{2\lambda} \leq \frac{|\beta_\nu(\xi_1)|}{\|h_{\nu-1}(\xi_1)\|} = \frac{\|h_\nu(\xi_1)\|}{\|h_{\nu-1}(\xi_1)\|} \leq \frac{2d_\nu(\xi_0)}{\lambda},$$

where $a_\nu(\xi_1)$ and $\beta_\nu(\xi_1)$ are the real numbers introduced in S.A.4.

The bounds given in (a) and (b) are used to estimate the loss of expansiveness at the returns. In Chapter 6 we shall prove that I.H.5 holds at time n.

I.H.6. *If $\nu \leq k$ is a return of a point ξ_0 bound to C_k, then its binding period satisfies $p \leq 2c^{-1}\alpha\nu < \nu$. Moreover,*

(a) There exists a constant τ_1, which only depends on K, α and β, such that for each $0 \leq j \leq p$ we have

$$\frac{1}{\tau_1} \leq \frac{\|h_{\nu+j}(\xi_1)\|}{|\beta_\nu(\xi_1)|\,\|h_j(\zeta_1)\|} \leq \tau_1,$$

where ζ_0 is the binding point associated to ξ_ν.

(b) There exists a constant $\tau_2 > 0$, which only depends on λ, K, α and β, such that

$$\frac{\|h_{\nu+p}(\xi_1)\|\,d_\nu(\xi_0)}{\|h_\nu(\xi_1)\|} \geq \tau_2 e^{\frac{1}{3}c(p+1)} > 1.$$

Denoting by $F_k(a; z_0)$ the whole number of free iterates of $z_0 \in C_k$ in $[1, k]$, we remove the parameter values for which the following condition does not hold

$$F_k(a; z_0) \geq (1 - \alpha) k \quad \text{for every } z_0 \in C_k. \quad \text{(FA)}$$

Notice that, for the remaining parameters, $F_k(a; \xi_0) \geq (1 - \alpha) k$ for every ξ_0 bound to C_k.

Statement (a) guarantees that during the binding period we have exponential growth of the derivatives. From statement (b) and I.H.5(b) this exponential growth compensates the small factor introduced at the return. In Chapter 7 we shall prove that C_n satisfies I.H.6. Now, we shall prove that the slope of $\omega_j(\xi_1)$ is small outside the folding periods.

Proposition 5.13. *For every $0 \leq \nu \leq k \leq n - 1$ and for every ξ_0 bound to C_k, it follows that*

$$|slope\ h_\nu(\xi_1)| \leq const\sqrt{b}.$$

Proposition 5.14. *For every $0 \leq k \leq n - 1$ and for every ξ_0 bound to C_k, it follows that*

$$\|h_k(\xi_1)\| \geq e^{2ck}.$$

The proofs of both results have to be developed simultaneously, because each one of them follows as soon as the other result is proved for previous iterates. Since for $n < N$ these propositions hold by themselves, we may obtain both of them in a recurrent way.

Proof of Proposition 5.14. Let us write

$$\|h_k(\xi_1)\| = \prod_{i=1}^{k} \frac{\|h_i(\xi_1)\|}{\|h_{i-1}(\xi_1)\|}.$$

Let $1 < \nu_1 < \nu_2 < ... < \nu_s \leq k$ be the free returns of ξ_0 and p_i be the lengths of the respective binding periods. Then, from I.H.5(b) and I.H.6(b), we obtain

$$\prod_{i=\nu_j}^{\nu_j+p_j} \frac{\|h_i(\xi_1)\|}{\|h_{i-1}(\xi_1)\|} = \frac{\|h_{\nu_j+p_j}(\xi_1)\|}{\|h_{\nu_j}(\xi_1)\|} \frac{\|h_{\nu_j}(\xi_1)\|}{\|h_{\nu_j-1}(\xi_1)\|} \geq \frac{\|h_{\nu_j+p_j}(\xi_1)\|}{\|h_{\nu_j}(\xi_1)\|} d_{\nu_j}(\xi_0) \geq 1. \quad (5.8)$$

On the other hand, if $\bar{z}_0$ is the point in $C_{\nu_{j+1}}$ which has a free return at time ν_{j+1}, then $|\bar{x}_{j+1} - c_\lambda| < \delta$ and $\left\|\bar{z}_{\nu_{j+1}} - \xi_{\nu_{j+1}}\right\| \le h_{\nu_{j+1}} e^{-\beta\nu_{j+1}}$, where $\bar{z}_{\nu_{j+1}} = \left(\bar{x}_{j+1}, \bar{y}_{j+1}\right) = T_{\lambda,a,b}^{\nu_{j+1}}(\bar{z}_0)$. Since ν_{j+1} is arbitrarily large whenever Ω_0 is close enough to $a(\lambda)$, it follows that $\left\|\bar{z}_{\nu_{j+1}} - \xi_{\nu_{j+1}}\right\| << \delta$. So, setting $\xi_{\nu_{j+1}} = (x_{j+1}, y_{j+1})$, we have $x_{j+1} \in W$, where W is the neighbourhood of c_λ given in Proposition 2.2. Then, repeating the arguments in the proof of Proposition 4.4 (replacing δ by $\frac{\delta}{2}$), we obtain

$$\prod_{i=\nu_j+p_j+1}^{\nu_{j+1}-1} \frac{\|h_i(\xi_1)\|}{\|h_{i-1}(\xi_1)\|} = \frac{\left\|\omega_{\nu_{j+1}-1}(\xi_1)\right\|}{\left\|\omega_{\nu_j+p_j}(\xi_1)\right\|} \ge C_0 e^{c_0(\nu_{j+1}-\nu_j-p_j-1)}.$$

Notice that Proposition 4.4 was proved for vectors $\mathbf{v}$ with $|\text{slope } \mathbf{v}| \le \sqrt[4]{b}$ and we cannot assume this condition for $\omega_{\nu_j+p_j}(\xi_1) = h_{\nu_j+p_j}(\xi_1)$, unless Proposition 5.13 is proved for $\nu_j + p_j < n$.

Let us now suppose that $k > \nu_s + p_s$. From Proposition 4.4, it follows that

$$\|h_k(\xi_1)\| \ge C_0^{s+1} e^{c_0 F_n(a;\xi_0)} \min\left\{\left|f_{\lambda,a}'(\eta)\right| : \eta \notin V_{\frac{\delta}{2}}\right\}$$

and (FA) leads to $\|h_k(\xi_1)\| \ge \frac{1}{2}C_0^{s+1} e^{c_0(1-\alpha)k}\delta$. Let us fix $c < \frac{1}{2}c_0$ and take α small enough so that $c_0(1-\alpha) > 2c + \alpha$. Then, since k can be chosen arbitrarily large, we get

$$\|h_k(\xi_1)\| \ge C_0^{s+1} e^{\alpha k} e^{2ck} \frac{\delta}{2} \ge C_0^{s+1} e^{\frac{1}{2}\alpha k} e^{2ck}. \tag{5.9}$$

From Propositions 5.9 and 5.23,

$$p_i \ge \frac{\log d_{\nu_i}(z_0)^{-1}}{4(\log K + 2\beta)} \ge \frac{\log(2\delta)^{-1}}{4(\log K + 2\beta)}.$$

So,

$$s + 1 \le \frac{4k(\log K + 2\beta)}{\log(2\delta)^{-1}}$$

and

$$\|h_k(\xi_1)\| \ge \left\{C_0^{\frac{4(\log K + 2\beta)}{\log(2\delta)^{-1}}} e^{\frac{1}{2}\alpha}\right\}^k e^{2ck}.$$

If $\delta = \delta(K, \beta, \alpha)$ is small enough, then

$$C_0^{\frac{4(\log K + 2\beta)}{\log(2\delta)^{-1}}} e^{\frac{1}{2}\alpha} > 1$$

and consequently, $\|h_k(\xi_1)\| \ge e^{2ck}$.

Let us suppose that $k \leq \nu_s + p_s$ and write

$$\|h_k(\xi_1)\| = \frac{\|h_{\nu_s}(\xi_1)\|}{\|h_{\nu_s-1}(\xi_1)\|} \frac{\|h_k(\xi_1)\|}{\|h_{\nu_s}(\xi_1)\|} \prod_{i=1}^{\nu_s-1} \frac{\|h_i(\xi_1)\|}{\|h_{i-1}(\xi_1)\|}.$$

From I.H.5(b),

$$\frac{\|h_{\nu_s}(\xi_1)\|}{\|h_{\nu_s-1}(\xi_1)\|} \geq \frac{d_{\nu_s}(\xi_0)}{2\lambda} > d_{\nu_s}(\xi_0)$$

and, from I.H.6(a), it follows that

$$\|h_k(\xi_1)\| \geq \frac{1}{\tau_1} |\beta_{\nu_s}(\xi_1)| \|h_{k-\nu_s}(\zeta_1)\| = \frac{1}{\tau_1} \|h_{\nu_s}(\xi_1)\| \|h_{k-\nu_s}(\zeta_1)\|,$$

where ζ_1 is the binding point associated to ξ_{ν_s}. Now, by induction hypothesis, we obtain

$$\frac{\|h_k(\xi_1)\|}{\|h_{\nu_s}(\xi_1)\|} \geq \frac{1}{\tau_1} e^{2c(k-\nu_s)}. \tag{5.10}$$

The arguments used in the previous case lead to

$$\prod_{i=1}^{\nu_s-1} \frac{\|h_i(\xi_1)\|}{\|h_{i-1}(\xi_1)\|} \geq C_0^s e^{c_0 F_n(a;\xi_0)}$$

and, from (FA), we conclude that $\|h_k(\xi_1)\| \geq C_0^s \tau_1^{-1} d_{\nu_s}(\xi_0) e^{2c(k-\nu_s)} e^{c_0(1-\alpha)k}$.

Let $\overline{z}_0$ be the point in C_{ν_s} which has a free return at time ν_s. Then, Remark 5.22 and (BA) imply $d_{\nu_s}(\xi_0) \geq \frac{6}{7} d_{\nu_s}(\overline{z}_0) \geq \frac{6}{7} e^{-\alpha \nu_s}$ and consequently,

$$\|h_k(\xi_1)\| \geq \frac{C_0^s}{2\tau_1} e^{-\alpha k} e^{c_0(1-\alpha)k}.$$

Let us fix $c < \frac{1}{2} c_0$ and take α sufficiently small so that $c_0(1-\alpha) > 2(\alpha + c)$. For k large enough we have $\|h_k(\xi_1)\| \geq C_0^s e^{\frac{1}{2}\alpha k} e^{2ck}$. Therefore, this case follows as the previous one does from (5.9). $\square$

Proof of Proposition 5.13. We shall inductively prove that

$$|\text{slope } h_\nu(\xi_1)| \leq \text{const} \sum_{i=1}^{\nu} \left(\sqrt{b}\right)^i \leq \text{const}\sqrt{b}.$$

Clearly, this claim holds for $\nu = 0$ and for every return ν of ξ_0. Assume that the statement holds for iterates μ of every point bound to C_k, with $\mu < \nu \leq k \leq n-1$. We consider the two possible cases:

1. No folding period ends at time ν. Let $(1, s)$ be a vector on the direction of $h_{\nu-1}$. Therefore, by induction hypothesis,

$$|s| \leq \text{const} \sum_{i=1}^{\nu-1} \left(\sqrt{b}\right)^i.$$

Then, by applying S.A.3, we obtain

$$|\text{slope } h_\nu(\xi_1)| \leq \frac{|C| + |D| \, |s|}{|A| - |B| \, |s|}.$$

Let $\xi_\nu = (x_\nu, y_\nu)$. Since ν is not a return of ξ_0, from Proposition 5.23, we get $|x_\nu - c_\lambda| = d_\nu(\xi_0) > \frac{1}{2}\delta$ and $|A| > \frac{3}{2}\delta$. Hence, from Proposition 1.3, we conclude that

$$|\text{slope } h_\nu(\xi_1)| \leq const \sum_{i=1}^{\nu} \left(\sqrt{b}\right)^i.$$

2. Let ν coincide with the end of a folding period $\mu + l$. Let μ be the minimum. Then, from the definitions of folding period and splitting algorithm, it follows that

$$h_\nu(\xi_1) = \tilde{h}_\nu(\xi_1) + \alpha_\mu(\xi_1) DT^l_{\lambda,a,b} e^{(l)}(\xi_{\mu+1}) =$$

$$= \beta_\mu(\xi_1) h_l(\xi_{\mu+1}) + \alpha_\mu(\xi_1) DT^l_{\lambda,a,b} e^{(l)}(\xi_{\mu+1}). \tag{5.11}$$

Let $\theta_1 = \text{ang}(h_\nu(\xi_1), (1,0))$, $\theta_2 = \text{ang}(h_l(\xi_{\mu+1}), (1,0))$ and notice that

$$|\text{tg } \theta_2| \leq const \sum_{i=1}^{l} \left(\sqrt{b}\right)^i.$$

From I.H.5(a) and Proposition 5.11, it follows that

$$\left\| \alpha_\mu(\xi_1) DT^l_{\lambda,a,b} e^{(l)}(\xi_{\mu+1}) \right\| \leq const \left(\sqrt{b}\right)^{l+1} \|h_{\mu-1}(\xi_1)\| \, d_\mu(\xi_0)^2.$$

Furthermore, I.H.5(b) leads to

$$\|h_{\mu-1}(\xi_1)\| \, d_\mu(\xi_0) \leq |\beta_\mu(\xi_1)|$$

and, from Propositions 5.14 and 5.23,

$$d_\mu(\xi_0) < 2\delta < e^{cl} < \|h_l(\xi_{\mu+1})\|.$$

Therefore,

$$\left\| \alpha_\mu(\xi_1) DT^l_{\lambda,a,b} e^{(l)}(\xi_{\mu+1}) \right\| \leq const \left(\sqrt{b}\right)^{l+1} \|\beta_\mu(\xi_1) h_l(\xi_{\mu+1})\|.$$

Now, from (5.11), we get $|\text{tg } (\theta_1 - \theta_2)| \leq const \left(\sqrt{b}\right)^{l+1}$. Then, the proposition follows from

$$|\text{tg } (\theta_1)| \leq 2 \left(|\text{tg } (\theta_2)| + |\text{tg } (\theta_1 - \theta_2)|\right) \leq const \sum_{i=1}^{\nu} \left(\sqrt{b}\right)^i. \quad \square$$

Proposition 5.14 and the next results are needed to get expansiveness for $\omega_k(\xi_1)$.

Proposition 5.15. *There exists a constant $C = C(C_0, \delta) > 0$ such that*

$$\|h_\nu(\xi_1)\| \geq C^{\nu-\mu} \min_{\mu < j \leq \nu} \left(\frac{\|h_j(\xi_1)\|}{\|h_{j-1}(\xi_1)\|} \right) \|h_\mu(\xi_1)\| \geq C^{\nu-\mu} \min_{\mu < j \leq \nu} (d_j(\xi_0)) \|h_\mu(\xi_1)\|$$

for $1 \leq \mu < \nu \leq k \leq n-1$ and for every ξ_0 bound to C_k.

Proof. Since

$$\frac{\|h_\nu(\xi_1)\|}{\|h_\mu(\xi_1)\|} = \prod_{j=\mu+1}^{\nu} \frac{\|h_j(\xi_1)\|}{\|h_{j-1}(\xi_1)\|},$$

we may assume that

$$\min_{\mu < j \leq \nu} \left(\frac{\|h_j(\xi_1)\|}{\|h_{j-1}(\xi_1)\|} \right) < 1.$$

Now we consider the different cases:

1. There exist neither returns nor ends of folding periods between μ and ν. Then, $\|h_\nu(\xi_1)\| = \left\| DT_{\lambda,a,b}^{\nu-\mu}(\xi_{\mu+1}) h_\mu(\xi_1) \right\|$ and, from Proposition 4.4(a),

$$\frac{\|h_\nu(\xi_1)\|}{\|h_\mu(\xi_1)\|} \geq C_0 e^{c_0(\nu-\mu)} \frac{\delta}{2} \geq C_0 \frac{\delta}{2} \geq \left(C_0 \frac{\delta}{2} \right)^{\nu-\mu} \min_{\mu < j \leq \nu} \left(\frac{\|h_j(\xi_1)\|}{\|h_{j-1}(\xi_1)\|} \right).$$

2. There are no returns between μ and ν but let us assume that there exist folding periods $[\mu_i + 1, \mu_i + l_i]$ such that every $\mu_i + l_i$ coincide with the same iterate $j \in [\mu, \nu]$. Let us denote by μ_1 the first μ_j and write

$$\frac{\|h_\nu(\xi_1)\|}{\|h_\mu(\xi_1)\|} = \frac{\|h_\nu(\xi_1)\|}{\|h_{\mu_1+l_1}(\xi_1)\|} \frac{\|h_{\mu_1+l_1}(\xi_1)\|}{\|\tilde{h}_{\mu_1+l_1}(\xi_1)\|} \frac{\|\tilde{h}_{\mu_1+l_1}(\xi_1)\|}{\|h_\mu(\xi_1)\|}.$$

From the previous case, we have

$$\frac{\|\tilde{h}_{\mu_1+l_1}(\xi_1)\|}{\|h_\mu(\xi_1)\|} \geq C_0 \frac{\delta}{2} \quad \text{and} \quad \frac{\|h_\nu(\xi_1)\|}{\|h_{\mu_1+l_1}(\xi_1)\|} \geq C_0 \frac{\delta}{2}.$$

Furthermore, since S.A.5 leads to

$$h_{\mu_1+l_1}(\xi_1) = \tilde{h}_{\mu_1+l_1}(\xi_1) + \alpha_{\mu_1}(\xi_1) DT_{\lambda,a,b}^{l_1}(\xi_{\mu_1+1}) e^{(l_1)}(\xi_{\mu_1+1}),$$

from I.H.5 and Proposition 5.11, we get

$$\frac{\|h_{\mu_1+l_1}(\xi_1)\|}{\|\tilde{h}_{\mu_1+l_1}(\xi_1)\|} \geq 1 - \frac{3K\sqrt{b} \|h_{\mu_1}(\xi_1)\| \left(\sqrt{b} \right)^{l_1} d_{\mu_1}(\xi_0)}{\|\tilde{h}_{\mu_1+l_1}(\xi_1)\|}.$$

Since Proposition 5.14 implies

$$\left\|\tilde{h}_{\mu_1+l_1}(\xi_1)\right\| = \|\beta_{\mu_1}(\xi_1)h_{l_1}(\xi_{\mu_1+1})\| \geq \|h_{\mu_1}(\xi_1)\| e^{cl_1},$$

it follows that

$$\frac{\|h_{\mu_1+l_1}(\xi_1)\|}{\left\|\tilde{h}_{\mu_1+l_1}(\xi_1)\right\|} \geq \frac{1}{2}. \tag{5.12}$$

Then, we conclude that

$$\frac{\|h_\nu(\xi_1)\|}{\|h_\mu(\xi_1)\|} \geq \frac{1}{2}\left(C_0\frac{\delta}{2}\right)^2 \geq \left(C_0\frac{\delta}{4}\right)^{\nu-\mu} \min_{\mu < j \leq \nu}\left(\frac{\|h_j(\xi_1)\|}{\|h_{j-1}(\xi_1)\|}\right).$$

Notice that, if there exist different ends of folding periods between μ and ν, then the proof follows in the same way.

For the remaining cases, that is, when there exist returns between μ and ν, we shall write

$$\frac{\|h_\nu(\xi_1)\|}{\|h_\mu(\xi_1)\|} = \left(\prod_{\substack{j=\mu+1 \\ j \neq i}}^{\nu} \frac{\|h_j(\xi_1)\|}{\|h_{j-1}(\xi_1)\|}\right) \frac{\|h_i(\xi_1)\|}{\|h_{i-1}(\xi_1)\|}.$$

Then, for a suitable choice of i we shall prove that $C^{\nu-\mu}$ is a lower bound for the first factor.

3. Assume that there exists only one return μ_1 between μ and ν. Then write

$$\frac{\|h_\nu(\xi_1)\|}{\|h_\mu(\xi_1)\|} = \frac{\|h_\nu(\xi_1)\|}{\|h_{\mu_1}(\xi_1)\|} \frac{\|h_{\mu_1}(\xi_1)\|}{\|h_{\mu_1-1}(\xi_1)\|} \frac{\|h_{\mu_1-1}(\xi_1)\|}{\|h_\mu(\xi_1)\|}.$$

Take $i = \mu_1$. From the previous case we obtain

$$\frac{\|h_\nu(\xi_1)\|}{\|h_{\mu_1}(\xi_1)\|} \frac{\|h_{\mu_1-1}(\xi_1)\|}{\|h_\mu(\xi_1)\|} \geq C^{\nu-\mu}.$$

4. Finally, suppose that there exists several returns of ξ_0 between μ and ν. Let $\mu_1, ..., \mu_s$ be all the returns between μ and ν whose binding periods are not contained in the binding period of other returns of ξ_0 taking place between μ and ν. If ν does not belong to any binding period of these returns, write

$$\frac{\|h_\nu(\xi_1)\|}{\|h_\mu(\xi_1)\|} = \frac{\|h_\nu(\xi_1)\|}{\|h_{\mu_s+p_s}(\xi_1)\|} \frac{\|h_{\mu_s+p_s}(\xi_1)\|}{\|h_{\mu_s-1}(\xi_1)\|} \cdots \frac{\|h_{\mu_1+p_1}(\xi_1)\|}{\|h_{\mu_1-1}(\xi_1)\|} \frac{\|h_{\mu_1-1}(\xi_1)\|}{\|h_\mu(\xi_1)\|}.$$

From Proposition 4.4 we obtain, as we did in (5.8),

$$\frac{\|h_\nu(\xi_1)\|}{\|h_\mu(\xi_1)\|} \geq C_0^{s+1} e^{c_0(\nu-\mu_s-p_s)} ... e^{c_0(\mu_2-\mu_1-p_1-1)} e^{c_0(\mu_1-\mu-1)}\frac{\delta}{2} \geq \left(C_0\frac{\delta}{2}\right)^{\nu-\mu}.$$

Otherwise, if $\nu \in [\mu_j, \mu_j + p_j]$ for some j, then take the minimum μ_j and write

$$\frac{\|h_\nu(\xi_1)\|}{\|h_\mu(\xi_1)\|} = \frac{\|h_\nu(\xi_1)\|}{\|h_i(\xi_1)\|} \frac{\|h_i(\xi_1)\|}{\|h_{i-1}(\xi_1)\|} \frac{\|h_{i-1}(\xi_1)\|}{\|h_\mu(\xi_1)\|},$$

where $i = \mu_j$. Now, repeating the arguments of the above case and those used to reach (5.10), we have

$$\frac{\|h_\nu(\xi_1)\|}{\|h_\mu(\xi_1)\|} \geq \frac{\|h_i(\xi_1)\|}{\|h_{i-1}(\xi_1)\|} C^{\nu-\mu}.$$

Finally, the proof ends by checking that

$$\frac{\|h_j(\xi_1)\|}{\|h_{j-1}(\xi_1)\|} \geq d_j(\xi_0),$$

for all j. When j is a return, this is a consequence of I.H.5(b). Otherwise, we distinguish two cases:

- No folding period ends at time j. Then $h_j(\xi_1) = DT_{\lambda,a,b}(\xi_j)h_{j-1}(\xi_1)$. Setting $(x_j, y_j) = \xi_j$, Lemma 4.1 leads to

$$\frac{\|h_j(\xi_1)\|}{\|h_{j-1}(\xi_1)\|} \geq \left(1 - \sqrt[5]{b}\right)\left|f'_{\lambda,a}(x_j)\right| > 3d_j(\xi_0).$$

- Some folding period ends at time j. Then, from (5.12),

$$\frac{\|h_j(\xi_1)\|}{\|h_{j-1}(\xi_1)\|} = \frac{\|h_j(\xi_1)\|}{\|\tilde{h}_j(\xi_1)\|} \frac{\|\tilde{h}_j(\xi_1)\|}{\|h_{j-1}(\xi_1)\|} > \frac{3}{2}d_j(\xi_0). \quad \square$$

Proposition 5.16. *There exists a constant $K_1 = K_1(K, C_0, \alpha, \delta)$ such that*

$$\tilde{\Lambda}K_1^{-5}e^{-\alpha\mu}\|h_\mu(\xi_1)\| \leq \|\omega_\mu(\xi_1)\| \leq \tilde{\Lambda}^{-1}K_1^5 e^{2\alpha\mu}\|h_\mu(\xi_1)\|$$

for every ξ_0 bound to C_k and for every $1 \leq \mu \leq k \leq n - 1$.

Proof. Clearly, we may assume that $\mu+1$ is not a fold-free iterate and so, there exists a return μ_1 of ξ_0 such that $\mu \in [\mu_1, \mu_1 + l_1)$. Take μ_1 minimum and let $\mu_2 = \mu_1 + l_1 + 1$. From the proof of Proposition 5.12, it follows that

$$\mu_2 - \mu_1 \leq \max\left\{\frac{10\alpha \log K}{c \log b^{-1}}\mu, 4\right\} + 1 \leq \frac{10\alpha \log K}{c \log b^{-1}}\mu + 5. \tag{5.13}$$

From Proposition 5.15, we obtain

$$\frac{\|h_\mu(\xi_1)\|}{\|h_{\mu_1}(\xi_1)\|} \geq C^{\mu-\mu_1} \min_{\mu_1 < j \leq \mu} (d_j(\xi_0))$$

and

$$\frac{\|h_{\mu_2-1}(\xi_1)\|}{\|h_\mu(\xi_1)\|} \geq C^{\mu_2-\mu-1} \min_{\mu<j<\mu_2} (d_j(\xi_0)).$$

On the other hand, if ζ_0 is the binding point associated to μ_1, from Propositions 5.21 and 5.19, $d_j(\xi_0) \geq \frac{6}{7}d_{j-\mu_1}(\zeta_0) \geq \frac{6}{7}\tilde{\Lambda}e^{-\alpha(j-\mu_1)}$ for all $j \in [\mu_1+1, \mu_1+l_1]$. Then,

$$\frac{\|h_\mu(\xi_1)\|}{\|h_{\mu_1}(\xi_1)\|} \geq \frac{6}{7}C^{\mu-\mu_1}\tilde{\Lambda}e^{-\alpha(\mu-\mu_1)}$$

and

$$\frac{\|h_{\mu_2-1}(\xi_1)\|}{\|h_\mu(\xi_1)\|} \geq \frac{6}{7}C^{\mu_2-\mu-1}\tilde{\Lambda}e^{-\alpha(\mu_2-\mu_1-1)}.$$

Furthermore, I.H.5(b), Remark 5.22 and (BA) lead to

$$\frac{\|h_{\mu_1}(\xi_1)\|}{\|h_{\mu_1-1}(\xi_1)\|} \geq d_{\mu_1}(\xi_0) > \frac{6}{7}e^{-\alpha\mu_1}.$$

Therefore,

$$\frac{\|h_\mu(\xi_1)\|}{\|h_{\mu_1-1}(\xi_1)\|} \geq \frac{\tilde{\Lambda}}{2}C^{\mu-\mu_1}e^{-\alpha\mu}.$$

Since μ_1 and μ_2 are fold-free iterates, we get $\|h_{\mu_1-1}(\xi_1)\| \geq K^{\mu_1-\mu-1}\|\omega_\mu(\xi_1)\|$ and $\|h_{\mu_2-1}(\xi_1)\| \leq K^{\mu_2-\mu-1}\|\omega_\mu(\xi_1)\|$. Hence,

$$\frac{\|\omega_\mu(\xi_1)\|}{\|h_\mu(\xi_1)\|} \geq \tilde{\Lambda}\left(\frac{C}{Ke^\alpha}\right)^{\mu_2-\mu_1}.$$

Let $K_1 = KC^{-1}e^\alpha$. From (5.13), we obtain

$$\frac{\|\omega_\mu(\xi_1)\|}{\|h_\mu(\xi_1)\|} \geq \tilde{\Lambda}K_1^{-5}K_1^{-\frac{10\alpha\log K}{c\log b^{-1}}\mu}.$$

Then,

$$\frac{\|\omega_\mu(\xi_1)\|}{\|h_\mu(\xi_1)\|} \geq \tilde{\Lambda}K_1^{-5}e^{-\alpha\mu},$$

provided that b is small enough.

In order to complete the proof of the proposition, notice that

$$\frac{\|h_\mu(\xi_1)\|}{\|\omega_\mu(\xi_1)\|} \geq \frac{\tilde{\Lambda}}{2}C^{\mu-\mu_1}e^{-\alpha\mu}K^{\mu_1-\mu-1} > \tilde{\Lambda}e^{-2\alpha\mu}K_1^{-5}. \quad \square$$

Corollary 5.17. *For $k \leq n-1$ and for every ξ_0 bound to C_k,*

$$\|\omega_k(\xi_1)\| \geq e^{ck}.$$

Proof. From Proposition 5.16,

$$\|\omega_k(\xi_1)\| \geq \tilde{\Lambda} K_1^{-5} e^{-\alpha k} \|h_k(\xi_1)\|$$

and, from Proposition 5.14,

$$\|\omega_k(\xi_1)\| \geq \tilde{\Lambda} K_1^{-5} e^{-\alpha k} e^{2ck}.$$

Then, for every $k > N$ we get $\|\omega_k(\xi_1)\| \geq e^{ck}$, provided that N is large enough. $\square$

We finish this chapter by proving a result which will be used in the construction of binding points.

Proposition 5.18. *There exists a constant $\tilde{c} > 0$ such that if ν is a free return for ξ_0, then*

$$\|\omega_{\nu-1}(\xi_1)\| \geq K^{-5} e^{\tilde{c}(\nu-\mu)} \|\omega_{\mu-1}(\xi_1)\|$$

for every ξ_0 bound to C_k and for every $1 \leq \mu < \nu \leq k \leq n-1$.

Proof. Let us assume that μ is a fold-free iterate. If there are no returns of ξ_0 between μ and ν, then the statement is an easy consequence of Proposition 4.4(b),

$$\frac{\|\omega_{\nu-1}(\xi_1)\|}{\|\omega_{\mu-1}(\xi_1)\|} \geq C_0 e^{c_0(\nu-\mu)} \geq K^{-5} e^{c_0(\nu-\mu)}.$$

Let us now assume that there exist returns of ξ_0 between μ and ν. Let $\nu_1 < ... < \nu_r$ be all the returns of ξ_0 between μ and ν whose binding periods are not contained in the binding period of other returns of ξ_0 taking place between μ and ν. Let $p_1, ..., p_r$ be the lengths of the respective binding periods and take $d_1 = \nu_1 - \mu - 1$, $d_2 = \nu_2 - \nu_1 - p_1 - 1, ..., d_r = \nu_r - \nu_{r-1} - p_{r-1} - 1$ and $d_{r+1} = \nu - \nu_r - p_r - 1$. Then, Proposition 4.4 and I.H.6(b) imply

$$\frac{\|\omega_{\nu-1}(\xi_1)\|}{\|\omega_{\mu-1}(\xi_1)\|} \geq C_0^{r+1} \tau_2^r e^{c_0(d_1+...+d_{r+1})} e^{\frac{1}{3}c(p_1+...+p_r+r)} \geq (C_0\tau_2)^{r+1} e^{\frac{1}{3}c(\nu-\mu)},$$

where $\tau_2 = \tau_2(\lambda, K, \beta, \alpha)$. From Propositions 5.9 and 5.23, we also get

$$p_i \geq \frac{\log(2\delta)^{-1}}{3\log K}$$

and consequently,

$$r < \frac{3(\nu-\mu)\log K}{\log(2\delta)^{-1}}.$$

From

$$(\nu - \mu) > \frac{\log(2\delta)^{-1}}{3 \log K},$$

it follows that

$$r + 1 < \frac{6(\nu - \mu) \log K}{\log(2\delta)^{-1}}.$$

So, we obtain

$$\frac{\|\omega_{\nu-1}(\xi_1)\|}{\|\omega_{\mu-1}(\xi_1)\|} \geq \left\{ (C_0 \tau_2)^{\frac{6 \log K}{\log(2\delta)^{-1}}} e^{\frac{1}{3}c} \right\}^{(\nu-\mu)} = e^{\bar{c}(\nu-\mu)}.$$

Finally, let us assume that μ belongs to a folding period. Take μ_1 and μ_2 as in Proposition 5.12(b), that is,

$$\mu_2 - \mu_1 \leq \frac{50 \log^2 K}{c \log b^{-1}} (\nu - \mu) + 5.$$

From the previous case, $\|\omega_{\nu-1}(\xi_1)\| \geq e^{\bar{c}(\nu-\mu_1)} \|\omega_{\mu_1-1}(\xi_1)\|$. Moreover,

$$\|\omega_{\mu-1}(\xi_1)\| \leq K^{\mu-\mu_1} \|\omega_{\mu_1-1}(\xi_1)\| \leq K^{\frac{50 \log^2 K}{c \log b^{-1}}(\nu-\mu)+5} \|\omega_{\mu_1-1}(\xi_1)\|$$

and hence, $\|\omega_{\mu-1}(\xi_1)\| \leq K^5 e^{\frac{1}{2}\bar{c}(\nu-\mu)} \|\omega_{\mu_1-1}(\xi_1)\|$, provided that b is small enough. Therefore, $\|\omega_{\nu-1}(\xi_1)\| \geq K^{-5} e^{\frac{1}{2}\bar{c}(\nu-\mu)} \|\omega_{\mu-1}(\xi_1)\|$. $\square$

5.5 Appendix: The function $d_\mu(\xi_0)$

For each natural number $k \leq n$ and for every ξ_0 bound to C_k, let us define

$$d_j(\xi_0) = \begin{cases} |x_j - c_\lambda|, & \text{if } j \text{ is not a return of } \xi_0 \\ \|\xi_j - \zeta_0\|, & \text{if } j \text{ is a return of } \xi_0 \end{cases}$$

for $1 \leq j \leq k$, where $\xi_j = (x_j, y_j) = T^j_{\lambda,a,b}(\xi_0)$ and ζ_0 is the binding point associated to ξ_j when j is a return of ξ_0.

Proposition 5.19. *For* $1 \leq j \leq k \leq n$ *and for every* $z_0 \in C_k$ *we have*

$$d_j(z_0) \geq \tilde{\Lambda} e^{-\alpha j}.$$

Proof. Let $N_0 \in \mathbf{N}$ large enough so that

$$e^{(\alpha-\beta)N_0} < \frac{4}{3}\tilde{\Lambda}^{-1}\left(1 - \tilde{\Lambda}\right) \quad \text{and} \quad e^{-\alpha N_0} < \frac{4}{7}\delta\tilde{\Lambda}^{-1}.$$

Take Ω_0 sufficiently close to $a(\lambda)$ and b small enough such that if $z_0 = (x_0, y_0)$ is a critical approximation, then $|x_j - c_\lambda| \geq \tilde{\Lambda}e^{-\alpha}$ for $1 \leq j \leq N_0$. So, the proposition is proved up to time N_0 for every critical approximation z_0.

Now, let us argue by induction. Let $j \in [N_0, k]$ and $z_0 \in C_k$. First, let us assume that j does not belong to the binding period of any return of z_0. Then we consider the two possible cases:

1. If $j = \nu$ is a return of z_0, let $\overline{z}_0 \in C_\nu$ be the critical approximation which has a free return at time ν. Then, (BA) leads to $d_\nu(\overline{z}_0) \geq e^{-\alpha\nu}$. Hence, from

$$e^{(\alpha-\beta)\nu} < e^{(\alpha-\beta)N_0} \leq \frac{4}{3}\tilde{\Lambda}^{-1}(1 - \tilde{\Lambda}),$$

it follows that

$$d_\nu(z_0) \geq d_\nu(\overline{z}_0) - \|z_\nu - \overline{z}_\nu\| \geq e^{-\alpha\nu} - \frac{3}{4}\tilde{\Lambda}e^{-\beta\nu} \geq \tilde{\Lambda}e^{-\alpha\nu}.$$

2. If j is not a return, let $\overline{z}_0 \in C_j$ be the point given in Definition 5.5. Since $d_j(\overline{z}_0) > \delta$, we obtain

$$d_j(z_0) \geq d_j(\overline{z}_0) - \frac{3}{4}\tilde{\Lambda}e^{-\beta j} \geq \delta - \frac{3}{4}\tilde{\Lambda}e^{-\alpha j} > \tilde{\Lambda}e^{-\alpha j},$$

provided that N_0 is large enough.

Now, let us assume that j belongs to the binding period of some return μ of z_0. Take μ maximum and let ζ_0 be the binding point associated to z_μ. Then, whether j is a return or not, we have

$$d_j(z_0) \geq d_{j-\mu}(\zeta_0) - \|z_j - \zeta_{j-\mu}\|.$$

Then, the binding condition implies

$$d_j(z_0) \geq d_{j-\mu}(\zeta_0) - he^{-\beta(j-\mu)}$$

and, by the induction hypothesis,

$$d_j(z_0) \geq \frac{7}{8}\tilde{\Lambda}e^{-\alpha(j-\mu)} > \tilde{\Lambda}e^{-\alpha j}$$

whenever N is large enough. $\square$

Remark 5.20. *In the same way as in Proposition 5.19, we obtain the following result: Let $k \geq N_0$. Then, $d_j(\xi_0) \geq \tilde{\Lambda} e^{-\alpha j}$ for every ξ_0 bound to C_k and for every $j \in [N_0, k]$.*

Proposition 5.21. *Let ν be a return of $z_0 \in C_n$ and let ζ_0 and $[\nu + 1, \nu + p]$ be the binding point and the binding period associated to z_ν, respectively. Then,*

$$\frac{6}{7} < \frac{d_{\nu+j}(z_0)}{d_j(\zeta_0)} < \frac{7}{6}$$

for every $1 \leq j \leq p$.

Proof. Notice that $d_{\nu+j}(z_0) \geq d_j(\zeta_0) - \|z_{\nu+j} - \zeta_j\|$ for every $1 \leq j \leq p$. So, Proposition 5.19 and the definition of binding period imply

$$\frac{d_{\nu+j}(z_0)}{d_j(\zeta_0)} \geq 1 - \frac{h}{\tilde{\Lambda}} e^{(\alpha-\beta)j} \geq \frac{7}{8}.$$

On the other hand, the same arguments lead to

$$\frac{d_j(\zeta_0)}{d_{\nu+j}(z_0)} \geq 1 - \frac{\|z_{\nu+j} - \zeta_j\|}{d_{\nu+j}(z_0)}$$

and

$$\frac{\|z_{\nu+j} - \zeta_j\|}{d_{\nu+j}(z_0)} \leq \frac{8}{7} \frac{h}{\tilde{\Lambda}} e^{(\alpha-\beta)j} \leq \frac{1}{7}.$$

Consequently,

$$\frac{d_j(\zeta_0)}{d_{\nu+j}(z_0)} > \frac{6}{7}. \quad \square$$

Remark 5.22. *In the same way as in Proposition 5.21, we prove the following result: Let $k \geq N_0$. Then, for every ξ_0 bound to C_k and for each $j \in [N_0, k]$, we have*

$$\frac{6}{7} < \frac{d_j(z_0)}{d_j(\xi_0)} < \frac{7}{6}$$

where $z_0 \in C_k$ is the point given by Definition 5.5.

Proposition 5.23. *For $1 \leq \nu \leq k \leq n$ and for every $z_0 \in C_k$, it follows that*
  (a) *If $d_\nu(z_0) < \frac{1}{2}\delta$, then ν is a return of z_0.*
  (b) *If ν is a return of z_0, then $d_\nu(z_0) < 2\delta$.*

Proof. Take N_1 sufficiently large so that

$$\frac{3}{4}\tilde{\Lambda}e^{-\beta N_1} < \frac{1}{8}\delta.$$

Choose Ω_0 close enough to $a(\lambda)$ and b small enough such that if z_0 is a critical approximation, then $d_j(z_0) > 10\delta$ for every $1 \leq j \leq N_1$. Then, we may take $\nu > N_1$. Let us assume that ν is not a return of z_0. Since $d_\nu(z_0) < \frac{1}{2}\delta$, ν has to be contained in the binding period of a return μ of z_0. Let μ be the largest return of z_0 whose binding period contains ν and let ζ_0 be the respective binding point. Then, $\nu - \mu$ is a free iterate of ζ_0, $\nu - \mu$ is not a return of ζ_0 and moreover, from Proposition 5.21,

$$\delta < d_{\nu-\mu}(\zeta_0) < \frac{7}{6}d_\nu(z_0).$$

Thus, $d_\nu(z_0) > \frac{6}{7}\delta$.

To prove statement (b), let us take the point $\overline{z}_0 \in C_\nu$ which z_0 is bound to. Then, by definition, ν is a return of $\overline{z}_0$. If ν is a free return, then, since all the critical approximations belong to B_L, we have

$$d_\nu(\overline{z}_0) = \|\overline{z}_\nu - \zeta_0\| < 2K\sqrt{b} + \lambda b^{\frac{1}{20}} + \delta < \frac{9}{8}\delta.$$

So,

$$d_\nu(z_0) \leq d_\nu(\overline{z}_0) + \frac{3}{4}\tilde{\Lambda}e^{-\beta N_1} < \frac{5}{4}\delta.$$

If ν is a bound return, then let μ be the largest return of $\overline{z}_0$ whose binding period contains ν and let ζ_0 be the respective binding point. From Proposition 5.21 and bearing in mind that $\nu - \mu$ is a free return of ζ_0, we obtain

$$d_\nu(\overline{z}_0) \leq \frac{7}{6}d_{\nu-\mu}(\zeta_0) < \frac{5}{3}\delta.$$

Therefore,

$$d_\nu(z_0) \leq \frac{5}{3}\delta + \frac{1}{8}\delta < 2\delta. \quad \square$$

Chapter 6

THE BINDING POINT

In Chapter 5 we showed that the assumption I.H.4 holds at time n provided that
I.H.5 and I.H.6 do, see Corollary 5.17. In this chapter, we shall prove that I.H.5 holds
actually at time n.

Let n be a free return of $z_0 \in C_n$. We need to find a binding point $\zeta_0 \in C_n$
associated to z_n satisfying the properties stated in I.H.5(a) and I.H.5(b). To this end,
we look for the afore-mentioned ζ_0 to be in *tangential position* with respect to z_n,
in the sense that $\mathrm{dist}(z_n, \gamma) << \|z_n - \zeta_0\|$ and $|\mathrm{ang}\,(\omega_{n-1}(z_1), t(\gamma; \eta))| << \|z_n - \zeta_0\|$,
where γ is a certain curve on $W_{a,m}^u$ which contains ζ_0 and $t(\gamma; \eta)$ is the tangent vector
to $W_{a,m}^u$ at $\eta \in \gamma$.

From now on, we fix the constant $\lambda_0 = \left(\frac{1}{2}\delta\right)^4$.

Definition 6.1. *Let $z_0 \in C_n$. We say that $\mu \in [1, n]$ is a semifavourable iterate for
z_0 if the following properties hold:*

*1. μ is neither a return of z_0 nor belongs to the folding period associated to any return
of z_0.*

*2. $z_\mu \in \tilde{V} = \{(x, y) : x \in I_2\}$, where I_2 is the interval given in the definition of folding
period.*

We say that μ is a favourable iterate for z_0 if, moreover,

3. $d_j(z_\mu) = d_{j+\mu}(z_0) \geq \lambda_0^{j+1}$ for $0 \leq j \leq n - \mu - 1$.

Lemma 6.2. *Let $I = [p, q)$ be a time interval such that p is a semifavourable iterate
for z_0 and*

$$\inf_I d_j(z_0) \geq \left(K^4 e^{8\beta}\right)^{-|I|}.$$

Then, there exists a semifavourable iterate $\nu \in \left[p, \frac{1}{2}(p+q)\right)$ such that

$$d_{\nu+j}(z_0) \geq \lambda_0^{j+1}$$

for every j satisfying $\nu \leq \nu + j < q$.

Proof. If $\left(K^4 e^{8\beta}\right)^{-|I|} \geq \lambda_0$ the statement follows by taking $\nu = p$. Therefore, we suppose that

$$|I| > \frac{\log \lambda_0^{-1}}{4\log K + 8\beta}. \tag{6.1}$$

By choosing δ small enough, we may assume that

$$\frac{\log \lambda_0^{-1}}{4\log K + 8\beta} \geq 100$$

and the result is proved for $|I| \leq 100$. Now, we proceed by induction on the length of I. Let us take $m \geq 12$ such that $8m - 7 \leq |I| \leq 8m$ and $J = [p, p+2m)$. We distinguish between the following cases:

1. If $\inf_J d_j(z_0) \geq \left(K^4 e^{8\beta}\right)^{-|J|}$, then there exists $\nu \in [p, p+m)$ such that $d_{\nu+j}(z_0) \geq \lambda_0^{j+1}$ for $\nu \leq \nu + j < p + 2m$. If $p + 2m \leq \nu + j < q$, then $j \geq m$. Hence, from $d_j(z_0) \geq \left(K^4 e^{8\beta}\right)^{-|I|} \geq \left(K^4 e^{8\beta}\right)^{-8m}$, we deduce $d_j(z_0) \geq \left(K^4 e^{8\beta}\right)^{-8j} \geq \lambda_0^{j+1}$.

2. Now, let us assume that $\inf_J d_j(z_0) < \left(K^4 e^{8\beta}\right)^{-|J|}$. Let $\overline{\nu} \in J$ be such that $d_{\overline{\nu}}(z_0) = \inf_J d_j(z_0)$. Then, $d_{\overline{\nu}}(z_0) \leq \left(K^4 e^{8\beta}\right)^{-|J|} \leq \left(K^4 e^{8\beta}\right)^{-\frac{1}{4}|I|}$ and, from (6.1), it follows that $d_{\overline{\nu}}(z_0) \leq \sqrt[4]{\lambda_0} = \frac{1}{2}\delta$. So, from Proposition 5.23, we deduce that $\overline{\nu}$ is a return of z_0. Let $\overline{p}$ be the length of its binding period. We claim that $\overline{p} \geq \frac{3}{2}m$. Indeed, let h be the constant given in the definition of primary binding period. Since h does not depend on δ, we may choose $\lambda_0 \leq h^{32}$. Now, (6.1) and the fact that $|I| \leq 8m$, imply

$$m \geq \frac{4\log h^{-1}}{4\log K + 8\beta}. \tag{6.2}$$

So, $\|z_{\overline{\nu}+j} - \zeta_j\| \leq K^j d_{\overline{\nu}}(z_0) < K^{4(j-2m)} e^{-16\beta m}$ for $1 \leq j \leq \frac{7}{4}m$. Now, from (6.2), $\left(K^4 e^{8\beta}\right)^{-\frac{1}{4}m} \leq h$ and hence $\|z_{\overline{\nu}+j} - \zeta_j\| \leq h e^{-\beta j}$ for $1 \leq j \leq \frac{7}{4}m$. Therefore, $\overline{p}_0 \geq \frac{7}{4}m$ and, from (5.6), $\overline{p} \geq (1 - 2c^{-1}\alpha)\overline{p}_0 \geq \frac{3}{2}m$, provided that α is much smaller than c.

Let $\overline{l}$ be the length of the folding period associated to the return $\overline{\nu}$ of z_0 and let $\overline{\mu} = \overline{\nu} + \overline{l} + 1$. Notice that, from the definition of folding period, it follows that

$$\overline{\mu} - \overline{\nu} = \overline{l} + 1 \leq \frac{10\log K}{c\log b^{-1}} \log d_{\overline{\nu}}(z_0)^{-1} + 5.$$

Then

$$\overline{\mu} - \overline{\nu} \leq \frac{80\left(8\beta + 4\log K\right)\log K}{c\log b^{-1}} m + 5.$$

Hence, since $m \geq 12$, we get $\overline{\mu} - \overline{\nu} \leq \frac{1}{2}m$ whenever b is small enough.

To finish the proof we argue as follows:

(a) If $\overline{\mu}$ is a semifavourable iterate for z_0, take $L = [\overline{\mu}, \overline{\mu} + m)$. Since $\overline{p} \geq \frac{3}{2}m$ and $\overline{\mu} \leq \overline{\nu} + \frac{1}{2}m$, we obtain $L \subset [\overline{\nu} + 1, \overline{\nu} + \overline{p}]$. So, if ζ_0 is the binding point associated to $\overline{\nu}$, then Propositions 5.21 and 5.19 imply

$$d_j(z_0) > \frac{6}{7} d_{j-\overline{\nu}}(\zeta_0) > \frac{6}{7}\tilde{\Lambda}e^{-\alpha(j-\overline{\nu})} > \frac{6}{7}\tilde{\Lambda}e^{-\frac{3}{2}\alpha m} > \left(K^4 e^{8\beta}\right)^{-m}$$

for every $j \in L$. Therefore L satisfies the hypothesis of the lemma and, by induction, there exists a semifavourable iterate $\nu \in \left[\overline{\mu}, \overline{\mu} + \frac{1}{2}m\right)$ such that $d_{\nu+j}(z_0) \geq \lambda_0^{j+1}$ for $\nu \leq \nu + j < \overline{\mu} + m$. Since $\nu - p < 3m$, we have $\nu \in \left[p, \frac{1}{2}(p+q)\right)$. Furthermore, if $\overline{\mu} + m \leq \nu + j < q$, then $m \leq 2j$ and thus $d_{\nu+j}(z_0) \geq \left(K^4 e^{8\beta}\right)^{-|I|} \geq \left(K^4 e^{8\beta}\right)^{-8m} \geq \lambda_0^{j+1}$.

(b) If $\overline{\mu}$ belongs to a folding period, then there exists a return $\tilde{\nu}$ of z_0 in (p, ν) such that $\overline{\nu} \in [\tilde{\nu} + 1, \tilde{\nu} + \tilde{p}]$ and consequently, $\tilde{p} \geq \overline{p} > \frac{3}{2}m$. Moreover, $\tilde{\mu} = \tilde{\nu} + \tilde{l} + 1$ is a semifavourable iterate for z_0 and $\tilde{\mu} - \tilde{\nu} \leq \frac{1}{2}m$. Therefore, we may proceed as in (a). □

Lemma 6.3. *Let n be a free return of $z_0 \in C_n$. Then, there exists a sequence $1 = m_1 < m_2 < ... < m_s \leq n$ such that $n - m_i$ is a favourable iterate for z_0. Furthermore, $\frac{5}{4}m_i \leq m_{i+1} \leq 3m_i$ for $1 \leq i \leq s - 1$ and $n \leq 3m_s$.*

Proof. Let μ be the return of z_0 such that if l is the length of its folding period, then all the iterates of z_0 between $\mu + l$ and n are fold-free. From Proposition 5.23, $d_\mu(z_0) < 2\delta$. Then, $n - \mu \geq 24$, provided that δ is sufficiently small and Ω_0 is close enough to $a(\lambda)$.

Let $\tilde{\mu} = \mu + l + 1$. Then there are neither returns nor folding periods in $[\tilde{\mu}, n)$. Therefore, for every $m_i \in [1, n - \tilde{\mu}]$, $n - m_i$ is a semifavourable iterate for z_0. Moreover, since Proposition 5.23 implies $d_j(z_{n-m_i}) = d_{n-m_i+j}(z_0) > \frac{1}{2}\delta > \lambda_0 > \lambda_0^{j+1}$ for $0 \leq j \leq m_i - 1$, it follows that each $n - m_i$ is a favourable iterate for z_0. Next, we shall proceed by induction. First, let us observe that Proposition 5.12 yields

$$\tilde{\mu} - \mu = l + 1 \leq \frac{50 \log^2 K}{c \log b^{-1}}(n - \mu) + 5 \leq \frac{1}{2}(n - \mu).$$

Thus, $\tilde{\mu} \leq \frac{1}{2}(n + \mu)$ and consequently, $m_r = n - \tilde{\mu} \geq \frac{1}{2}(n - \mu)$. So, we assume that we have already found a natural number $m_i \geq \frac{1}{2}(n - \mu) \geq 12$ satisfying the statement. If $n \leq 3m_i$ the lemma is proved. Otherwise, to obtain m_{i+1} from m_i, notice that, from

Proposition 5.12, there exists a fold-free iterate $\mu_i \geq n - 3m_i$ such that

$$\mu_i - (n - 3m_i) \leq \frac{150 \log^2 K}{c \log b^{-1}} m_i + 5 \leq \frac{1}{2} m_i. \tag{6.3}$$

Furthermore, since $\mu_i - 1$ coincides with the end of a folding period, we obtain that μ_i is a semifavourable iterate for z_0. Let $I = [\mu_i, n)$. We claim that $d_j(z_0) \geq \left(K^4 e^{8\beta}\right)^{-|I|}$ for every $j \in I$. So, from Lemma 6.2, there exists a semifavourable iterate for z_0, $n - m_{i+1} \in \left[\mu_i, \frac{1}{2}(\mu_i + n)\right)$, such that $d_j(z_{n-m_{i+1}}) \geq \lambda_0^{j+1}$ for $0 \leq j < m_{i+1}$. Therefore, $n - m_{i+1}$ is a favourable iterate for z_0 and, since $n - 3m_i \leq \mu_i \leq n - m_{i+1}$, we get $m_{i+1} \leq 3m_i$. Finally, (6.3) leads to $n - m_{i+1} < \frac{1}{2}(\mu_i + n) < n - \frac{5}{4} m_i$.

In order to prove that $d_j(z_0) \geq \left(K^4 e^{8\beta}\right)^{-|I|}$ for every $j \in I$, notice that, if j is a return of z_0 and p_j is the length of its binding period, then Proposition 5.9 gives

$$n - j > p_j > \frac{\log d_j(z_0)^{-1}}{4 \left(\log K + 2\beta\right)}.$$

So, $d_j(z_0) \geq \left(K^4 e^{8\beta}\right)^{j-n} \geq \left(K^4 e^{8\beta}\right)^{-|I|}$. If j is not a return, then, from Proposition 5.23, we have $d_j(z_0) \geq \frac{1}{4} d_\mu(z_0)$ and, from the previous case, $d_\mu(z_0) \geq \left(K^4 e^{8\beta}\right)^{\mu-n}$. So, since $m_i \geq \frac{1}{2}(n - \mu)$, it follows that $d_j(z_0) \geq \frac{1}{4} \left(K^4 e^{8\beta}\right)^{-2m_i}$. Finally, from (6.3), $d_j(z_0) \geq \left(K^4 e^{8\beta}\right)^{-|I|}$. This completes the proof. $\square$

Lemma 6.4. *Let n be a free return of $z_0 \in C_n$ and let $\{m_i\}_{i=1}^s$ be the sequence given in Lemma 6.3. Then z_{n-m_i} is expanding up to time m_i for every $i \in \{1, ..., s\}$.*

Proof. As in the proof of Proposition 5.12, for every $1 \leq j \leq m_i$, we get an iterate $k \in [j, m_i]$ such that $n - m_i + k$ is a fold-free iterate of z_0. Moreover,

$$k - j \leq \frac{10 \log K}{c \log b^{-1}} \log d_{\tilde{\mu}}(\overline{z}_0)^{-1} + 5,$$

where $\tilde{\mu} \in (n - m_i, n - m_i + j)$ is a free return of $\overline{z}_0 \in C_{\tilde{\mu}}$. From Remark 5.22 and the fact that $n - m_i$ is a favourable iterate for z_0, we obtain

$$k - j \leq \frac{20 \log K}{c \log b^{-1}} \log d_{\tilde{\mu}}(z_0)^{-1} + 5 \leq \frac{20 \log K \log \lambda_0^{-1}}{c \log b^{-1}} j + 5. \tag{6.4}$$

Now, since $n - m_i + k$ is a fold-free iterate, $\|\omega_{n-m_i+j-1}(z_1)\| \geq K^{j-k} \|h_{n-m_i+k-1}(z_1)\|$. So, setting

$$v_i(z_1) = \frac{\omega_{n-m_i-1}(z_1)}{\|\omega_{n-m_i-1}(z_1)\|},$$

Proposition 5.15 and the fact that $n - m_i$ is a favourable iterate for z_0, imply that

$$\left\| DT^j_{\lambda,a,b}(z_{n-m_i}) v_i(z_1) \right\| \geq K^{j-k} C^k \min_{0 \leq t \leq k-1} d_t(z_{n-m_i}) \geq \left(\frac{K}{C\lambda_0} \right)^{j-k} (C\lambda_0)^j .$$

Therefore, from (6.4), we get $\left\| DT^j_{\lambda,a,b}(z_{n-m_i}) v_i(z_1) \right\| \geq (CK^{-1}\lambda_0)^5 (C\lambda_0)^{2j}$, provided that b is small enough.

Hence, from Proposition 3.4, $\| \omega_j(z_{n-m_i}) \| = \left\| DT^j_{\lambda,a,b}(z_{n-m_i})(1,0) \right\| \geq (CK^{-1}\lambda_0)^{8j}$. Thus, z_{n-m_i} is $(CK^{-1}\lambda_0)^8$-expanding up to time m_i and the lemma is proved. $\square$

Lemma 6.4 allows us to apply the results stated in Chapter 4 related to the existence of contractive fields. Let us take there $z = z_{n-m_i}$, $m = m_i$ and $\xi = (CK^{-1}\lambda_0)^8$. There exist integral curves Γ^{m_i} which intercept G_1 in one point $\eta_0^{[i]}$. Let

$$\rho_0 = \left(\frac{(CK^{-1}\lambda_0)^8}{10K^2} \right)^2 ,$$

$\eta^{[i]} = T^{m_i}_{\lambda,a,b}\left(\eta_0^{[i]} \right)$, $\gamma_0^{[i]} = \gamma\left(\eta_0^{[i]}, \rho_0^{m_i} \right) \subset W^u_{a,m}$ and $\gamma^{[i]} = T^{m_i}_{\lambda,a,b}\left(\gamma_0^{[i]} \right)$. According to Proposition 4.13, we check below that each $\gamma_0^{[i]}$ is a $C^2(b)$-curve.

Lemma 6.5. *Let $I_0 = [Q_{a_0} + \delta_0, \xi_1(a_0) - \delta_0]$ be the interval given in Proposition 4.11(a). Then, $\gamma_0^{[i]} \subset I_0 \times \mathbf{R}$ for $1 \leq i \leq s$.*

Proof. Let $m = m_i$ and $z_{n-m} = (x_{n-m}, y_{n-m})$. For every x in the domain of f_a we have $|Q_{a_0} - f_a(x)| \leq 3\lambda^{-1}|Q_{a_0} - x| + \lambda^{-1}|a - a_0|$. Therefore, $|Q_{a_0} - x_1| \leq 3\lambda^{-1}|Q_{a_0} - x| + \lambda^{-1}|a - a_0| + K\sqrt{b}$, where $(x_j, y_j) = T^j_{\lambda,a,b}(x,y)$ and $(x,y) \in U_m$. Furthermore, since $n - m$ is a favourable iterate,

$$|Q_{a_0} - x_{n-m}| \geq \delta_2 = R \left(\frac{3}{2} + \frac{\sqrt{a^2(\lambda) - 1}}{\lambda} \right) F(\lambda, a_0) > \frac{3}{2} F(\lambda, a_0). \tag{6.5}$$

On the other hand, since n is a return of z_0, we also get $|Q_{a_0} - x_n| \geq \frac{1}{2}\tilde{\Lambda}$. Hence,

$$\frac{\tilde{\Lambda}}{2} \leq \frac{3}{\lambda}|Q_{a_0} - x_{n-1}| + \frac{1}{\lambda}|a - a_0| + K\sqrt{b} \leq \ldots \leq$$

$$\leq \left(\frac{3}{\lambda} \right)^m |Q_{a_0} - x_{n-m}| + 2\left(\frac{1}{\lambda}|a - a_0| + K\sqrt{b} \right)\left(\frac{3}{\lambda} \right)^m . \tag{6.6}$$

For b sufficiently small, (4.6) implies

$$\frac{\tilde{\Lambda}}{2} \leq \left(\frac{3}{\lambda} \right)^m |Q_{a_0} - x_{n-m}| + \frac{1}{2}F(\lambda, a_0)\left(\frac{3}{\lambda} \right)^m .$$

Then, from (6.5), it follows that $\frac{1}{2}\tilde{\Lambda} < (3\lambda^{-1})^{m+1}|Q_{a_0} - x_{n-m}|$ and consequently,

$$|Q_{a_0} - x_{n-m}| \geq \frac{\tilde{\Lambda}}{2}\left(\frac{\lambda}{3}\right)^{m+1}. \tag{6.7}$$

Let us set $\eta_0^{[i]} = (x_0, y_0)$ and check that $|Q_{a_0} - x_0| - \rho_0^m \geq \delta_0$. To this end, write $|Q_{a_0} - x_0| \geq |Q_{a_0} - x_{n-m}| - |x_{n-m} - x_0|$. Since Γ^m is a nearly vertical curve, we may assume that $|x_{n-m} - x_0| < \frac{1}{4}F(\lambda, a_0)$, provided that b is small enough. Moreover, by taking δ, and consequently ρ_0, sufficiently small, (6.7) gives $(1 - R^{-1})|Q_{a_0} - x_{n-m}| \geq \rho_0^m$. So, $\rho_0^m \leq (1 - R^{-1})|Q_{a_0} - x_{n-m}|$ and $|Q_{a_0} - x_{n-m}| - \rho_0^m \geq R^{-1}|Q_{a_0} - x_{n-m}|$. Finally, we get $|Q_{a_0} - x_0| - \rho_0^m \geq R^{-1}|Q_{a_0} - x_{n-m}| - \frac{1}{4}F(\lambda, a_0)$ and, from (6.5),

$$|Q_{a_0} - x_0| - \rho_0^m \geq \left(\frac{5}{4} + \frac{\sqrt{a^2(\lambda) - 1}}{\lambda}\right)F(\lambda, a_0) = \delta_0.$$

The proof of the lemma is completed by checking $|f_{a_0}(c_\lambda) - x_0| - \rho_0^m \geq \delta_0$. To this end, we claim that $|f_{a_0}(c_\lambda) - x_{n-m}| \geq \frac{1}{2}\tilde{\Lambda}\left(\frac{1}{3}\lambda\right)^{m+1}$. Then, this inequality will play the same role that (6.7) did in the above arguments. To prove the claim, notice that, as in (6.5), we get

$$|f_{a_0}(c_\lambda) - x_{n-m}| > \delta_2 > \frac{3}{2}F(\lambda, a_0). \tag{6.8}$$

On the other hand, replacing m by $m - 1$ in (6.6), we obtain

$$\frac{\tilde{\Lambda}}{2} \leq \left(\frac{3}{\lambda}\right)^m|f_{a_0}(c_\lambda) - x_{n-m}| + \left(F(\lambda, a_0) + \frac{2}{\lambda}|a - a_0| + 2K\sqrt{b}\right)\left(\frac{3}{\lambda}\right)^{m-1}.$$

So, as in the proof of (6.7), we conclude from (6.8) that

$$\frac{\tilde{\Lambda}}{2} \leq \left(\frac{3}{\lambda}\right)^{m+1}|f_{a_0}(c_\lambda) - x_{n-m}|. \quad \square$$

Lemma 6.6. $\gamma^{[i]}$ is a $C^2(b)$-curve for every $i \in \{1, ..., s\}$.

Proof. Let $\gamma_0^{[i]} = \gamma_0$, $\eta_0^{[i]} = \eta_0$, $\gamma^{[i]} = \gamma$, $\eta^{[i]} = \eta$ and $m_i = m$, for each $i \in [1, s]$. From Proposition 5.13 and Lemma 6.5, the slope of the vectors $\omega_{n-m-1}(z_1)$ and $t(\gamma_0; \eta_0)$ are less than $\frac{1}{10}$. Therefore, by applying (4.18) for $\nu = m$ and $z_r = \eta_0$, we get $|\text{ang}(t(\gamma; \eta), \omega_{n-1}(z_1))| \leq \left(K_7\sqrt[4]{b}\right)^{m+1} \leq const\sqrt{b}$. Since n is also a fold-free iterate, from Proposition 5.13, it follows that $\omega_{n-1}(z_1)$ is a nearly horizontal vector. Hence, $|\text{slope } t(\gamma; \eta)| \leq const\sqrt{b}$.

Let $\Xi(\gamma)$ be the curvature of γ and let us assume that $\Xi(\gamma) \leq const\sqrt{b}$. Then $|slope\ t(\gamma;\xi)| \leq const\sqrt{b}$ for every $\xi \in \gamma$ and the lemma follows by taking the parametrization of γ given by $x \in [a,b] \to z(x) = (x, y(x))$, with $|y'(x)| = |slope\ t(\gamma;z(x))| < \sqrt[4]{b}$ and $|y''(x)| = \Xi(\gamma; z(x))(1 + y'^2(x))^{\frac{3}{2}} < \sqrt[4]{b}$.

Therefore, let us prove that $\Xi(\gamma) \leq const\sqrt{b}$. Let γ_0 be parametrized by arc length in $(-\tau, \tau)$ with $\gamma_0(0) = \eta_0$ and let $\gamma_j = T^j_{\lambda,a,b}(\gamma_0) = T^j(\gamma_0)$ for $1 \leq j \leq m$. Write $\gamma_j = \gamma_j(s) = (\gamma_{j_1}, \gamma_{j_2})$. Then, $\gamma'_{j+1} = DT(\gamma_j)\gamma'_j = \left(A\gamma'_{j_1} + B\gamma'_{j_2}, C\gamma'_{j_1} + D\gamma'_{j_2}\right)$ and

$$\gamma''_{j+1} = \left(\begin{array}{c} A\gamma''_{j_1} + B\gamma''_{j_2} \\ C\gamma''_{j_1} + D\gamma''_{j_2} \end{array} \right) + \left(\begin{array}{c} M_{11}\gamma'^2_{j_1} + M_{12}\gamma'_{j_1}\gamma'_{j_2} + M_{22}\gamma'^2_{j_2} \\ N_{11}\gamma'^2_{j_1} + N_{12}\gamma'_{j_1}\gamma'_{j_2} + N_{22}\gamma'^2_{j_2} \end{array} \right), \qquad (6.9)$$

where $(M_{11}, N_{11}) = \partial^2_x T(\gamma_j)$, $(M_{12}, N_{12}) = \partial^2_{xy} T(\gamma_j)$ and $(M_{22}, N_{22}) = \partial^2_y T(\gamma_j)$. Since $\Xi(\gamma_j) = \left\|\gamma'_j\right\|^{-3} \left|\det\left(\gamma'_j, \gamma''_j\right)\right|$, we get $\Xi(\gamma_{j+1}) \leq K_j \left(|\det DT(\gamma_j)|\,\Xi(\gamma_j) + L_j\right)$ with

$$t_j = \frac{\gamma'_j}{\left\|\gamma'_j\right\|}, \quad K_j = \left(\frac{\left\|\gamma'_j\right\|}{\left\|\gamma'_{j+1}\right\|}\right)^3 \text{ and } L_j = \left|\det\left(DT(\gamma_j)t_j, D^2T(\gamma_j)(t_j, t_j)\right)\right|.$$

Now, from (6.9) and Proposition 1.3, we have $L_j \leq 12K^2\sqrt{b}$ and $|\det DT(\gamma_j)| \leq Kb$. Therefore, $\Xi(\gamma_{j+1}) \leq K_j\left(Kb\,\Xi(\gamma_j) + 12K^2\sqrt{b}\right)$ and recurrently,

$$\Xi(\gamma) = \Xi(\gamma_m) \leq K_{m-1}\left(Kb\,\Xi(\gamma_{m-1}) + 12K^2\sqrt{b}\right) \leq ... \leq$$

$$\leq (Kb)^m K_{m-1}K_{m-2}...K_0\,\Xi(\gamma_0) + 12K^2\sqrt{b}\sum_{j=1}^{m} K_{m-1}...K_{j-1}(Kb)^{m-j}. \qquad (6.10)$$

Next, we shall bound the terms $K_{m-1}...K_{j-1}$. From Lemma 6.4, z_{n-m} is $(CK^{-1}\lambda_0)^8$-expanding up to time m and, from (4.17), we get

$$\frac{1}{2} \leq \frac{\left\|\gamma'_j(0)\right\|}{\left\|\omega_j(z_{n-m})\right\|} \leq 2.$$

Hence, $\gamma'_0(0)$ is also $(CK^{-1}\lambda_0)^8$-expanding up to time m. Furthermore,

$$\left\|\gamma_0(s) - \gamma_0(0)\right\| \leq \rho_0^m = \left(\frac{(CK^{-1}\lambda_0)^8}{10K^2}\right)^{2m}$$

and $\left\|\gamma'_0(s) - \gamma'_0(0)\right\| \leq \sqrt[4]{b}\left(1 + \sqrt[4]{b}\right)\left\|\gamma_0(s) - \gamma_0(0)\right\| \leq \rho_0^m$. So, we may apply Proposition 3.6 to obtain

$$\frac{1}{2} \leq \frac{\left\|\gamma'_j(s)\right\|}{\left\|\gamma'_j(0)\right\|} \leq 2.$$

Since $\omega_{n-m-1}(z_1)$ is a nearly horizontal vector, (4.17) implies

$$\frac{1}{2} \leq \frac{\|\omega_{n-m+j-1}(z_1)\|}{\|\omega_{n-m-1}(z_1)\| \, \|\gamma_j'(0)\|} \leq 2.$$

Thus,

$$\frac{1}{4}\left\|\gamma_j'(s)\right\| \leq \frac{1}{2}\left\|\gamma_j'(0)\right\| \leq \frac{\|\omega_{n-m+j-1}(z_1)\|}{\|\omega_{n-m-1}(z_1)\|} \leq 2\left\|\gamma_j'(0)\right\| \leq 4\left\|\gamma_j'(s)\right\|$$

for each s and consequently,

$$\frac{\|\gamma_j'\|}{\|\gamma_m'\|} \leq 16 \frac{\|\omega_{n-m+j-1}(z_1)\|}{\|\omega_{n-1}(z_1)\|},$$

for every $1 \leq j \leq m$. From Proposition 5.18,

$$\frac{\|\gamma_j'\|}{\|\gamma_m'\|} \leq const \, e^{-\tilde{c}(m-j)}$$

and $K_{m-1}...K_{j-1} \leq const \, e^{-3\tilde{c}(m-j+1)}$. Finally, since γ_0 is a $C^2(b)$-curve, from (6.10) it follows that $\Xi(\gamma_m) \leq const\sqrt{b}$. $\square$

Corollary 6.7. $\left\|\left(\gamma^{[i]}\right)'\right\| \geq \frac{1}{4}C_0$ for every $i \in \{1,...,s\}$.

Proof. Let $\gamma = \gamma^{[i]}$. From the proof of Lemma 6.6,

$$\|\gamma'\| = \|\gamma_m'\| \geq \frac{1}{4} \frac{\|\omega_{n-1}(z_1)\|}{\|\omega_{n-m-1}(z_1)\|}.$$

Thus, from Proposition 5.18, $\|\gamma'\| \geq \frac{1}{4}K^{-5}e^{\tilde{c}m}$. So, if $m \geq \tilde{c}^{-1}\log\left(4K^5\right)$, then $\|\gamma'\| \geq 1$.

On the other hand, for δ sufficiently small and Ω_0 close enough to $a(\lambda)$, it follows that if $m < \tilde{c}^{-1}\log\left(4K^5\right)$, then there are no returns between $n-m$ and n. Thus, we use Proposition 4.4(b) to obtain $\|\gamma'\| \geq \frac{1}{4}C_0 e^{cm} > \frac{1}{4}C_0$. $\square$

As a consequence of Corollary 6.7, we have that

$$\gamma^{[i]} \supset \gamma\left(\eta^{[i]}, \frac{C_0}{4}\rho_0^{m_i}\right) \supset \gamma\left(\eta^{[i]}, 5\rho_0^{m_i+1}\right),$$

provided that ρ_0 is small enough. Since $\gamma^{[0]} \subset G_1$, we also get $\gamma^{[i]} \subset G_{g_i}$ where $g_i = m_i + 1$. Furthermore, Lemma 6.3 implies that $g_{i+1} \leq 3g_i$. We complete the family of $C^2(b)$-curves $\left\{\gamma^{[i]}\right\}_{i=1}^{s}$, with the curve $\gamma^{[0]} = \gamma\left(\eta^{[0]}, \frac{1}{4}\Lambda\right)$, where $\eta^{[0]} = G_1 \cap \{x = x_n\}$ and $z_n = (x_n, y_n)$. This curve is also $C^2(b)$ and has generation $g_0 = 1$. Moreover,

$\gamma^{[0]} \supset \gamma \left(\eta^{[0]}, 5\rho_0^{m_0+1} \right).$

Definition 6.8. *Let $k \geq 0$ be the largest integer such that $\gamma \left(\eta^{[k]}, \rho_0^{g_k} \right)$, or $\gamma \left(\eta^{[0]}, \frac{1}{20}\Lambda \right)$ if $k = 0$, contains some $\zeta_{0,k} \in C_n$. $\zeta_0 = \zeta_{0,k}$ is the binding point associated to z_n.*

Proposition 6.9. *There exists a unique binding point associated to z_n.*

Proof. According to (4.12), $\gamma \left(\eta^{[0]}, \frac{1}{20}\Lambda \right)$ contains all the critical approximations $\omega_0^{(i-1)}$ of $G_1 \cap C_i$, for $2 \leq i \leq n$. Thus, the existence of $\zeta_{0,k}$ is obvious. If $k = 0$, then the uniqueness is also immediate. Otherwise, we proceed as in the proof of Lemma 5.2: It suffices to check that $\zeta_0 = \zeta_{0,k} \in \gamma \left(\eta^{[k]}, \rho_0^{g_k} \right) \cap C_n$ is the unique point of C_n in $\gamma \left(\zeta_0, 2\rho_0^{g_k} \right)$. First, notice that $\gamma \left(\zeta_0, 2\rho_0^{g_k} \right) \subset \gamma \left(\eta^{[k]}, 5\rho_0^{g_k} \right) \subset \gamma^{[k]}$ is a $C^2(b)$-curve. Let $\tilde{\zeta}_0 = \tilde{\zeta}_0^{(n-1)}$ be a point in $\gamma \left(\zeta_0, 2\rho_0^{g_k} \right) \cap C_n$ and let g_k be the generation of ζ_0. Since $g_k \leq \theta n$, from Lemma 5.2, we get $\sigma_0^n < \rho_0^{\theta n} < 3\rho_0^{\theta n} < 3\rho_0^{g_k}$. Moreover, unlike the proof of Lemma 5.2, since $g_k > 1$, we may assume that $\sigma_0 > 3\rho_0^{g_k}$. Therefore, there exists $l < n$ such that $\sigma_0^{l+1} \leq 3\rho_0^{g_k} < \sigma_0^l$. Then, by means of algorithm A we may find critical approximations $\zeta_0^{(l)}$ and $\tilde{\zeta}_0^{(l)}$ such that $\left| \zeta_0^{(l)} - \tilde{\zeta}_0^{(l)} \right| < 3\rho_0^{g_k} < \sigma_0^l$. Hence, as in Lemma 5.2, we conclude that $\zeta_0^{(l)} = \tilde{\zeta}_0^{(l)}$ and consequently, $\zeta_0^{(n-1)} = \tilde{\zeta}_0^{(n-1)}$. $\square$

From now on, we only consider parameter values a satisfying the basic assumption (BA) introduced in Chapter 5.

Proposition 6.10. *(Tangential position). Let n be a free return of $z_0 \in C_n$ and let $\zeta_0 \in C_n$ be the binding point associated to z_n. Then,*

$$\left\| z_n - \eta^{[k]} \right\| \leq b^{\frac{3}{10}} d_n(z_0) \quad and \quad \left| ang \left(h_{n-1}(z_1), t \left(\gamma^{[k]}; \eta^{[k]} \right) \right) \right| \leq b^{\frac{3}{10}} d_n(z_0).$$

Consequently, there exists a $C^2(b)$-curve $\tilde{\gamma}$ containing z_n and ζ_0, tangent to $\gamma^{[k]}$ at ζ_0 and to $h_{n-1}(z_1)$ at z_n.

Proof. From (4.16) and (4.18), it follows that $\left\| z_n - \eta^{[k]} \right\| \leq const \sqrt{b} \, (const b)^{m_k}$ and $\left| ang \left(h_{n-1}(z_1), t \left(\gamma^{[k]}; \eta^{[k]} \right) \right) \right| \leq \left(const \sqrt[4]{b} \right)^{g_k}$, respectively. The first inequality leads to

$$\left\| z_n - \eta^{[k]} \right\| \leq b^{\frac{3}{10}} b^{\frac{1}{20} g_k}, \tag{6.11}$$

whenever b is small enough. If $g_k \geq 2$, then $\left| ang \left(h_{n-1}(z_1), t \left(\gamma^{[k]}; \eta^{[k]} \right) \right) \right| \leq b^{\frac{3}{10}} b^{\frac{1}{20} g_k}$. If $g_k = 1$ the same follows from Propositions 4.13 and 5.13.

Hence, to prove the proposition it suffices to show that $d_n(z_0) \geq b^{\frac{1}{20}g_k}$. We shall proceed by contradiction. Let us assume that

$$d_n(z_0) < b^{\frac{1}{20}g_k}. \tag{6.12}$$

Then, from (BA), we obtain $g_k \leq \frac{1}{3}\theta n$, provided that α is small enough. Therefore, since $g_{k+1} < 3g_k \leq \theta n$, there exists $\mu < n$ such that $g_k \leq \theta\mu < g_{k+1} \leq \theta(\mu+1)$. This fact implies that, in the construction of C_n, we could apply algorithm B just to construct $C_{\mu+1}$. We shall see that, in fact, this has to be the case if $d_n(z_0) < b^{\frac{1}{20}g_k}$. Thus, we could find a point of C_n in $\gamma\left(\eta^{[k+1]}, \rho_0^{g_{k+1}}\right)$, which contradicts the definition of binding point. We shall develop the proof in three stages:

First stage. Since $\theta << 1 < g_{k+1} \leq 3g_k \leq \theta n$, there exists $\mu < n$ such that $g_k \leq \theta\mu < g_{k+1} \leq \theta(\mu+1)$. We construct, by applying algorithm A to $\zeta_0^{(n-1)}$, a sequence of critical approximations $\zeta_0^{(n-2)}, ..., \zeta_0^{(\mu)}$. We claim that $\zeta_0^{(\mu)} \in C_{\mu+1}$. Indeed, for any i with $\mu \leq i \leq n-1$, from (6.11) and (6.12) we get $\left\|\zeta_0^{(i)} - \eta^{[k]}\right\| \leq 3b^{\frac{1}{20}g_k}$, whenever θ is small enough so that $(Kb)^i \leq (Kb)^\mu \leq (Kb)^{\frac{1}{\theta}g_k} \leq b^{\frac{1}{20}g_k}$. Therefore, taking b sufficiently small, we have

$$\left|\zeta_0^{(i)} - \eta^{[k]}\right| \leq \rho_0^{5g_k} < 5\rho_0^{g_k}. \tag{6.13}$$

So, $\zeta_0^{(i)} \in \gamma\left(\eta^{[k]}, 5\rho_0^{g_k}\right) \subset G_{g_k}$ for each $i \in \{\mu, ..., n-1\}$ and $\zeta_0^{(i)}$ has generation $g_k \leq \theta\mu$. Then, by construction, $\zeta_0^{(\mu)} \in C_{\mu+1}$.

Second stage. Now, let us see how to construct, from $\zeta_0^{(\mu)}$, a critical approximation $\bar{\zeta}_0^{(\mu)}$ in $C_{\mu+1} \cap \gamma\left(\eta^{[k+1]}, \rho_0^{g_{k+1}}\right)$. Let $\zeta_0^{(\mu)} = (x_0, y_0)$. Recall that $\gamma\left(\eta^{[k+1]}, 5\rho_0^{g_{k+1}}\right)$ and $\gamma\left(\eta^{[k]}, 5\rho_0^{g_k}\right)$ are $C^2(b)$-curves. Next, we shall prove that both of them are defined on $\left(x_0 - \rho_0^{g_{k+1}}, x_0 + \rho_0^{g_{k+1}}\right)$. Let $\eta^{[k]} = \left(\eta_x^{[k]}, \eta_y^{[k]}\right)$ and $\eta^{[k+1]} = \left(\eta_x^{[k+1]}, \eta_y^{[k+1]}\right)$. From (6.13), we have $\left|x_0 - \eta_x^{[k]}\right| \leq \frac{1}{8}\rho_0^{g_{k+1}}$. On the other hand, from (4.16), we obtain $\left|\eta_x^{[k]} - \eta_x^{[k+1]}\right| \leq \left(const\sqrt{b}\right)^{g_k} < \frac{1}{8}\rho_0^{g_{k+1}}$ and consequently, $\left|x_0 - \eta_x^{[k+1]}\right| \leq \frac{1}{4}\rho_0^{g_{k+1}}$. Then, we let $z_0 = \gamma\left(\eta^{[k+1]}, \frac{1}{2}\rho_0^{g_{k+1}}\right) \cap \{x = x_0\}$.

Now, we shall show that algorithm B can be applied to construct a critical approximation of order μ in $\gamma\left(\eta^{[k+1]}, \rho_0^{g_{k+1}}\right)$. That is, we shall check that $d = \left\|z_0 - \zeta_0^{(\mu)}\right\| \leq \frac{1}{4}\sigma_0^{2\mu}$ and $\sqrt{d} < l = \rho_0^{g_{k+1}}$. We argue as follows: Let $\Gamma^k = \Gamma^{m_k}$ be the integral curve of the field defined by $e^{(m_k)}$ containing z_{n-m_k}. Let $U^k = \bigcup_{z \in \Gamma^k} [z - \rho_0^{m_k}, z + \rho_0^{m_k}]$, where $[z - \rho_0^{m_k}, z + \rho_0^{m_k}]$ is the horizontal segment of radius $\rho_0^{m_k}$ centered at z. According to Chapter 4, every point in U^k is expanding up to time m_k. We firstly show that $T_{\lambda,a,b}^{m_{k+1}-m_k}\left(\gamma_0^{[k+1]}\right) \subset U^k$. To this end, let $\eta_{m_{k+1}-m_k}^{[k+1]} = T_{\lambda,a,b}^{m_{k+1}-m_k}\left(\eta_0^{[k+1]}\right)$. It suffices to prove that $\mathrm{length}\left(T_{\lambda,a,b}^{m_{k+1}-m_k}\left(\gamma_0^{[k+1]}\right)\right) + \mathrm{dist}\left(\eta_{m_{k+1}-m_k}^{[k+1]}, \Gamma^k\right) < \frac{1}{5}\rho_0^{m_k}$. In fact, $\mathrm{length}\left(T_{\lambda,a,b}^{m_{k+1}-m_k}\left(\gamma_0^{[k+1]}\right)\right) \leq K^{m_{k+1}-m_k}\rho_0^{m_{k+1}} << \frac{1}{10}\rho_0^{m_k}$. On the other hand, from Lemma 6.3, it follows that $\frac{5}{4}m_k \leq m_{k+1}$. Then, from (4.16), we also get $\mathrm{dist}\left(\eta_{m_{k+1}-m_k}^{[k+1]}, \Gamma^k\right) \leq \left(const\sqrt{b}\right)^{m_{k+1}-m_k} \leq \left(const\sqrt[8]{b}\right)^{m_k} << \frac{1}{10}\rho_0^{m_k}$. Now, let $z_{0,0}$ be the point in $T_{\lambda,a,b}^{m_{k+1}-m_k}\left(\gamma_0^{[k+1]}\right)$ such that $T_{\lambda,a,b}^{m_k}(z_{0,0}) = z_0$. Since $z_{0,0} \in U_k$, there exists the integral curve Γ_0^k of the vector field $e^{(m_k)}$ containing $z_{0,0}$. From Proposition 3.10 and Gronwall's lemma, it follows that Γ_0^k intercepts $\gamma_0^{[k]}$ in one point $\bar{\zeta}_{0,0}$. Hence, since $\gamma^{[k]}$ and $\gamma^{[k+1]}$ are $C^2(b)$-curves, (4.16) leads to

$$\left\| z_0 - \zeta_0^{(\mu)} \right\| < 2\left\| z_0 - T_{\lambda,a,b}^{m_k}\left(\bar{\zeta}_{0,0}\right) \right\| \leq const\sqrt{b}\,(const\,b)^{m_k}. \tag{6.14}$$

Therefore, $4\left\| z_0 - \zeta_0^{(\mu)} \right\| < \left(const\sqrt{b}\right)^{g_k} \leq b^{\frac{1}{10}g_{k+1}}$ and, since $g_{k+1} > \theta\mu$, the definition of θ implies that $4\left\| z_0 - \zeta_0^{(\mu)} \right\| \leq \sigma_0^{2\mu}$.

Let us now take b small enough so that $\left\| z_0 - \zeta_0^{(\mu)} \right\|^{\frac{1}{2}} < \left(const\sqrt{b}\right)^{\frac{1}{2}g_k} < \frac{1}{2}\rho_0^{g_{k+1}}$. By means of algorithm B, we find a critical approximation $\bar{\zeta}_0^{(\mu)} \in \gamma\left(z_0, \frac{1}{2}\rho_0^{g_{k+1}}\right) \subset \gamma\left(\eta^{[k+1]}, \rho_0^{g_{k+1}}\right) \subset \gamma^{[k+1]}$. Finally, we claim that $\bar{\zeta}_0^{(\mu)} \in C_{\mu+1}'''$. Indeed:

a. $\gamma\left(\bar{\zeta}_0^{(\mu)}, \rho_0^{\theta(\mu+1)}\right) \subset \gamma^{[k+1]}$ is a $C^2(b)$-curve.

b. From Lemma 6.5, $T_{\lambda,a,b}^{1-g_{k+1}}\left(\gamma\left(\bar{\zeta}_0^{(\mu)}, \rho_0^{\theta(\mu+1)}\right)\right) \subset \gamma^{[k+1]} \subset G_1 \cap (I_0 \times \mathbf{R})$ and, from Corollary 6.7, its tangent vectors are $\frac{1}{4}C_0$ expanded by $DT_{\lambda,a,b}^{g_{k+1}-1}$.

c. From (6.14) and (4.15) we have $\left\| \bar{\zeta}_0^{(\mu)} - \zeta_0^{(\mu)} \right\| < const\sqrt[4]{b}\left(const\sqrt{b}\right)^{m_k} \leq b^{\frac{1}{10}g_{k+1}}$.

Third stage. We apply algorithm A to $\bar{\zeta}_0^{(\mu)} \in C_{\mu+1}$ to obtain a point $\bar{\zeta}_0^{(n-1)} \in C_n$. Furthermore, since $\left\| \bar{\zeta}_0^{(n-1)} - \bar{\zeta}_0^{(\mu)} \right\| < (Kb)^\mu < \frac{1}{8}\rho_0^{g_{k+1}}$ and, since we already proved $\left\| \bar{\zeta}_0^{(\mu)} - \zeta_0^{(\mu)} \right\| \leq b^{\frac{1}{10}g_{k+1}}$, $\left\| \zeta_0^{(\mu)} - \eta^{[k]} \right\| \leq 3b^{\frac{1}{20}g_k}$ and $\left\| \eta^{[k]} - \eta^{[k+1]} \right\| \leq \frac{1}{8}\rho_0^{g_{k+1}}$, we conclude $\bar{\zeta}_0^{(n-1)} \in \gamma\left(\eta^{[k+1]}, \rho_0^{g_{k+1}}\right) \cap C_n$.

Finally, to finish the proof, we shall outline how to obtain a $C^2(b)$-curve $\tilde{\gamma}$ satisfying the statement. Let $\zeta_0 = (\zeta_{0,x}, \zeta_{0,y})$ and $z_n = (x_n, y_n)$. From (6.11), we obtain $\left\| z_n - \eta^{[k]} \right\| << \rho_0^{m_k}$. Let us take $z_0 = (x_0, y_0) = \gamma\left(\eta^{[k]}, \rho_0^{m_k}\right) \cap \{x = x_n\}$ and let $z(x) = (x, y(x))$ be a parametrization of $\gamma^{[k]}$ with $z(x_0) = z_0$. Without loss of generality, let us assume that $x_0 < \zeta_{0,x}$. Then, there exists a map

$$\tilde{y} : [x_0, \zeta_{0,x}] \to \mathbf{R}$$

satisfying the following conditions: $\tilde{y}(x_0) = y_0 - y_n$, $\tilde{y}(\zeta_{0,x}) = 0$, $\tilde{y}'(\zeta_{0,x}) = 0$ and $\tilde{y}'(x_0) = y'(x_0) - t$, where $(1,t)$ is colinear to $h_{n-1}(z_1)$. Moreover, $\tilde{y}$ may be chosen such that $|\tilde{y}'(x)| < \sqrt[4]{b}$ and $|\tilde{y}''(x)| < \sqrt[4]{b}$, for every $x \in [x_0, \zeta_{0,x}]$. $\square$

Next, we shall estimate the loss of exponential growth on returns. We start by estimating this loss on free returns. Notice that the bounds obtained in the following proposition are better than the ones established in I.H.5.

Proposition 6.11. *Let n be a free return of $z_0 \in C_n$. Then,*

(a)

$$|\alpha_n(z_1)| \leq 2K\sqrt{b}\,\|h_{n-1}(z_1)\|.$$

(b)

$$\frac{5}{6}\lambda^{-1}d_n(z_0) \leq \frac{|\beta_n(z_1)|}{\|h_{n-1}(z_1)\|} \leq \frac{6}{5}\lambda^{-1}d_n(z_0).$$

Proof. Let $\zeta_0 = (x_0, y_0)$ be the binding point associated to z_n and let $\tilde{\gamma}$ be the $C^2(b)$-curve given by Proposition 6.10 parametrized by $s \in (-\tau, \tau) \to z(s) = (x_0 + s, y(s))$. Let $s_0 \in (-\tau, \tau)$ be such that $z(s_0) = z_n$ and assume that $s_0 > 0$. First, we shall prove that if $[n+1, n+l]$ is the folding period associated to n, then $z_1(s) = T_{\lambda,a,b}(z(s))$ is e^c-expanding up to time l for every $s \in [0, s_0]$. To this end, since (BA) leads to

$$l \leq \frac{10\alpha \log K}{c \log b^{-1}}n + 4 < \frac{n}{2},$$

it suffices to prove that $z(s)$ is bound to C_l. In fact, according to Proposition 5.7, it is sufficient to check that $z(s)$ is bound to C_n up to time l. Now, since $\tilde{\gamma}$ is a $C^2(b)$-curve, it follows that $\|z(s) - \zeta_0\| < 2d_n(z_0)$ for every $s \in [0, s_0]$. Then, $\|z_j(s) - \zeta_j\| < 2K^j d_n(z_0)$ for all j. Furthermore, since $l \leq l_0$, we get $d_n(z_0) < \left(\frac{1}{4}K^{-1}e^{-\beta}\tilde{\Lambda}\right)^l$, provided that δ and b are small enough. Hence, $\|z_j(s) - \zeta_j\| < \frac{1}{2}\tilde{\Lambda}e^{-\beta j} < h_l e^{-\beta j}$ for $1 \leq j \leq l$ and $z(s)$ is bound to C_n up to time l.

Therefore, we may take a vector $e^{(l)}(z_1(s)) = (q_l(s), 1)$ on the maximally contracting direction at $z_1(s)$. Let

$$t(s) = (A(z(s)) + B(z(s))y'(s), C(z(s)) + D(z(s))y'(s)) \qquad (6.15)$$

be the tangent vector to $T_{\lambda,a,b}(\tilde{\gamma})$ at $z_1(s)$. Write $t(s) = \alpha_t(s)e^{(l)}(z_1(s)) + \beta_t(s)(1,0) = (\alpha_t(s)q_l(s) + \beta_t(s), \alpha_t(s))$. Then, we obtain that $\alpha_t(s) = C(z(s)) + D(z(s))y'(s)$ and

$\beta_t(s) = A(z(s)) + B(z(s))y'(s) - \alpha_t(s)q_1(s)$. From Proposition 1.3 and the fact that $\tilde{\gamma}$ is $C^2(b)$, it follows that

$$|\alpha_t(s)| < 2K\sqrt{b} \text{ and } |\alpha_t'(s)| < 3K\sqrt{b}. \tag{6.16}$$

Furthermore, $\beta_t' = A'(z)(1,y') + B'(z)(1,y')y' + B(z)y'' - \alpha_t'q_1 - \alpha_t q_1'$ and, from Proposition 1.3, (6.16), (4.9) and (4.10), we conclude that $|\beta_t' - A'(z)(1,y')| \le 4K\sqrt{b}$. Moreover, since $A'(z) = A'(z(s)) = (-\lambda^{-1}(1 + \text{tg}^2(x_0 + s)), 0)$ and

$$\left|\lambda^{-1}(1 + \text{tg}^2 c_\lambda) - \lambda^{-1}(1 + \text{tg}^2(x_0 + s))\right| \le const\ \delta,$$

we get $|\beta_t' + \lambda^{-1}(1 + \lambda^2)| \le 4K\sqrt{b} + const\ \delta$. Thus, by applying the mean value theorem we obtain

$$\left(\lambda^{-1}(1 + \lambda^2) - 4K\sqrt{b} - const\ \delta\right)|s| \le$$

$$\le |\beta_t(s) - \beta_t(0)| \le \left(\lambda^{-1}(1 + \lambda^2) + 4K\sqrt{b} + const\ \delta\right)|s|. \tag{6.17}$$

On the other hand, since $e^{(l)}$ is a nearly vertical vector, we have that $|\beta_t(0)| \le 2\|t(0)\| \left|\text{ang}\left(t(0), e^{(l)}(\zeta_1)\right)\right|$. So, from (6.15) and Proposition 1.3, we obtain $\|t(0)\|^2 \le |A(z(0))|^2 + const\ b < 16\lambda^{-2}$. Furthermore, since ζ_0 is a critical approximation of order $n-1$, $t(0)$ is colinear to $e^{(n-1)}(\zeta_1)$. Hence, Proposition 3.3 gives $\left|\text{ang}\left(t(0), e^{(l)}(\zeta_1)\right)\right| \le 3K(Kb)^l$. Therefore,

$$|\beta_t(0)| \le 24K\lambda^{-1}(Kb)^l. \tag{6.18}$$

Now, from Proposition 5.11, $\left(K\sqrt{b}\right)^l < d_n(z_0)^2$ and, since $d_n(z_0) < 2\delta$, we get

$$|\beta_t(0)| \le \left(\sqrt{b}\right)^l d_n(z_0). \tag{6.19}$$

Finally, write

$$\frac{h_{n-1}(z_1)}{\|h_{n-1}(z_1)\|} = \chi\ t(\tilde{\gamma}, z_n) = \chi(1, y'(s_0))$$

and notice that

$$|\chi| = \frac{1}{\sqrt{1 + y'^2(s_0)}} \in \left(\frac{19}{20}, 1\right) \tag{6.20}$$

and

$$s_0 \le d_n(z_0) \le \frac{11}{10}s_0. \tag{6.21}$$

Let

$$\hat{\alpha}_n = \frac{\alpha_n}{\|h_{n-1}(z_1)\|} \text{ and } \hat{\beta}_n = \frac{\beta_n}{\|h_{n-1}(z_1)\|}. \tag{6.22}$$

Then, from (6.16), $|\hat{a}_n(z_1)| < |\alpha_t(s_0)| < 2K\sqrt{b}$ and statement (a) is proved.

To prove statement (b) we use (6.17) and (6.20) to obtain $\left|\hat{\beta}_n(z_1)\right| < |\beta_t(s_0)| \leq |\beta_t(0)| + \left(\lambda^{-1}(1 + \lambda^2) + 4K\sqrt{b} + const\delta\right) s_0$. Therefore, from (6.19) and (6.21), we deduce that $\left|\hat{\beta}_n(z_1)\right| \leq \left(\sqrt{b}\right)^l d_n(z_0) + \left(\lambda^{-1}(1 + \lambda^2) + 4K\sqrt{b} + const\delta\right) d_n(z_0)$. Therefore, $\left|\hat{\beta}_n(z_1)\right| \leq \frac{6}{5}\lambda^{-1}d_n(z_0)$, whenever $\lambda < \frac{1}{3}$ and δ and b are sufficiently small.

On the other hand, from (6.20) and (6.17), we obtain

$$\left|\hat{\beta}_n(z_1)\right| \geq \frac{19}{20}|\beta_t(s_0)| \geq \frac{19}{20}\left\{\left(\lambda^{-1}(1 + \lambda^2) - 4K\sqrt{b} - const\delta\right) s_0 - |\beta_t(0)|\right\}.$$

From (6.19) and (6.21), $\left|\hat{\beta}_n(z_1)\right| \geq \frac{5}{6}\lambda^{-1}d_n(z_0)$, provided that δ and b are small enough. $\square$

Remark 6.12. *Proposition 6.11 still holds if we suppose that $\tilde{\gamma}$ is a curve which can be parametrized by $(x, y(x))$ with $|y'(x)| < \frac{1}{10}$ and $|y''(x)| < \frac{1}{10}$.*

To extend Proposition 6.11 to free returns of points bound to C_n, we need results which will be stated in Chapter 7. Therefore, we have to postpone the proof of the following result:

Proposition 6.13. *Let n be a free return of $z_0 \in C_n$. Then, for every ξ_0 bound to z_0 up to time n, we have*

(a)
$$|\alpha_n(\xi_1)| \leq 2K\sqrt{b}\,\|h_{n-1}(\xi_1)\|.$$

(b)
$$\frac{5}{6}\lambda^{-1}d_n(\xi_0) \leq \frac{|\beta_n(\xi_1)|}{\|h_{n-1}(\xi_1)\|} \leq \frac{6}{5}\lambda^{-1}d_n(\xi_0).$$

Now, by using Proposition 6.13 for iterates $j < n$, we prove the following result for bound returns of points bound to C_n.

Proposition 6.14. *Let n be a bound return of $z_0 \in C_n$. Then, for every ξ_0 bound to z_0 up to time n, we have*

(a)
$$|\alpha_n(\xi_1)| \leq 3K\sqrt{b}\,\|h_{n-1}(\xi_1)\|.$$

(b)
$$\frac{1}{2}\lambda^{-1}d_n(\xi_0) \leq \frac{|\beta_n(\xi_1)|}{\|h_{n-1}(\xi_1)\|} \leq 2\lambda^{-1}d_n(\xi_0).$$

Proof. If n is a bound return of ξ_0, then there exists a return $\nu < n$ of ξ_0 such that its binding period, $[\nu + 1, \nu + p_\nu]$, contains n. Take ν maximum. Then, $k = n - \nu$ is a free return of $\tilde{\zeta}_0$, where $\tilde{\zeta}_0 \in \overset{\nu}{\underset{j=1}{\cup}} C_j$ is the binding point associated to ξ_ν. Furthermore, from the definition of binding period, ξ_ν is bound to $\tilde{\zeta}_0$ up to time k, with $k \le p_\nu \le d$ and $d \in \{1, ..., \nu\}$ such that $\tilde{\zeta}_0 \in C_d$. Therefore, by induction, it follows that

$$|\hat{a}_k(\xi_{\nu+1})| = \frac{|\alpha_k(\xi_{\nu+1})|}{\|h_{k-1}(\xi_{\nu+1})\|} \le 2K\sqrt{b} \qquad (6.23)$$

and

$$\frac{5}{6}\lambda^{-1}d_k(\xi_\nu) \le |\hat{\beta}_k(\xi_{\nu+1})| = \frac{|\beta_k(\xi_{\nu+1})|}{\|h_{k-1}(\xi_{\nu+1})\|} \le \frac{6}{5}\lambda^{-1}d_k(\xi_\nu). \qquad (6.24)$$

We consider the two possible cases:

1. There exist no folding periods $[\mu + 1, \mu + l]$ associated to returns of ξ_0 such that $\mu \le \nu < \mu + l < n$. Then, we shall prove that $\hat{a}_n(\xi_1) = \hat{a}_k(\xi_{\nu+1})$ and $\hat{\beta}_n(\xi_1) = \hat{\beta}_k(\xi_{\nu+1})$, where $\hat{a}_n(\xi_1)$ and $\hat{\beta}_n(\xi_1)$ are given in (6.22). To this end, since $h_\nu(\xi_1) = \beta_\nu(\xi_1)(1,0) = \beta_\nu(\xi_1)h_0(\xi_{\nu+1})$, we get

$$\frac{|\beta_n(\xi_1)|}{|\beta_k(\xi_{\nu+1})|} = \frac{\|h_n(\xi_1)\|}{\|h_k(\xi_{\nu+1})\|} = \frac{\|\tilde{h}_n(\xi_1)\|}{\|\tilde{h}_k(\xi_{\nu+1})\|} = \frac{\|h_{n-1}(\xi_1)\|}{\|h_{k-1}(\xi_{\nu+1})\|} = |\beta_\nu(\xi_1)|$$

and consequently, $\hat{\beta}_n(\xi_1) = \hat{\beta}_k(\xi_{\nu+1})$. In the same way, we may obtain $\hat{a}_n(\xi_1) = \hat{a}_k(\xi_{\nu+1})$. Thus, since $d_k(\xi_\nu) = d_{k+\nu}(\xi_0) = d_n(\xi_0)$, the proposition holds.

2. There exist returns of ξ_0, $\mu_1, ..., \mu_s$, satisfying that $\mu_1 < ... < \mu_s = \nu < \nu + l = \mu_s + l_s \le ... \le \mu_1 + l_1 < n$. Then, $h_{\mu_1+l_1}(\xi_1) = \beta_{\mu_1}(\xi_1)h_{l_1}(\xi_{\mu_1+1}) + \alpha_{\mu_1}(\xi_1)DT_{\lambda,a,b}^{l_1}e^{(l_1)}(\xi_{\mu_1+1})$. Moreover, if there exists a return of ξ_0 between $\mu_1 + l_1$ and n, then its folding period ends before n. Therefore,

$$h_{n-1}(\xi_1) = \beta_{\mu_1}(\xi_1)h_{n-\mu_1-1}(\xi_{\mu_1+1}) + \alpha_{\mu_1}(\xi_1)DT_{\lambda,a,b}^{n-\mu_1-1}e^{(l_1)}(\xi_{\mu_1+1}). \qquad (6.25)$$

Now, since ξ_{μ_1} is bound to $C_{d'}$ up to time $p_{\mu_1} = p_1$, with $p_1 < d' \le \mu_1 < n$, Proposition 5.7 implies that ξ_{μ_1} is bound to C_{p_1}. Furthermore, $\nu \in [\mu_1 + 1, \mu_1 + l_1] \subset [\mu_1 + 1, \mu_1 + p_1]$ and $n \in [\nu + 1, \nu + p_\nu]$. So, from the stair structure of binding periods, we get $n \in [\mu_1 + 1, \mu_1 + p_1]$. Therefore, ξ_{μ_1} is bound to $C_{n-\mu_1}$ and, from Proposition 5.14, $\|h_{n-\mu_1-1}(\xi_{\mu_1+1})\| \ge e^{c(n-\mu_1-1)}$. Then, the induction hypotheses yield

$$\|\beta_{\mu_1}(\xi_1)h_{n-\mu_1-1}(\xi_{\mu_1+1})\| \ge \frac{1}{2}\lambda^{-1}d_{\mu_1}(\xi_0)\|h_{\mu_1-1}(\xi_1)\|e^{c(n-\mu_1-1)}. \qquad (6.26)$$

On the other hand, from the induction hypotheses and Proposition 5.11,

$$\left\| \alpha_{\mu_1}(\xi_1) DT_{\lambda,a,b}^{n-\mu_1-1} e^{(l_1)}(\xi_{\mu_1+1}) \right\| \leq 3K\sqrt{b} \left\| h_{\mu_1-1}(\xi_1) \right\| K^{n-\mu_1-l_1-1} \left(\sqrt{b} \right)^{l_1} d_{\mu_1}(\xi_0)^2.$$

(6.27)

Now, writing $h_{n-1}(\xi_1)$ as in (6.25) and using (6.26) and (6.27) we get

$$\left| \text{ang}\left(h_{n-1}(\xi_1), h_{n-\mu_1-1}(\xi_{\mu_1+1}) \right) \right| \leq const\sqrt{b} \left(\sqrt{b} \right)^{l_1} K^k,$$

where $k = n - \nu \geq n - \mu_1 - l_1 - 1$.

Repeating the arguments for each $i \in \{1, ..., s-1\}$ and adding all the respective inequalities we obtain

$$\left| \text{ang}\left(h_{n-1}(\xi_1), h_{k-1}(\xi_{\nu+1}) \right) \right| \leq const\sqrt{b} K^k \left(\sqrt{b} \right)^{l_s}.$$

(6.28)

Next, we shall prove that $K^k \left(\sqrt{b} \right)^{l_s} < d_n(\xi_0)$. Since ν is a return of ξ_0, there exist $\overline{\mu} \leq \nu$ and $\overline{z}_0 \in C_{\overline{\mu}}$ such that $\overline{\mu}$ is a free return of $\overline{z}_0$. From I.H.6, it follows that $k \leq p_\nu = p_{\overline{\mu}} < 2c^{-1} \log d_{\overline{\mu}}(\overline{z}_0)^{-1}$. Furthermore, Proposition 5.10 leads to

$$l_s > \frac{1}{2}l_{0,s} \geq \frac{5\log K}{c\log b^{-1}} \log d_{\overline{\mu}}(\overline{z}_0)^{-1}$$

and consequently,

$$K^k \left(\sqrt{b} \right)^{l_s} \leq \left\{ K \left(\sqrt{b} \right)^{\frac{5\log K}{2\log b^{-1}}} \right\}^k.$$

Now, Remark 5.22 and Proposition 5.21 imply $d_n(\xi_0) > \frac{7}{6}d_n(z_0) > \frac{3}{4}d_k(\widetilde{\zeta}_0)$ and, from (BA), $d_n(\xi_0) > \frac{1}{2}e^{-\alpha k} > \left(\frac{1}{2}e^{-\alpha} \right)^k$. Then, we obtain $K^k \left(\sqrt{b} \right)^{l_s} < d_n(\xi_0)$ by taking K large enough so that

$$K \left(\sqrt{b} \right)^{\frac{5\log K}{2\log b^{-1}}} < \frac{1}{2}e^{-\alpha},$$

Hence, from (6.28), $\left| \text{ang}\left(h_{n-1}(\xi_1), h_{k-1}(\xi_{\nu+1}) \right) \right| \leq const\sqrt{b} d_n(\xi_0)$. Therefore,

$$\left\| \frac{h_{n-1}(\xi_1)}{\|h_{n-1}(\xi_1)\|} - \frac{h_{k-1}(\xi_{\nu+1})}{\|h_{k-1}(\xi_{\nu+1})\|} \right\| \leq const\sqrt{b} d_n(\xi_0)$$

and consequently,

$$\left\| \frac{\widetilde{h}_n(\xi_1)}{\|h_{n-1}(\xi_1)\|} - \frac{\widetilde{h}_k(\xi_{\nu+1})}{\|h_{k-1}(\xi_{\nu+1})\|} \right\| \leq constK\sqrt{b} d_n(\xi_0).$$

Let $\theta = \text{ang}\left((1,0), e^{(l_n)}(\xi_{n+1}) \right)$. Then Proposition 3.4 leads to $\left| \widehat{\alpha}_n(\xi_1) - \widehat{\alpha}_k(\xi_{\nu+1}) \right| \leq constK\sqrt{b} d_n(\xi_0) < K\sqrt{b}$ and $\left| \widehat{\beta}_n(\xi_1) - \widehat{\beta}_k(\xi_{\nu+1}) \right| \leq constK\sqrt{b} d_n(\xi_0) < \frac{1}{4}\lambda^{-1}d_n(\xi_0)$. Finally, from (6.23), it follows that $\left| \widehat{\alpha}_n(\xi_1) \right| \leq 3K\sqrt{b}$ and, from (6.24),

$$\frac{1}{2}\lambda^{-1}d_n(\xi_0) \leq \left| \widehat{\beta}_n(\xi_1) \right| \leq 2\lambda^{-1}d_n(\xi_0). \quad \square$$

Chapter 7

THE BINDING PERIOD

In this chapter we shall prove that I.H.6 holds at time n. Let n be a return of a point ξ_0 bound to C_n and let $[n+1, n+p]$ be its binding period. As was said in Chapter 5, we need to obtain exponential growth during $[n+1, n+p]$ to compensate for the loss of derivative on the return. To this end, we shall state a distortion result. We start by proving an auxiliary lemma which will be used several times along this chapter:

Lemma 7.1. *Let* v_1, v_2, w_1 *and* w_2 *be vectors such that, for some* $\Upsilon > 0$, $\|v_i + w_i\| \geq \Upsilon \|v_i\|$ *for* $i = 1, 2$. *If*

$$\chi = |ang(v_1, v_2)| \frac{\|w_1\|}{\|v_1\|} + \frac{\|w_1 - w_2\|}{\|v_1\|} + \left| \frac{\|v_2\|}{\|v_1\|} - 1 \right| \frac{\|w_2\|}{\|v_2\|},$$

then

$$(a) \quad \frac{\|v_1 + w_1\|}{\|v_2 + w_2\|} \leq \frac{\|v_1\|}{\|v_2\|} \left(1 + \frac{\chi}{\Upsilon} \right).$$

$$(b) \quad |ang(v_1 + w_1, v_2 + w_2)| \leq |ang(v_1, v_2)| + \frac{2\chi}{\Upsilon}.$$

Proof. First, let us assume that $ang(v_1, v_2) = 0$ and define $E_1 = \|v_2\| \|v_1\|^{-1} w_1$ and $d = E_1 - w_2$. Then,

$$\frac{\|v_1 + w_1\|}{\|v_1\|} = \frac{\|v_2 + E_1\|}{\|v_2\|} \leq \frac{\|v_2 + w_2\|}{\|v_2\|} \left(1 + \frac{\|d\|}{\|v_2 + w_2\|} \right)$$

and consequently,

$$\frac{\|v_1 + w_1\|}{\|v_1\|} \leq \frac{\|v_2 + w_2\|}{\|v_2\|} \left(1 + \frac{\|d\|}{\Upsilon \|v_2\|} \right).$$

Furthermore, notice that

$$\frac{\|d\|}{\|v_2\|} = \left\| \frac{w_1}{\|v_1\|} - \frac{w_2}{\|v_2\|} \right\| \leq \frac{1}{\|v_1\|} \|w_1 - w_2\| + \left| \frac{\|v_2\|}{\|v_1\|} - 1 \right| \frac{\|w_2\|}{\|v_2\|} = \chi_0.$$

So statement (a) is proved. To prove (b) let $\alpha = \text{ang}(\mathbf{v}_1 + \mathbf{w}_1, \mathbf{v}_2 + \mathbf{w}_2)$. Then,

$$|\sin \alpha| \leq \frac{\|\mathbf{d}\|}{\|\mathbf{v}_2 + \mathbf{w}_2\|}$$

and, repeating the above argument, $|\sin \alpha| \leq \chi_0 \Upsilon^{-1}$. Hence, $|\alpha| \leq 2\chi_0 \Upsilon^{-1}$.

If $\text{ang}(\mathbf{v}_1, \mathbf{v}_2) \neq 0$, then this case is reduced to the previous one by rotating $\mathbf{v}_1$ through an angle $\beta = \text{ang}(\mathbf{v}_1, \mathbf{v}_2)$. Notice that $\mathbf{w}_1$ becomes a new vector $\mathbf{w}_1'$ such that $\|\mathbf{w}_1\| = \|\mathbf{w}_1'\|$ and $\text{ang}(\mathbf{w}_1', \mathbf{w}_1) = \beta$. The proof finishes by replacing $\mathbf{w}_1$ by $\mathbf{w}_1' - \mathbf{w}_1 + \mathbf{w}_1$ in the above expression of χ_0. $\quad\square$

In what follows, we shall consider $\theta_k = \theta_k(\eta_0, \zeta_0) = \sum_{j=1}^{k} \left(\sqrt[4]{b}\right)^{k-j} \|\eta_j - \zeta_j\|$ to be a distance between the orbits of the points η_0 and ζ_0 up to time k.

Proposition 7.2. *(Bounded distortion). Let η_0 and ζ_0 be two points bound to the same $z_0 \in C_s$ up to time q, with $q \leq s \leq n$. Then,*

(a)

$$\frac{\|h_\nu(\eta_1)\|}{\|h_\nu(\zeta_1)\|} \leq \exp\left(8K \sum_{k=1}^{\nu} \frac{\theta_k}{d_k(\zeta_0)}\right) \text{ for } 0 \leq \nu \leq q.$$

(b)

$$|\text{ang}\left(h_\nu(\eta_1), h_\nu(\zeta_1)\right)| \leq 2\sqrt[4]{b}\theta_\nu \text{for } 0 \leq \nu \leq q.$$

Proof. We shall inductively prove that

$$\text{(a)} \quad \frac{\|h_\nu(\eta_1)\|}{\|h_\nu(\zeta_1)\|} \leq \exp\left(4K \sum_{i=0}^{\nu} \left(\sqrt[4]{b}\right)^i \sum_{k=1}^{\nu} \frac{\theta_k}{d_k(\zeta_0)}\right),$$

$$\text{(b)} \quad |\text{ang}\left(h_\nu(\eta_1), h_\nu(\zeta_1)\right)| \leq \sum_{i=1}^{\nu} \left(\sqrt[4]{b}\right)^i \theta_\nu.$$

To this end, let us assume that (a) and (b) hold for every $\mu \leq \nu - 1$ and for every pair of points bound to the same point of C_s. We consider the three possible cases:

A. Let us assume that ν is a return of z_0. Let $[\nu + 1, \nu + l]$ be its folding period and $(q, 1)$ a vector on the l-th contractive direction. Write

$$\frac{\tilde{h}_\nu(\eta_1)}{\left\|\tilde{h}_\nu(\eta_1)\right\|} = \hat{\alpha}_\nu(\eta_1)\left(q(\eta_{\nu+1}), 1\right) + \hat{\beta}_\nu(\eta_1)(1, 0)$$

and

$$\frac{\tilde{h}_\nu(\zeta_1)}{\left\|\tilde{h}_\nu(\zeta_1)\right\|} = \hat{\alpha}_\nu(\zeta_1)\left(q(\zeta_{\nu+1}), 1\right) + \hat{\beta}_\nu(\zeta_1)(1, 0).$$

Then, from I.H.5,

$$|\hat{\alpha}_\nu(\eta_1)| = \frac{|\alpha_\nu(\eta_1)|}{\|h_{\nu-1}(\eta_1)\|} \le 3K\sqrt{b}, \tag{7.1}$$

$$\frac{1}{2\lambda}d_\nu(\eta_0) \le |\hat{\beta}_\nu(\eta_1)| = \frac{|\beta_\nu(\eta_1)|}{\|h_{\nu-1}(\eta_1)\|} \le \frac{2}{\lambda}d_\nu(\eta_0) \tag{7.2}$$

and the same holds by replacing η by ζ. Let

$$\frac{h_{\nu-1}(\eta_1)}{\|h_{\nu-1}(\eta_1)\|} = (u_\eta, v_\eta) \ \text{ and } \ \frac{h_{\nu-1}(\zeta_1)}{\|h_{\nu-1}(\zeta_1)\|} = (u_\zeta, v_\zeta).$$

Since $\tilde{h}_\nu(\eta_1) = DT_{\lambda,a,b}(\eta_\nu)h_{\nu-1}(\eta_1)$, we have

$$\begin{cases} \hat{\alpha}_\nu(\eta_1) = C(\eta_\nu)u_\eta + D(\eta_\nu)v_\eta \\ \\ \hat{\beta}_\nu(\eta_1) = A(\eta_\nu)u_\eta + B(\eta_\nu)v_\eta - \hat{\alpha}_\nu(\eta_1)q(\eta_{\nu+1}) \end{cases} \tag{7.3}$$

and the same holds for ζ. So, $|\hat{\alpha}_\nu(\eta_1) - \hat{\alpha}_\nu(\zeta_1)| \le |C(\eta_\nu)(u_\eta - u_\zeta) + D(\eta_\nu)(v_\eta - v_\zeta)|$
$+ |C(\eta_\nu) - C(\zeta_\nu)||u_\zeta| + |D(\eta_\nu) - D(\zeta_\nu)||v_\zeta|$.

Since (u_η, v_η) and (u_ζ, v_ζ) are unit vectors, we have

$$|u_\eta - u_\zeta| + |v_\eta - v_\zeta| \le \frac{3}{2}|\text{ang}(h_{\nu-1}(\eta_1), h_{\nu-1}(\zeta_1))|. \tag{7.4}$$

Hence, from Proposition 1.3 and the induction, it follows that

$$|\hat{\alpha}_\nu(\eta_1) - \hat{\alpha}_\nu(\zeta_1)| \le 2K\sqrt{b}\left(2\sqrt[4]{b}\theta_{\nu-1} + \|\eta_\nu - \zeta_\nu\|\right). \tag{7.5}$$

Furthermore,

$$2\sqrt[4]{b}\theta_{\nu-1} + \|\eta_\nu - \zeta_\nu\| < 2\theta_\nu. \tag{7.6}$$

Therefore, we conclude that

$$|\hat{\alpha}_\nu(\eta_1) - \hat{\alpha}_\nu(\zeta_1)| = |C(\eta_\nu)u_\eta + D(\eta_\nu)v_\eta - (C(\zeta_\nu)u_\zeta + D(\zeta_\nu)v_\zeta)| \le 4K\sqrt{b}\theta_\nu. \tag{7.7}$$

On the other hand, (7.3) leads to

$$\left|\hat{\beta}_\nu(\eta_1) - \hat{\beta}_\nu(\zeta_1)\right| \le |A(\eta_\nu)u_\eta + B(\eta_\nu)v_\eta - (A(\zeta_\nu)u_\zeta + B(\zeta_\nu)v_\zeta)| +$$

$$+ |\hat{\alpha}_\nu(\zeta_1)q(\zeta_{\nu+1}) - \hat{\alpha}_\nu(\eta_1)q(\eta_{\nu+1})|. \tag{7.8}$$

Now, from (7.4) and Proposition 1.3, $|A(\eta_\nu)u_\eta + B(\eta_\nu)v_\eta - (A(\zeta_\nu)u_\zeta + B(\zeta_\nu)v_\zeta)| \le \frac{3}{2}K|\text{ang}(h_{\nu-1}(\eta_1), h_{\nu-1}(\zeta_1))| + K(1 + \sqrt{b})\|\eta_\nu - \zeta_\nu\|$. So, from (7.6) and the induction,

$$|A(\eta_\nu)u_\eta + B(\eta_\nu)v_\eta - (A(\zeta_\nu)u_\zeta + B(\zeta_\nu)v_\zeta)| \le 3K\theta_\nu. \tag{7.9}$$

Furthermore, (7.1), (7.5) and Propositions 3.4 and 3.10 imply

$$|\hat{a}_\nu(\zeta_1)q(\zeta_{\nu+1}) - \hat{a}_\nu(\eta_1)q(\eta_{\nu+1})| \leq \frac{1}{2}K\left(2\sqrt[4]{b}\theta_{\nu-1} + \|\eta_\nu - \zeta_\nu\|\right) \leq K\theta_\nu. \qquad (7.10)$$

Therefore, from (7.9), (7.10) and (7.8), we obtain

$$\left|\hat{\beta}_\nu(\eta_1) - \hat{\beta}_\nu(\zeta_1)\right| \leq 4K\theta_\nu \qquad (7.11)$$

and, since I.H.5(b) leads to

$$1 \leq \frac{|\beta_\nu(\zeta_1)|}{\|h_{\nu-1}(\zeta_1)\| \, d_\nu(\zeta_0)} = \frac{|\hat{\beta}_\nu(\zeta_1)|}{d_\nu(\zeta_0)},$$

we get

$$\left|\hat{\beta}_\nu(\eta_1) - \hat{\beta}_\nu(\zeta_1)\right| \leq \frac{4K\theta_\nu}{d_\nu(\zeta_0)}\left|\hat{\beta}_\nu(\zeta_1)\right|. \qquad (7.12)$$

Finally, since ν is a return of z_0, it follows that $\|h_\nu(\zeta_1)\| = \|h_{\nu-1}(\zeta_1)\|\left|\hat{\beta}_\nu(\zeta_1)\right|$ and $\|h_\nu(\eta_1)\| = \|h_{\nu-1}(\eta_1)\|\left|\hat{\beta}_\nu(\eta_1)\right|$. Hence, from (7.12),

$$\frac{\|h_\nu(\eta_1)\|}{\|h_\nu(\zeta_1)\|} = \frac{\|h_{\nu-1}(\eta_1)\|\left|\hat{\beta}_\nu(\eta_1)\right|}{\|h_{\nu-1}(\zeta_1)\|\left|\hat{\beta}_\nu(\zeta_1)\right|} \leq \frac{\|h_{\nu-1}(\eta_1)\|}{\|h_{\nu-1}(\zeta_1)\|}\left(1 + \frac{4K\theta_\nu}{d_\nu(\zeta_0)}\right).$$

Therefore,

$$\frac{\|h_\nu(\eta_1)\|}{\|h_\nu(\zeta_1)\|} \leq \exp\left(4K\sum_{i=0}^\nu \left(\sqrt[4]{b}\right)^i \sum_{k=1}^\nu \frac{\theta_k}{d_k(\zeta_0)}\right). \qquad (7.13)$$

So statement (a) is proved. Notice that statement (b) is evident when ν is a return.

B. Assume that ν is neither a return of z_0 nor the end of the folding period associated to any return of z_0. Write again

$$\frac{h_{\nu-1}(\eta_1)}{\|h_{\nu-1}(\eta_1)\|} = (u_\eta, v_\eta) \text{ and } \frac{h_{\nu-1}(\zeta_1)}{\|h_{\nu-1}(\zeta_1)\|} = (u_\zeta, v_\zeta).$$

We also have

$$\frac{h_\nu(\eta_1)}{\|h_{\nu-1}(\eta_1)\|} = (A(\eta_\nu)u_\eta + B(\eta_\nu)v_\eta, C(\eta_\nu)u_\eta + D(\eta_\nu)v_\eta)$$

and

$$\frac{h_\nu(\zeta_1)}{\|h_{\nu-1}(\zeta_1)\|} = (A(\zeta_\nu)u_\zeta + B(\zeta_\nu)v_\zeta, C(\zeta_\nu)u_\zeta + D(\zeta_\nu)v_\zeta).$$

Furthermore, from Propositions 1.3 and 5.13, $|A(\eta_\nu)u_\eta + B(\eta_\nu)v_\eta| \geq \frac{3}{4}|A(\eta_\nu)| - K\sqrt{b}$. Hence, since ν is not a return, $|A(\eta_\nu)u_\eta + B(\eta_\nu)v_\eta| \geq \delta$ and the same holds by replacing η by ζ. So, from (7.7), it follows that $|\text{ang}\,(h_\nu(\eta_1), h_\nu(\zeta_1))| \leq 4K\delta^{-1}\sqrt{b}\theta_\nu \leq$

$\sum_{i=1}^{\nu} \left(\sqrt[4]{b} \right)^i \theta_\nu$ and statement (b) is proved. To prove (a) write $DT_{\lambda,a,b} = DT$ and notice that

$$\frac{\|h_\nu(\eta_1)\|}{\|h_\nu(\zeta_1)\|} \leq \frac{\|h_{\nu-1}(\eta_1)\|}{\|h_{\nu-1}(\zeta_1)\|} \left(\frac{\|DT(\eta_\nu)(u_\eta, v_\eta) - DT(\zeta_\nu)(u_\zeta, v_\zeta)\|}{\|DT(\zeta_\nu)(u_\zeta, v_\zeta)\|} + 1 \right). \tag{7.14}$$

Now, from (7.7) and (7.9), we obtain $\|DT(\eta_\nu)(u_\eta, v_\eta) - DT(\zeta_\nu)(u_\zeta, v_\zeta)\|^2 \leq (4K\theta_\nu)^2$ and, since ν is not a return, $\|DT(\zeta_\nu)(u_\zeta, v_\zeta)\|^2 \geq \frac{1}{2}|A(\zeta_\nu)|^2 > d_\nu(\zeta_0)^2$. Thus,

$$\frac{\|DT(\eta_\nu)(u_\eta, v_\eta)\|}{\|DT(\zeta_\nu)(u_\zeta, v_\zeta)\|} \leq 1 + \frac{4K\theta_\nu}{d_\nu(\zeta_0)}.$$

From (7.14), we eventually obtain

$$\frac{\|h_\nu(\eta_1)\|}{\|h_\nu(\zeta_1)\|} \leq \exp\left(4K \sum_{i=0}^{\nu} \left(\sqrt[4]{b} \right)^i \sum_{k=1}^{\nu} \frac{\theta_k}{d_k(\zeta_0)} \right).$$

C. Finally, let us assume that a folding period $[\mu+1, \mu+l]$ ends at time ν. Take μ minimum. Then, $h_\nu(\eta_1) = \|h_{\mu-1}(\eta_1)\| (v(\eta_1) + w(\eta_1))$, where $v(\eta_1) = \hat{\beta}_\mu(\eta_1)h_l(\eta_{\mu+1})$ and $w(\eta_1) = \hat{\alpha}_\mu(\eta_1)DT^l e^{(l)}(\eta_{\mu+1})$. We use the same notation for ζ. We shall try to apply Lemma 7.1 with $v_1 = v(\eta_1)$, $w_1 = w(\eta_1)$, $v_2 = v(\zeta_1)$ and $w_2 = w(\zeta_1)$. First, we shall prove that the assumption of Lemma 7.1 holds. To this end, notice that ζ_μ and η_μ are bound to the same point $\tilde{\zeta}_0 \in C_j$ with $l < j \leq \mu < \nu$, where $\tilde{\zeta}_0$ is the binding point associated to the return μ of z_0. Then, by induction,

$$\frac{\|h_l(\zeta_{\mu+1})\|}{\|h_l(\eta_{\mu+1})\|} \leq \exp\left(4K \sum_{i=0}^{l} \left(\sqrt[4]{b} \right)^i \sum_{k=1}^{l} \frac{\theta_k'}{d_k(\eta_\mu)} \right) \tag{7.15}$$

and

$$|\text{ang}\,(h_l(\zeta_{\mu+1}), h_l(\eta_{\mu+1}))| \leq \sum_{i=1}^{l} \left(\sqrt[4]{b} \right)^i \theta_l', \tag{7.16}$$

where

$$\theta_k' = \theta_k(\eta_\mu, \zeta_\mu) = \sum_{i=1}^{k} \left(\sqrt[4]{b} \right)^{k-i} \|\eta_{\mu+i} - \zeta_{\mu+i}\|. \tag{7.17}$$

Since μ is a return of z_0, we may apply (7.7) and (7.12) to get, respectively,

$$|\hat{\alpha}_\mu(\eta_1) - \hat{\alpha}_\mu(\zeta_1)| \leq 4K\sqrt{b}\theta_\mu \tag{7.18}$$

and

$$\left| \hat{\beta}_\mu(\eta_1) - \hat{\beta}_\mu(\zeta_1) \right| \leq \frac{4K\theta_\mu}{d_\mu(\eta_0)} \left| \hat{\beta}_\mu(\eta_1) \right|. \tag{7.19}$$

Then, (7.15) gives

$$\frac{\|v(\zeta_1)\|}{\|v(\eta_1)\|} \leq \left(1 + \frac{4K\theta_\mu}{d_\mu(\eta_0)} \right) \exp\left(8K \sum_{k=1}^{l} \frac{\theta_k'}{d_k(\eta_\mu)} \right).$$

Now, from Remark 5.22,

$$\frac{1}{2} \le \frac{d_k(\eta_\mu)}{d_k(\zeta_\mu)} = \frac{d_{k+\mu}(\eta_0)}{d_{k+\mu}(\zeta_0)} \le 2$$

for $0 \le k \le l$. Therefore,

$$\frac{\|\mathbf{v}(\zeta_1)\|}{\|\mathbf{v}(\eta_1)\|} \le \left(1 + \frac{8K\theta_\mu}{d_\mu(\zeta_0)}\right) \exp\left(16K \sum_{k=1}^{l} \frac{\theta_k'}{d_k(\zeta_\mu)}\right). \tag{7.20}$$

Furthermore, from (7.17) and Definition 5.5, it follows that

$$\theta_k' \le \frac{3\tilde{\Lambda}}{2} \sum_{i=1}^{k} \left(\sqrt[4]{b}\right)^{k-i} e^{-\beta i}$$

and Propositions 5.21 and 5.19 yield $d_k(\zeta_\mu) > \frac{1}{2}d_k(\tilde{\zeta}_0) > \frac{1}{2}\tilde{\Lambda}e^{-\alpha k}$. Thus,

$$\sum_{k=1}^{l} \frac{\theta_k'}{d_k(\zeta_\mu)} \le 4 \sum_{k=1}^{\infty} e^{(\alpha-\beta)k}$$

and consequently,

$$\exp\left(16K \sum_{k=1}^{l} \frac{\theta_k'}{d_k(\zeta_\mu)}\right) \le \exp\left(64K \sum_{k=1}^{\infty} e^{(\alpha-\beta)k}\right). \tag{7.21}$$

On the other hand, the same arguments lead to

$$\theta_\mu \le \frac{3\tilde{\Lambda}}{2} \sum_{j=1}^{\mu} \left(\sqrt[4]{b}\right)^{\mu-j} e^{-\beta j}$$

and $d_\mu(\zeta_0) > \frac{6}{7}d_\mu(z_0) > \frac{6}{7}\tilde{\Lambda}e^{-\alpha\mu}$. Therefore,

$$\frac{\theta_\mu}{d_\mu(\zeta_0)} \le 2e^{\alpha\mu} \sum_{j=1}^{\mu} \left(\sqrt[4]{b}\right)^{\mu-j} e^{-\beta j} \le \sum_{j=0}^{\infty} \left(\sqrt[4]{b}e^{\beta}\right)^{j} < \frac{3}{2}. \tag{7.22}$$

From (7.21) and (7.22), there exists $\tau' = \tau'(K,\alpha,\beta) > 0$ such that

$$\left(1 + \frac{8K\theta_\mu}{d_\mu(\zeta_0)}\right) \exp\left(16K \sum_{k=1}^{l} \frac{\theta_k'}{d_k(\zeta_\mu)}\right) \le 1 + \frac{8K\theta_\mu}{d_\mu(\zeta_0)} + \tau' \sum_{k=1}^{l} \frac{\theta_k'}{d_k(\zeta_\mu)}. \tag{7.23}$$

Hence, from (7.23) and (7.20),

$$\left|\frac{\|\mathbf{v}(\zeta_1)\|}{\|\mathbf{v}(\eta_1)\|} - 1\right| \le 8K \frac{\theta_\mu}{d_\mu(\zeta_0)} + \tau' \sum_{k=1}^{l} \frac{\theta_k'}{d_k(\zeta_\mu)}. \tag{7.24}$$

Now (7.1), (7.18), (3.5) and Proposition 3.15 give $\|\mathbf{w}(\eta_1) - \mathbf{w}(\zeta_1)\| \le 4K\sqrt{b}\theta_\mu (Kb)^l + 3K^2\sqrt{b}(K_{15}b)^l \|\eta_\mu - \zeta_\mu\|$. Recall that in the proof of Proposition 5.11 we proved that

$(Kb)^l < \left(\sqrt{b}\right)^l d_\mu(\zeta_0)^2$ and, in the same way, $(K_{15}b)^l < \left(\sqrt{b}\right)^l d_\mu(\zeta_0)^2$. Then, since $\|\eta_\mu - \zeta_\mu\| < \theta_\mu$, we have $\|\mathbf{w}(\eta_1) - \mathbf{w}(\zeta_1)\| \leq const \left(\sqrt{b}\right)^{l+1} \theta_\mu d_\mu(\zeta_0)^2$. Furthermore, since I.H.5(b) and Proposition 5.14 yield

$$\|\mathbf{v}(\eta_1)\| = \frac{|\beta_\mu(\eta_1)|}{\|h_{\mu-1}(\eta_1)\|} \|h_l(\eta_{\mu+1})\| \geq \frac{d_\mu(\eta_0)}{2\lambda} > \frac{1}{2}d_\mu(\zeta_0), \qquad (7.25)$$

we get

$$\frac{\|\mathbf{w}(\eta_1) - \mathbf{w}(\zeta_1)\|}{\|\mathbf{v}(\eta_1)\|} \leq const \left(\sqrt{b}\right)^{l+1} \theta_\mu d_\mu(\zeta_0). \qquad (7.26)$$

By means of (7.1), (7.25) and Proposition 5.11 we also obtain

$$\frac{\|\mathbf{w}(\eta_1)\|}{\|\mathbf{v}(\eta_1)\|} \leq \frac{3K\sqrt{b}\left(\sqrt{b}\right)^l d_\mu(\eta_0)^2}{d_\mu(\eta_0)} \leq const \left(\sqrt{b}\right)^{l+1} d_\mu(\zeta_0). \qquad (7.27)$$

Hence, $\|\mathbf{v}(\eta_1) + \mathbf{w}(\eta_1)\| > \frac{1}{2}\|\mathbf{v}(\eta_1)\|$ and the same holds by replacing η by ζ. Therefore, we may apply Lemma 7.1 with $\Upsilon = \frac{1}{2}$ and

$$\chi = |\text{ang}\,(\mathbf{v}(\eta_1), \mathbf{v}(\zeta_1))| \frac{\|\mathbf{w}(\eta_1)\|}{\|\mathbf{v}(\eta_1)\|} + \frac{\|\mathbf{w}(\eta_1) - \mathbf{w}(\zeta_1)\|}{\|\mathbf{v}(\eta_1)\|} + \left|\frac{\|\mathbf{v}(\zeta_1)\|}{\|\mathbf{v}(\eta_1)\|} - 1\right| \frac{\|\mathbf{w}(\zeta_1)\|}{\|\mathbf{v}(\zeta_1)\|}.$$

From (7.16), (7.27), (7.26) and (7.24),

$$\chi \leq const \left(\sqrt{b}\right)^{l+1} \left(\theta_\mu + d_\mu(\zeta_0) \left(\sqrt[4]{b}\theta_l' + \theta_\mu + \sum_{k=1}^{l} \frac{\theta_k'}{d_k(\zeta_\mu)}\right)\right)$$

and, since $d_l(\zeta_\mu)\sqrt[4]{b} < 2\Lambda\sqrt[4]{b} < 1$, we have

$$4\chi \leq const \left(\sqrt{b}\right)^{l+1} \left(\theta_\mu + d_\mu(\zeta_0) \sum_{k=1}^{l} \frac{\theta_k'}{d_k(\zeta_\mu)}\right),$$

where the constant only depends on K, α and β.

Now, from the definition of l, $d_k(\zeta_\mu) > \frac{1}{2}d_k(\tilde{\zeta}_0) \geq \frac{1}{2}\tilde{\Lambda}e^{-\alpha l} \geq \frac{1}{2}\tilde{\Lambda}e^{-4\alpha}\{d_\mu(z_0)\}^{\frac{1}{2}}$ for every $k \in \{1,...,l\}$. Thus, since $d_\mu(z_0) < 2\delta$, it follows that $d_k(\zeta_\mu) > 2d_\mu(z_0) > d_\mu(\zeta_0)$ for every $k \in \{1,...,l\}$ and consequently, $4\chi \leq const \left(\sqrt{b}\right)^{l+1} \left(\theta_\mu + \sum_{k=1}^{l} \theta_k'\right)$. The definition of θ_k' given in (7.17) implies $\sum_{k=1}^{l} \theta_k' = \sum_{j=1}^{l} \|\eta_{\mu+j} - \zeta_{\mu+j}\| \sum_{k=j}^{l} \left(\sqrt[4]{b}\right)^{k-j}$. Hence, $const \left(\sqrt{b}\right)^{l+1} \sum_{k=1}^{l} \theta_k' \leq \left(\sqrt[4]{b}\right)^{l+1} \theta_l'$ and thus $4\chi \leq \left(\sqrt[4]{b}\right)^{l+1} \left(\left(\sqrt[4]{b}\right)^l \theta_\mu + \theta_l'\right)$. Finally, since the definitions of θ_μ and θ_l' lead to

$$\left(\sqrt[4]{b}\right)^l \theta_\mu + \theta_l' = \theta_\nu, \qquad (7.28)$$

we obtain

$$4\chi \leq \left(\sqrt[4]{b}\right)^{l+1} \theta_\nu. \qquad (7.29)$$

From Lemma 7.1(b), (7.16) and (7.29), it follows that

$$|\text{ang}\,(h_\nu(\eta_1), h_\nu(\zeta_1))| \leq \sum_{i=1}^{l} \left(\sqrt[4]{b}\right)^i \theta_l' + \left(\sqrt[4]{b}\right)^{l+1} \theta_\nu.$$

Since (7.28) implies $\theta_l' < \theta_\nu$, we conclude that $|\text{ang}\,(h_\nu(\eta_1), h_\nu(\zeta_1))| \leq \sum_{i=1}^{\nu} \left(\sqrt[4]{b}\right)^i \theta_\nu$ and statement (b) is proved.

To prove (a), notice that Lemma 7.1(a) leads to

$$\frac{\|h_\nu(\eta_1)\|}{\|h_\nu(\zeta_1)\|} \leq \frac{\|h_{\mu-1}(\eta_1)\|\,\|\mathbf{v}(\eta_1)\|}{\|h_{\mu-1}(\zeta_1)\|\,\|\mathbf{v}(\zeta_1)\|}\,(1+2\chi)\,.$$

Therefore, in the same way as (7.20) was reached, we deduce that

$$\frac{\|\mathbf{v}(\eta_1)\|}{\|\mathbf{v}(\zeta_1)\|} \leq \left(1 + \frac{4K\theta_\mu}{d_\mu(\zeta_0)}\right) \exp\left(4K \sum_{i=0}^{l} \left(\sqrt[4]{b}\right)^i \sum_{k=1}^{l} \frac{\theta_k'}{d_k(\zeta_\mu)}\right)\,.$$

Furthermore, since $\theta_k' \leq \theta_{\mu+k}$, the inductive method and (7.29) give

$$\log\left(\frac{\|h_\nu(\eta_1)\|}{\|h_\nu(\zeta_1)\|}\right) \leq 4K \sum_{i=0}^{\mu-1} \left(\sqrt[4]{b}\right)^i \sum_{k=1}^{\mu-1} \frac{\theta_k}{d_k(\zeta_0)} +$$

$$+ \frac{4K\theta_\mu}{d_\mu(\zeta_0)} + 4K \sum_{i=0}^{l} \left(\sqrt[4]{b}\right)^i \sum_{k=1}^{l} \frac{\theta_{\mu+k}}{d_{k+\mu}(\zeta_0)} + \left(\sqrt[4]{b}\right)^{l+1} \theta_\nu.$$

Hence,

$$\log\left(\frac{\|h_\nu(\eta_1)\|}{\|h_\nu(\zeta_1)\|}\right) \leq 4K \sum_{i=0}^{\nu} \left(\sqrt[4]{b}\right)^i \sum_{k=1}^{\nu} \frac{\theta_k}{d_k(\zeta_0)}$$

and the proposition is proved. $\square$

Next, we shall prove a result which was stated in Chapter 6:

Proof of Proposition 6.13. Let n be a free return of $z_0 \in C_n$ and let ζ_0 be its binding point. Let ξ_0 be bound to z_0 up to time n. From Proposition 7.2, $|\text{ang}\,(h_{n-1}(\xi_1), h_{n-1}(z_1))| \leq 2\sqrt[4]{b}\theta_{n-1}$. Furthermore, since $\theta_{n-1} \leq 2e^{-\beta n}$, it follows that $|\text{ang}\,(h_{n-1}(\xi_1), h_{n-1}(z_1))| \leq 4\sqrt[4]{b}e^{-\alpha n}$. Hence, Proposition 6.10 and (BA) lead to $\left|\text{ang}\left(h_{n-1}(\xi_1), t\left(\gamma^{[k]}; \eta^{[k]}\right)\right)\right| \leq 10\sqrt[4]{b}d_n(\xi_0)$. Furthermore, from (BA), we have $\|\xi_n - z_n\| \leq e^{-\beta n} < \frac{1}{50}d_n(\xi_0)$. Using once again Proposition 6.10, $\left\|\xi_n - \eta^{[k]}\right\| \leq \frac{1}{40}d_n(\xi_0)$. Therefore, by repeating the argument used in Proposition 6.10, we may construct a curve $\tilde{\gamma}$ containing ξ_n and ζ_0, tangent to $\gamma^{[k]}$ at ζ_0 and to $h_{n-1}(\xi_1)$ at ξ_n. Moreover, $\tilde{\gamma}$ can be parametrized by $(x, y(x))$ such that $|y'(x)| \leq \frac{1}{10}$ and $|y''(x)| \leq \frac{1}{10}$. The proof follows from Proposition 6.11 and Remark 6.12. $\square$

Proposition 7.3. *Let n be a return of $z_0 \in C_n$. Let ζ_0 and $[n+1, n+p]$ be its binding point and binding period, respectively. Let $\overline{z}_0 \in C_\nu$ be the critical approximation which has a free return at ν. Then, there exists a constant $\tau_1 = \tau_1(K, \alpha, \beta) > 1$ such that*

$$\frac{1}{\tau_1} \leq \frac{\|h_{n+k}(\xi_1)\|}{|\beta_n(\xi_1)| \, \|h_k(\zeta_1)\|} \leq \tau_1$$

for every $0 \leq k \leq \min\{p, 2c^{-1}\log d_\nu(\overline{z}_0)^{-1}\}$ and for every ξ_0 bound to z_0 up to time $n + k$.

Proof. From the definition of binding period, we have $\|\xi_{n+i} - \zeta_i\| \leq \|\xi_{n+i} - z_{n+i}\| + \|z_{n+i} - \zeta_i\| \leq h_n e^{-\beta(n+i)} + h e^{-\beta i} \leq h_\nu e^{-\beta i}$ for $1 \leq i \leq k$. Thus, ξ_n is bound to ζ_0 up to time k and, since $k \leq 2c^{-1}\log d_\nu(\overline{z}_0)^{-1} \leq 2c^{-1}\alpha\nu < \nu$, from Proposition 7.2 we get

$$\frac{\|h_k(\xi_{n+1})\|}{\|h_k(\zeta_1)\|} \leq \exp\left(8K \sum_{j=1}^{k} \frac{\theta_j}{d_j(\zeta_0)}\right),$$

with $\theta_j \leq \frac{1}{2}\tilde{\Lambda} \sum_{i=1}^{j} \left(\sqrt[4]{b}\right)^{j-i} e^{-\beta i}$. Furthermore, from Proposition 5.19, $d_j(\zeta_0) \geq \tilde{\Lambda} e^{-\alpha j}$ and so

$$8K \sum_{j=1}^{k} \frac{\theta_j}{d_j(\zeta_0)} \leq 5K \sum_{j=1}^{\infty} e^{(\alpha-\beta)j}.$$

Hence,

$$\frac{\|h_k(\xi_{n+1})\|}{\|h_k(\zeta_1)\|} \leq \exp\left(5K \sum_{j=1}^{\infty} e^{(\alpha-\beta)j}\right).$$

In the same way we obtain

$$\frac{\|h_k(\zeta_1)\|}{\|h_k(\xi_{n+1})\|} \leq \exp\left(10K \sum_{j=1}^{\infty} e^{(\alpha-\beta)j}\right).$$

Therefore, if we define $\tau = \exp\left(10K \sum_{j=1}^{\infty} e^{(\alpha-\beta)j}\right)$, then

$$\frac{1}{\tau} \leq \frac{\|h_k(\xi_{n+1})\|}{\|h_k(\zeta_1)\|} \leq \tau. \tag{7.30}$$

Now, let us check the relationship between $h_{n+k}(\xi_1)$ and $\beta_n(\xi_1)h_k(\xi_{n+1})$. If there are no returns μ of z_0 such that $\mu \leq n < \mu + l_\mu < n + k$, then the above vectors coincide and (7.30) implies the proposition.

Let us assume that there exist returns $\mu_1, ..., \mu_s$ of z_0 such that

$$\mu_1 < ... < \mu_s = n < \mu_s + l_s \leq ... \leq \mu_1 + l_1 \leq n + k.$$

Then,

$$h_{n+k}(\xi_1) = \beta_{\mu_1}(\xi_1)h_{n+k-\mu_1}(\xi_{\mu_1+1}) + \alpha_{\mu_1}(\xi_1)DT^{n+k-\mu_1}e^{l_1}(\xi_{\mu_1+1})$$

and, as in (6.26) and (6.27), we obtain

$$\frac{\|h_{n+k}(\xi_1)\|}{|\beta_{\mu_1}(\xi_1)|\,\|h_{n+k-\mu_1}(\xi_{\mu_1+1})\|} \le 1 + const\sqrt{b}\left(\sqrt{b}\right)^{l_1}K^k. \tag{7.31}$$

Furthermore, since $\mu_2 - \mu_1$ is a return of ξ_{μ_1}, we have

$$h_{\mu_2-\mu_1}(\xi_{\mu_1+1}) = \beta_{\mu_2-\mu_1}(\xi_{\mu_1+1})h_0(\xi_{\mu_2+1}) + \alpha_{\mu_2-\mu_1}(\xi_{\mu_1+1})e^{l_2}(\xi_{\mu_2+1})$$

and

$$h_{n+k-\mu_1}(\xi_{\mu_1+1}) = \beta_{\mu_2-\mu_1}(\xi_{\mu_1+1})h_{n+k-\mu_2}(\xi_{\mu_2+1}) + \alpha_{\mu_2-\mu_1}(\xi_{\mu_1+1})DT^{n+k-\mu_2}e^{l_2}(\xi_{\mu_2+1}).$$

Hence,

$$\frac{\|h_{n+k-\mu_1}(\xi_{\mu_1+1})\|}{|\beta_{\mu_2-\mu_1}(\xi_{\mu_1+1})|\,\|h_{n+k-\mu_2}(\xi_{\mu_2+1})\|} \le 1 + const\sqrt{b}\left(\sqrt{b}\right)^{l_2}K^k. \tag{7.32}$$

Therefore, from (7.31) and (7.32), we get

$$\frac{\|h_{n+k}(\xi_1)\|}{|\beta_{\mu_1}(\xi_1)|\,|\beta_{\mu_2-\mu_1}(\xi_{\mu_1+1})|\,\|h_{n+k-\mu_2}(\xi_{\mu_2+1})\|} \le \prod_{i=1}^{2}\left(1 + const\sqrt{b}\left(\sqrt{b}\right)^{l_i}K^k\right).$$

Now, notice that the folding period associated to the return μ_2 of ξ_0 coincides with the one associated to the return $\mu_2 - \mu_1$ of ξ_{μ_1}. So, $|\beta_{\mu_1}(\xi_1)|\,|\beta_{\mu_2-\mu_1}(\xi_{\mu_1+1})| = |\beta_{\mu_2}(\xi_1)|$ and therefore,

$$\frac{\|h_{n+k}(\xi_1)\|}{|\beta_{\mu_2}(\xi_1)|\,\|h_{n+k-\mu_2}(\xi_{\mu_2+1})\|} \le \prod_{i=1}^{2}\left(1 + const\sqrt{b}\left(\sqrt{b}\right)^{l_i}K^k\right).$$

In the same way as (7.31) was obtained, we also get

$$\frac{\|h_{n+k}(\xi_1)\|}{|\beta_{\mu_1}(\xi_1)|\,\|h_{n+k-\mu_1}(\xi_{\mu_1+1})\|} \ge 1 - const\sqrt{b}\left(\sqrt{b}\right)^{l_1}K^k$$

and so,

$$\frac{\|h_{n+k}(\xi_1)\|}{|\beta_{\mu_2}(\xi_1)|\,\|h_{n+k-\mu_2}(\xi_{\mu_2+1})\|} \ge \prod_{i=1}^{2}\left(1 - const\sqrt{b}\left(\sqrt{b}\right)^{l_i}K^k\right).$$

Repeating the arguments up to $\mu_s = n$, we deduce

$$\prod_{i=1}^{s}\left(1 - const\sqrt{b}\left(\sqrt{b}\right)^{l_i}K^k\right) \le$$

$$\le \frac{\|h_{n+k}(\xi_1)\|}{|\beta_n(\xi_1)|\,\|h_k(\xi_{n+1})\|} \le \prod_{i=1}^{s}\left(1 + const\sqrt{b}\left(\sqrt{b}\right)^{l_i}K^k\right). \tag{7.33}$$

On the other hand, since $0 < s - i < \mu_s - \mu_i \leq l_i - l_s$, it follows that $l_i > l_s + s - i$. Hence, $K^k \left(\sqrt{b}\right)^{l_i} < K^k \left(\sqrt{b}\right)^{l_s} \left(\sqrt{b}\right)^{s-i}$. Furthermore, since $k \leq 2c^{-1} \log d_\nu(\bar{z}_0)^{-1}$ and $l_s > \frac{1}{2}l_{0,s}$ lead to $K^k \left(\sqrt{b}\right)^{l_s} < 1$, it follows that

$$\prod_{i=1}^{s} \left(1 + const\sqrt{b} \left(\sqrt{b}\right)^{l_i} K^k\right) \leq \prod_{i=0}^{\infty} \left(1 + const \left(\sqrt{b}\right)^{i+1}\right) < 2$$

and

$$\prod_{i=1}^{s} \left(1 - const\sqrt{b} \left(\sqrt{b}\right)^{l_i} K^k\right) \geq \frac{1}{2}.$$

Therefore, from (7.33),

$$\frac{1}{2} \leq \frac{\|h_{n+k}(\xi_1)\|}{|\beta_n(\xi_1)| \|h_k(\xi_{n+1})\|} \leq 2. \tag{7.34}$$

Finally, (7.30) and (7.34) imply

$$\frac{\|h_{n+k}(\xi_1)\|}{|\beta_n(\xi_1)| \|h_k(\zeta_1)\|} \leq 2\tau \quad \text{and} \quad \frac{\|h_{n+k}(\xi_1)\|}{|\beta_n(\xi_1)| \|h_k(\zeta_1)\|} \geq \frac{1}{2\tau}.$$

Then, the proposition is proved by taking $\tau_1 = 2\tau$. $\square$

Proposition 7.4. *Let η_0 and ζ_0 be bound to the same $z_0 \in C_s$ up to time q, with $q \leq s \leq n$. Then,*

(a)

$$\frac{\|\omega_\nu(\eta_1)\|}{\|\omega_\nu(\zeta_1)\|} \leq exp\left(8Ke^{2\alpha\nu} \sum_{k=1}^{\nu} \frac{\theta_k}{d_k(\zeta_0)}\right) \text{ for } 0 \leq \nu \leq q.$$

(b)

$$|ang\left(\omega_\nu(\eta_1), \omega_\nu(\zeta_1)\right)| \leq 3\sqrt[4]{b}e^{2\alpha\nu} \sum_{k=1}^{\nu} \frac{\theta_k}{d_k(\zeta_0)} \text{for } 0 \leq \nu \leq q.$$

Proof. From Proposition 5.16 it follows that $\|\omega_\nu\| \geq \tilde{\Lambda} K_1^{-5} e^{-\alpha\nu} \|h_\nu\| = \Upsilon \|h_\nu\|$, where $K_1 = K_1(K, C_0, \alpha, \delta)$ is a positive constant. Hence, we may apply Lemma 7.1 to $\mathbf{v}_1 = h_\nu(\eta_1)$, $\mathbf{v}_2 = h_\nu(\zeta_1)$, $\mathbf{w}_1 = \sigma_\nu(\eta_1)$ and $\mathbf{w}_2 = \sigma_\nu(\zeta_1)$.

If $\nu + 1$ is a fold-free iterate, then the proposition is an immediate consequence of Proposition 7.2. Therefore, we assume that there exist returns $\mu_1, ..., \mu_s$ whose folding periods contain $\nu + 1$. According to the splitting algorithm,

$$\sigma_\nu(\eta_1) = \sum_{i=1}^{s} \alpha_{\mu_i}(\eta_1) DT^{\nu - \mu_i} e^{l_i}(\eta_{\mu_i+1}). \tag{7.35}$$

From I.H.5 and Proposition 3.3 we get

$$\left\|\alpha_\mu(\eta_1) DT^{\nu-\mu} e^l(\eta_{\mu+1})\right\| \leq const\sqrt{b} |\beta_\mu(\eta_1)| d_\mu(\eta_0)^{-1} (K_3 b)^{\nu-\mu}. \tag{7.36}$$

Furthermore, if $\tilde{\zeta}_0$ is the binding point associated to μ, then Propositions 7.3 and 5.14 give

$$\|h_\nu(\eta_1)\| \geq \frac{1}{\tau_1} \left\|h_{\nu-\mu}(\tilde{\zeta}_1)\right\| |\beta_\mu(\eta_1)| \geq \frac{1}{\tau_1} |\beta_\mu(\eta_1)|. \tag{7.37}$$

Hence, from (7.35), (7.36) and (7.37),

$$\frac{\|\sigma_\nu(\eta_1)\|}{\|h_\nu(\eta_1)\|} \leq \sum_\mu const\sqrt{b}\,(K_3 b)^{\nu-\mu}\,d_\mu(\eta_0)^{-1}$$

and, from (BA),

$$\frac{\|\sigma_\nu(\eta_1)\|}{\|h_\nu(\eta_1)\|} \leq const\sqrt{b}\sum_\mu (K_3 b)^{\nu-\mu}\,e^{\alpha\mu} < const\sqrt{b}e^{\alpha\nu}\sum_{i=0}^\infty (K_3 b)^i < \sqrt[3]{b}e^{\alpha\nu}. \tag{7.38}$$

All the previous estimates hold by replacing η by ζ.

On the other hand,

$$|\alpha_\mu(\eta_1) - \alpha_\mu(\zeta_1)| \leq$$

$$\leq \|h_{\mu-1}(\eta_1)\| |\widehat{\alpha}_\mu(\eta_1) - \widehat{\alpha}_\mu(\zeta_1)| + |\widehat{\alpha}_\mu(\zeta_1)| \,|\|h_{\mu-1}(\eta_1)\| - \|h_{\mu-1}(\zeta_1)\||. \tag{7.39}$$

Now, from (7.7),

$$|\widehat{\alpha}_\mu(\eta_1) - \widehat{\alpha}_\mu(\zeta_1)| \leq 4K\sqrt{b}\theta_\mu \leq 4K\sqrt{b}\frac{\theta_\mu}{d_\mu(\zeta_0)} \tag{7.40}$$

and by I.H.5(a),

$$|\widehat{\alpha}_\mu(\zeta_1)| \leq 3K\sqrt{b}. \tag{7.41}$$

Finally, from Proposition 7.2,

$$\frac{|\|h_{\mu-1}(\eta_1)\| - \|h_{\mu-1}(\zeta_1)\||}{\|h_{\mu-1}(\eta_1)\|} \leq \exp\left(8K\sum_{k=1}^{\mu-1}\frac{\theta_k}{d_k(\eta_0)}\right) + 1 \leq const\sum_{k=1}^{\mu-1}\frac{\theta_k}{d_k(\zeta_0)}. \tag{7.42}$$

Therefore,

$$|\|h_{\mu-1}(\eta_1)\| - \|h_{\mu-1}(\zeta_1)\|| \leq const\,\|h_{\mu-1}(\eta_1)\|\sum_{k=1}^{\mu-1}\frac{\theta_k}{d_k(\zeta_0)}. \tag{7.43}$$

Then, (7.39), (7.40), (7.41) and (7.43) yield

$$|\alpha_\mu(\eta_1) - \alpha_\mu(\zeta_1)| \leq const\sqrt{b}\,\|h_{\mu-1}(\eta_1)\|\sum_{k=1}^\mu\frac{\theta_k}{d_k(\zeta_0)}.$$

Hence, since

$$\|\eta_\mu - \zeta_\mu\| < \sum_{k=1}^\mu\frac{\theta_k}{d_k(\zeta_0)},$$

from Propositions 3.3 and 3.15, we get

$$\left\| \alpha_\mu(\eta_1)DT^{\nu-\mu}e^{(l)}(\eta_{\mu+1}) - \alpha_\mu(\zeta_1)DT^{\nu-\mu}e^{(l)}(\zeta_{\mu+1}) \right\| \leq$$

$$\leq const\sqrt{b} \, \|h_{\mu-1}(\eta_1)\| \, (K_M b)^{\nu-\mu} \sum_{k=1}^{\mu} \frac{\theta_k}{d_k(\zeta_0)},$$

where $K_M = \max\{K_3, K_{15}\}$. Then, from (7.35),

$$\frac{\|\sigma_\nu(\eta_1) - \sigma_\nu(\zeta_1)\|}{\|h_\nu(\eta_1)\|} \leq \frac{1}{\|h_\nu(\eta_1)\|} \sum_\mu const\sqrt{b} \, \|h_{\mu-1}(\eta_1)\| \, (K_M b)^{\nu-\mu} \sum_{k=1}^{\mu} \frac{\theta_k}{d_k(\zeta_0)}.$$

Furthermore, from Proposition 5.15, there exists $C = C(C_0, \delta) > 0$ such that

$$\frac{\|h_\nu(\eta_1)\|}{\|h_{\mu-1}(\eta_1)\|} \geq C^{\nu-\mu+1} \min_{\mu \leq j \leq \nu} (d_j(\eta_0)) \geq const C^{\nu-\mu+1} e^{-\alpha\nu}.$$

Hence,

$$\frac{\|\sigma_\nu(\eta_1) - \sigma_\nu(\zeta_1)\|}{\|h_\nu(\eta_1)\|} \leq const\sqrt{b} \sum_{k=1}^{\nu} \frac{\theta_k}{d_k(\zeta_0)} e^{\alpha\nu} \sum_{j=1}^{\infty} \left(\frac{K_M b}{C}\right)^j < \sqrt[3]{b} e^{\alpha\nu} \sum_{k=1}^{\nu} \frac{\theta_k}{d_k(\zeta_0)}. \quad (7.44)$$

Now, let χ be the constant in Lemma 7.1. From Proposition 7.2, (7.38), (7.42) and (7.44), we obtain

$$\chi \leq const\sqrt[3]{b} e^{\alpha\nu} \sum_{k=1}^{\nu} \frac{\theta_k}{d_k(\zeta_0)}$$

and we may apply Lemma 7.1 and Proposition 7.2 to conclude that

$$\frac{\|\omega_\nu(\eta_1)\|}{\|\omega_\nu(\zeta_1)\|} \leq \frac{\|h_\nu(\eta_1)\|}{\|h_\nu(\zeta_1)\|} \left(1 + \tilde{\Lambda}^{-1} K_1^5 e^{\alpha\nu} \chi\right) \leq \exp\left(8K e^{2\alpha\nu} \sum_{k=1}^{\nu} \frac{\theta_k}{d_k(\zeta_0)}\right)$$

and

$$\left| ang\left(\omega_\nu(\eta_1), \omega_\nu(\zeta_1)\right)\right| \leq 3\sqrt[4]{b} e^{2\alpha\nu} \sum_{k=1}^{\nu} \frac{\theta_k}{d_k(\zeta_0)}.$$

The proof is complete. $\square$

Proposition 7.5. *Let $\zeta_0 = (x_0, y_0) \in C_k$ be a critical approximation of order $k - 1$, with $k \leq n$. Let $s \to \eta_0(s) = (x_0 \pm s, y(x_0 \pm s))$ be a $C^2(b)$-curve such that $\eta_0(0) = \zeta_0$. Assume that there exist $q \leq k - 1$ and $0 < \sigma < 2\delta$ such that*

$$\|\eta_\nu(s) - \zeta_\nu\| = \left\|T_{\lambda,a,b}^\nu(\eta_0(s)) - T_{\lambda,a,b}^\nu(\zeta_0)\right\| \leq h e^{-\beta\nu} \quad (7.45)$$

for every $1 \leq \nu \leq q$ and for each $0 \leq s \leq \sigma$. Then:

(a) $\|\omega_\nu(\zeta_1)\| \sigma^2 \leq e^{-\beta\nu}$ *for $0 \leq \nu \leq q - 1$.*

(b) *Moreover, if $\|\omega_q(\zeta_1)\| \sigma^2 \leq h^2 e^{-2\beta q}$, then $\|\eta_{q+1}(s) - \zeta_{q+1}\| \leq h e^{-2\beta(q+1)}$.*

Proof. We shall prove (a) by induction on ν. Notice that if $\nu = 0$, then the statement holds for $4\delta^2 < 1$.

Since $\zeta_0 \in C_k$ with $k \le n$, ζ_1 is e^c-expanding up to time $k - 1$. Furthermore, from (7.45), it follows that $\eta_0(s)$ is bound to ζ_0 up to time $q \le k - 1$ for $0 \le s \le \sigma$. Then I.H.4 implies that $\eta_1(s)$ is also e^c-expanding up to time q. We may split the tangent vector $t(s) = \alpha_t(s)e(s) + \beta_t(s)(1,0)$, where $e(s) = (\bar{e}(s), 1)$ is colinear to the q-th contractive direction at $\eta_1(s)$. The arguments used to get (6.17) and (6.18) yield

$$|\beta_t(0)| \le const\,(Kb)^q \tag{7.46}$$

and

$$\frac{1}{2\lambda}|s| \le |\beta_t(s) - \beta_t(0)| \le \frac{2}{\lambda}|s|, \tag{7.47}$$

provided that δ and b are small enough.

Let us assume that statement (a) has been proved up to time $\nu - 1$. Let

$$t_\nu(s) = DT_{\lambda,a,b}^\nu(t(s)) = \alpha_t(s)DT_{\lambda,a,b}^\nu(e(s)) + \beta_t(s)\omega_\nu(\eta_1(s))$$

and

$$\kappa(s) = \frac{\|\omega_\nu(\eta_1(s))\|}{\|\omega_\nu(\eta_1(0))\|} = \frac{\|\omega_\nu(\eta_1(s))\|}{\|\omega_\nu(\zeta_1)\|}$$

for all s. Take $\epsilon_\nu(s)$ such that $\omega_\nu(\eta_1(s)) = \kappa(s)\,(\omega_\nu(\zeta_1) + \epsilon_\nu(s))$. From Proposition 7.4 we get

$$\kappa(s) \le \exp\left(8Ke^{2\alpha\nu}\sum_{k=1}^\nu \frac{\theta_k}{d_k(\zeta_0)}\right) \tag{7.48}$$

and

$$\kappa^{-1}(s) \le \exp\left(8Ke^{2\alpha\nu}\sum_{k=1}^\nu \frac{\theta_k}{d_k(\eta_0(s))}\right) \le \exp\left(16Ke^{2\alpha\nu}\sum_{k=1}^\nu \frac{\theta_k}{d_k(\zeta_0)}\right). \tag{7.49}$$

On the other hand, since $\|\omega_\nu(\zeta_1)\| = \|\omega_\nu(\zeta_1) + \epsilon_\nu(s)\|$, Proposition 7.4 leads to

$$\|\epsilon_\nu(s)\| \le \|\omega_\nu(\zeta_1)\|\,3\sqrt[4]{b}e^{2\alpha\nu}\sum_{k=1}^\nu \frac{\theta_k}{d_k(\zeta_0)}. \tag{7.50}$$

Next, we shall check that

$$e^{2\alpha\nu}\sum_{k=1}^\nu \frac{\theta_k}{d_k(\zeta_0)} \le \frac{4}{\Lambda}\sum_{j=1}^\nu e^{-\alpha j}. \tag{7.51}$$

To this end, notice that, from Proposition 5.19 and the definition of θ_k, it follows that

$$e^{2\alpha\nu}\sum_{k=1}^\nu \frac{\theta_k}{d_k(\zeta_0)} \le \frac{2}{\Lambda}\sum_{j=1}^\nu e^{2\alpha\nu+\alpha j}\,\|\eta_j(s) - \zeta_j\|. \tag{7.52}$$

If $j \geq 2\alpha(\beta - 2\alpha)^{-1}\nu$, then, from (7.45), $e^{2\alpha\nu + \alpha j}\|\eta_j(s) - \zeta_j\| \leq 2e^{-\alpha j}$ and (7.51) is proved. Now, we assume that $j < 2\alpha(\beta - 2\alpha)^{-1}\nu$ and take $\alpha = \alpha(K, \beta)$ sufficiently small so that

$$\alpha - \left(\frac{\beta + c - 4\alpha}{2}\right)\frac{\beta - 2\alpha}{2\alpha} < -\log K - \alpha. \tag{7.53}$$

Since $\nu \leq q - 1 \leq k - 2 < n$, from I.H.4, we inductively obtain

$$e^{c(\nu - 1)}s^2 \leq \|\omega_{\nu-1}(\zeta_1)\|\sigma^2 \leq e^{-\beta(\nu - 1)}$$

and consequently, $|s| \leq \exp\left(-\frac{1}{2}(\beta + c)(\nu - 1)\right)$. Therefore,

$$\|\eta_j(s) - \zeta_j\| \leq K^j\|\eta_0(s) - \zeta_0\| \leq \frac{11}{10}K^j|s| \leq 2K^j\exp\left(-\frac{1}{2}(\beta + c)\nu\right)$$

and, from (7.53), we have

$$e^{2\alpha\nu + \alpha j}\|\eta_j(s) - \zeta_j\| \leq 2K^j\exp\left(\alpha - \left(\frac{\beta + c - 4\alpha}{2}\right)\frac{\beta - 2\alpha}{2\alpha}\right)j \leq 2e^{-\alpha j}.$$

So, (7.51) follows from (7.52).

Now, from (7.48), (7.49) and (7.51), we have

$$\tau^{-1} \leq \kappa(s) \leq \tau, \tag{7.54}$$

with

$$\tau = \exp\left(64K\tilde{\Lambda}^{-1}\sum_{j=1}^{\infty}e^{-\alpha j}\right) = \tau(\lambda, K, \alpha).$$

Furthermore, (7.50) and (7.51) give

$$\|\epsilon_\nu(s)\| < \frac{1}{2}\|\omega_\nu(\zeta_1)\|. \tag{7.55}$$

Finally, let us write

$$\eta_{\nu+1}(\sigma) - \zeta_{\nu+1} = \int_0^\sigma t_\nu(s)ds = \int_0^\sigma \alpha_t(s)DT_{\lambda,a,b}^\nu(e(s))ds + \int_0^\sigma \beta_t(0)\omega_\nu(\eta_1(s))ds +$$

$$+ \int_0^\sigma \omega_\nu(\zeta_1)\kappa(s)\left(\beta_t(s) - \beta_t(0)\right)ds + \int_0^\sigma \kappa(s)\left(\beta_t(s) - \beta_t(0)\right)\epsilon_\nu(s)ds. \tag{7.56}$$

Notice that, from Proposition 3.3 and (6.16), we get

$$\left\|\int_0^\sigma \alpha_t(s)DT_{\lambda,a,b}^\nu(e(s))ds\right\| \leq 2K\sqrt{b}\,(K_3b)^\nu\sigma < \left(\sqrt{b}\right)^\nu \tag{7.57}$$

and (7.46) yields

$$\left\|\int_0^\sigma \beta_t(0)\omega_\nu(\eta_1(s))ds\right\| \leq const\,(Kb)^q\,K^\nu\sigma < \left(\sqrt{b}\right)^\nu. \tag{7.58}$$

On the other hand, from (7.55),

$$\left\| \int_0^\sigma \kappa(s) \left(\beta_t(s) - \beta_t(0) \right) \epsilon_\nu(s) ds \right\| \leq \frac{1}{2} \left\| \omega_\nu(\zeta_1) \int_0^\sigma \kappa(s) \left(\beta_t(s) - \beta_t(0) \right) ds \right\| \quad (7.59)$$

and from (7.47) and (7.54)

$$\frac{\sigma^2}{4\lambda\tau} \leq \left| \int_0^\sigma \kappa(s) \left(\beta_t(s) - \beta_t(0) \right) ds \right| \leq \frac{2\tau\sigma^2}{\lambda}. \quad (7.60)$$

Therefore, (7.60) and (7.59) imply

$$\frac{\sigma^2}{8\lambda\tau} \left\| \omega_\nu(\zeta_1) \right\| \leq \frac{1}{2} \left\| \omega_\nu(\zeta_1) \int_0^\sigma \kappa(s) \left(\beta_t(s) - \beta_t(0) \right) ds \right\| \leq$$

$$\leq \left\| \omega_\nu(\zeta_1) \int_0^\sigma \kappa(s) \left(\beta_t(s) - \beta_t(0) \right) ds + \int_0^\sigma \kappa(s) \left(\beta_t(s) - \beta_t(0) \right) \epsilon_\nu(s) ds \right\|.$$

Then, from (7.56), (7.57) and (7.58) we obtain by induction

$$\frac{\sigma^2}{8\lambda\tau} \left\| \omega_\nu(\zeta_1) \right\| \leq h e^{-\beta(\nu+1)} + 2 \left(\sqrt{b} \right)^\nu \leq 2h e^{-\beta(\nu+1)}$$

and consequently, $\left\| \omega_\nu(\zeta_1) \right\| \sigma^2 \leq 16\lambda\tau h e^{-\beta\nu}$. Thus, if we take $h = h(\lambda, K, \alpha) = \frac{1}{16}\lambda\tau^{-1} < \frac{1}{8}\tilde{\Lambda}$, then $\left\| \omega_\nu(\zeta_1) \right\| \sigma^2 \leq \lambda^2 e^{-\beta\nu} < e^{-\beta\nu}$ and statement (a) is proved.

To prove (b) notice that if we put $\nu = q$ in (7.56) and (7.59), then, using also (7.58) and (7.60), we obtain $\left\| \eta_{q+1}(\sigma) - \zeta_{q+1} \right\| \leq 3\tau\lambda^{-1}\sigma^2 \left\| \omega_q(\zeta_1) \right\| + 2 \left(\sqrt{b} \right)^q$. Therefore, if $\left\| \omega_q(\zeta_1) \right\| \sigma^2 \leq h^2 e^{-2\beta q}$, then $\left\| \eta_{q+1}(\sigma) - \zeta_{q+1} \right\| \leq h e^{-2\beta(q+1)}$. $\square$

Remark 7.6. *From the proof of Proposition 7.5(a), we obtain*

$$\left\| \omega_\nu(\eta_1(s)) \right\| = \kappa(s) \left\| \omega_\nu(\zeta_1) \right\| \leq \tau \left\| \omega_\nu(\zeta_1) \right\|$$

for $0 \leq s \leq \sigma$. Therefore, if we take $h = \frac{1}{16}\lambda\tau^{-2}$, then Proposition 7.5(a) also holds for every $0 \leq s \leq \sigma$. That is, $\left\| \omega_\nu(\eta_1(s)) \right\| \sigma^2 \leq \tau \left\| \omega_\nu(\zeta_1) \right\| \sigma^2 \leq 16\lambda\tau^2 h e^{-\beta\nu} = \lambda^2 e^{-\beta\nu} < e^{-\beta\nu}$.

To bound the length of the binding periods we have to modify the definition of primary binding period given in Chapter 5. Let n be a free return of $z_0 \in C_n$. Let $s \to \eta_0(s)$ be the curve constructed in the proof of Proposition 6.10, with $z_n = \eta_0(\sigma)$ and $\zeta_0 = \eta_0(0)$ the binding point associated to z_n. Then, the primary binding period associated to n is $[n+1, n+p_0]$, where p_0 is the largest natural number such that

$$\left\| \eta_j(s) - \zeta_j \right\| \leq h e^{-\beta j} \text{ for every } 1 \leq j \leq p_0 \text{ and for every } 0 \leq s \leq \sigma.$$

The binding period $[n+1, n+p]$ remains defined from the primary binding period as in Chapter 5.

Notice that the lower bound of the length of the binding periods, obtained in Proposition 5.9, is still true. In fact, it suffices to replace there

$$\|z_{n+p_0+1} - \zeta_{p_0+1}\| \geq he^{-\beta(p_0+1)} \text{ by } \|\eta_{p_0+1}(s) - \zeta_{p_0+1}\| \geq he^{-\beta(p_0+1)}$$

and

$$\|z_{n+p+1} - \zeta_{p+1}\| \leq K^{p+1}d_n(z_0) \text{ by } \|\eta_{p+1}(s) - \zeta_{p+1}\| \leq K^{p+1}d_n(z_0)$$

for some $0 \leq s \leq \sigma$.

Proposition 7.7. *Let n be a return of $z_0 \in C_n$ and let p be the length of its binding period. Let $\bar{z}_0 \in C_\nu$ be the critical approximation which has a return at ν. Then*

(a) $p \leq 2c^{-1} \log d_\nu(\bar{z}_0)^{-1}$.

(b) There exists $\tau_2 = \tau_2(\lambda, K, \alpha, \beta)$ such that

$$\|h_{n+p}(\xi_1)\| d_n(\xi_0) \geq \tau_2 e^{\frac{1}{3}c(p+1)} \|h_n(\xi_1)\| \geq \|h_n(\xi_1)\|$$

for every ξ_0 bound to z_0 up to time $n+p$.

Proof. It suffices to prove the statement (a) when $\nu = n$ is a free return of $z_0 = \bar{z}_0$. Let $\zeta_0 \in C_n$ be the binding point associated to z_n and let $s \to \eta_0(s)$ be the curve constructed in the proof of Proposition 6.10, with $z_n = \eta_0(\sigma)$ and $\zeta_0 = \eta_0(0)$. Notice that this curve satisfies the hypotheses of Proposition 7.5. Furthermore, since ζ_1 is e^c-expanding up to time $n-1$, (BA) implies that ζ_1 is e^c-expanding up to time $4c^{-1} \log d_n(z_0)^{-1}$, provided that $4\alpha < c$. Therefore, if $\nu > 2c^{-1} \log d_n(z_0)^{-1}$, then $\|\omega_\nu(\zeta_1)\| \sigma^2 \geq \|\omega_\nu(\zeta_1)\| \frac{9}{10} d_n(z_0)^2 \geq \frac{9}{10} e^{c\nu} d_n(z_0)^2 > \frac{9}{10} > e^{-\beta\nu}$. Hence, from Proposition 7.5(a), there exists $s \in [0, \sigma]$ such that $\|\eta_\nu(s) - \zeta_\nu\| > he^{-\beta\nu}$ and consequently, $p_0 \leq 2c^{-1} \log d_n(z_0)^{-1}$.

To prove (b) notice that, from (5.7), there exists $s \in [0, \sigma]$ such that

$$\|\eta_{p+1}(s) - \zeta_{p+1}\| \geq he^{-2\beta(p+1)}.$$

Then, from Proposition 7.5(b),

$$\|\omega_p(\zeta_1)\| \sigma^2 \geq h^2 e^{-2\beta p}. \tag{7.61}$$

Since $p + 1$ is a free iterate for ζ_0, we get $\omega_p(\zeta_1) = h_p(\zeta_1)$. Hence, from Proposition 7.3 and (7.61),

$$\|h_{n+p}(\xi_1)\| \geq \tau_1^{-1} |\beta_n(\xi_1)| \|h_p(\zeta_1)\| \geq h^2 \tau_1^{-1} \sigma^{-2} \|h_n(\xi_1)\| e^{-2\beta p}.$$

Now, (a) leads to $e^{\frac{1}{2}cp} < d_\nu(\overline{z}_0)^{-1} < \sigma^{-1}$ and thus

$$\|h_{n+p}(\xi_1)\| \, d_n(\xi_0) \geq \frac{1}{2} \|h_{n+p}(\xi_1)\| \, \sigma \geq \tau_2 e^{\frac{1}{3}c(p+1)} \|h_n(\xi_1)\|,$$

where $\tau_2 = \tau_2(\lambda, K, \alpha, \beta) = \frac{1}{2}h^2\tau_1^{-1}$. Finally, from Propositions 5.9 and 5.23, it follows that $\tau_2 e^{\frac{1}{3}c(p+1)} > 1$, whenever $\delta = \delta(K, c, \alpha, \beta)$ is small enough. $\square$

Chapter 8

THE EXCLUSION OF PARAMETERS

Since I.H.5 and I.H.6 hold at time n, the proof of the main theorem ends if we prove the existence of a positive Lebesgue measure set of parameters for which every critical approximation satifies (BA) and (FA) at all times. This set will be constructed, as in the unidimensional case, by means of an inductive exclusion of parameters: Once a critical set C_n has been constructed for each $a \in E_{n-1}$, we take the values of the parameters for which the image of every point in C_n is e^c-expanding up to time n. Then, we try to construct C_{n+1} from C_n by using the algorithms described in Chapter 4. We remove the parameters which do not satisfy either (BA) or (FA). Notice that (BA) was formulated in terms of the distance between the respective return and its binding point. This distance depends on a and will be shown to vary slightly with respect to a.

Let $a \in E_{n-1}$ and $z_0^{(n-1)}(a) \in C_n(a)$. Recall that $z_0^{(n-1)}(a)$ is the solution of

$$\langle t(a, T_{\lambda,a,b}(z)), f^{(n-1)}(a, T_{\lambda,a,b}(z)) \rangle = 0, \tag{8.1}$$

where $t(a, T_{\lambda,a,b}(z))$ is the tangent vector to $W_{a,m}^u$ at $T_{\lambda,a,b}(z)$ and $f^{(n-1)}(a, T_{\lambda,a,b}(z))$ is the maximally contracting vector up to time $n - 1$. Using the results stated in Chapter 3 and by means of the implicit function theorem, we shall check that there exists $\tau > 0$ such that

$$\left\| D_a \left(z_0^{(n-1)}(a) \right) \right\| \leq b^\tau.$$

First, we shall prove this statement for the critical approximations $z_0^{(n-1)} \in G_0$ and $\omega_0^{(n-1)} \in G_1$, defined in Chapter 4 for $2 \leq n \leq N - 1$. Notice that we may also take $E_n = \Omega_0$ where $\Omega_0 = \Omega_0(\lambda) = (a_0(\lambda), a_M(\lambda))$ is given by Proposition 4.11.

Proposition 8.1. *The critical approximations $z_0^{(n-1)}$ and $\omega_0^{(n-1)}$ are defined for all $a \in \Omega_0$ and for every $2 \leq n \leq N - 1$. Moreover,*

$$\left\| D_a(z_0^{(n-1)}(a)) \right\| \leq \sqrt[3]{b} \ \text{ and } \ \left\| D_a(\omega_0^{(n-1)}(a)) \right\| \leq \sqrt[3]{b}.$$

Proof. Let I_0 be the interval given by Proposition 4.11. From Proposition 4.13, there exists a parametrization $z : (a, x) \in \Omega_0 \times I_0 \to z(a, x) = (x, y(a, x)) \in \mathbf{R}^2$ such that $z(a, I_0) = G_0(a) \cap I_0$ and

$$\|y\|_{C^2(a,x)} \leq const\sqrt{b}. \tag{8.2}$$

Then, we have

$$z_0^{(n-1)}(a) = (x(a), y(a, x(a))), \tag{8.3}$$

where $x(a)$ is the solution of

$$H(a, x) = \left\langle t\left(a, T(a, x, y(a, x))\right), f^{(n-1)}\left(a, T(a, x, y(a, x))\right) \right\rangle = 0.$$

From Proposition 1.3 and (8.2), we get $\|D_a t\left(a, T(a, x, y(a, x))\right)\| \leq \sqrt{b}$. Furthermore, Proposition 3.10 gives

$$\left\| D_a f^{(n-1)}(a, T(a, x, y(a, x))) \right\| \leq const\sqrt{b}$$

and consequently, $\|\partial_a H(a, x)\| \leq const\sqrt{b}$.

To bound $D_x t\left(a, T(a, x, y(a, x))\right)$ notice that, from Proposition 1.3, we obtain $|\partial_x A(a, x, y(a, x))| > 3$. Thus, from (8.2) and Proposition 1.3,

$$\left\| D_x t\left(a, T(a, x, y(a, x))\right) \right\| > \frac{2}{5}.$$

Moreover, from Proposition 3.10,

$$\left\| D_x f^{(n-1)}(a, T(a, x, y(a, x))) \right\| \leq const\sqrt{b}$$

and, since $f^{(n-1)}\left(a, T(a, x, y(a, x))\right)$ is nearly horizontal, we get $\|\partial_x H(a, x)\| > 2$.

Then, by applying the implicit function theorem, for each $a \in \Omega_0$ we find a unique $x = x(a)$ solution of $H(a, x) = 0$ with $|x'(a)| \leq const\sqrt{b}$. Therefore, (8.3) yields $\left\| D_a(z_0^{(n-1)}(a)) \right\| \leq const\sqrt{b} < \sqrt[3]{b}$. The proof of $\left\| D_a(\omega_0^{(n-1)}(a)) \right\| \leq \sqrt[3]{b}$ follows in the same way. □

Once the parameters not satisfying (BA) and (FA) are removed from E_{n-1} in order to construct E_n, for two parameters a and a' in the same connected component of E_n, each of the critical sets $C_{n+1}(a)$ and $C_{n+1}(a')$ may not be an analytic continuation of the other because of algorithm B. Actually, this fact permits the distance $d_{n+1}(a)$ in (BA) fail to be continuous for some critical approximation. Therefore, we shall only consider critical approximations which have an analytic continuation to the whole connected component. It will be necessary to check that these approximations are sufficient to define the binding point in tangential position on each return.

Let us fix a parameter a_0 and a critical approximation $z_0(a_0)$. We shall find an interval of parameters ω containing a_0 such that, for every $a \in \omega$, there exists a critical approximation $z_0(a)$ with the same order and generation as $z_0(a_0)$. The function $z_0(a)$, defined on ω, is the analytic continuation given by equation (8.1). The algorithm used to construct $z_0^{(n)}(a_0)$ from $z_0^{(n-1)}(a_0)$ also allows us to construct $z_0^{(n)}(a)$ from $z_0^{(n-1)}(a)$ for every $a \in \omega$. On each ω associated to $z_0(a_0) = z_0^{(k-1)}(a_0)$, we construct a positive Lebesgue measure set $E_k(z_0^{(k-1)})$ such that if $a \in E_k(z_0^{(k-1)})$, then $z_1^{(k-1)}(a) = T_{\lambda,a,b}(z_0^{(k-1)}(a))$ is e^c-expanding up to time k. Next, setting

$$E_k = \bigcap_{z_0^{(k-1)} \in C_k} E_k\left(z_0^{(k-1)}\right)$$

and using (5.1), we shall check that $E = \bigcap_{k \in \mathbb{N}} E_k$ has positive Lebesgue measure. The sets E_k are inductively constructed. From Proposition 4.5, we may take $E_k = \Omega_0 \subset \Omega_N$ for $k \leq N - 1$, where Ω_0 is the interval given by Proposition 4.11.

We have already pointed out that, regardless of the algorithm used to construct $z_0^{(k-1)}(a_0)$ from $z_0^{(k-2)}(a_0)$, $z_0^{(k-1)}(a)$ is bound to $z_0^{(k-2)}(a)$ for every $a \in \omega$. Therefore, from I.H.4, $z_0^{(k-1)}(a)$ is e^c-expanding up to time $k-1$. Hence, to obtain expansiveness up to time k, it seems sufficient to study the relative position between $z_k^{(k-1)}(\omega)$ and B_δ, according to (BA). Nevertheless, the expansiveness of $z_0^{(k-1)}$ also depends on the free assumption (FA). With respect to this assumption, we remove, as we did in Chapter 2, not only the parameters for which the orbit of a critical approximation spends too much time in binding periods, but also the parameters for which this orbit spends too much time outside escape periods. The key fact to controling the measure of the removed set is the persistence of escape situations (see Lemma 2.17). Therefore, we need to start the process under the assumption that a certain escape period exists, which according to Definition 2.23, is equivalent to the existence of a certain escape situation for $\nu \leq k + 1$. The following result ensures this fact for $1 \leq k \leq N - 1$:

Proposition 8.2. *Let $\Omega_0 = \Omega_0(\lambda) = (a_0(\lambda), a_M(\lambda))$ be the interval given by Proposition 4.11. Let $z_0^{(k-1)}$ be the critical approximation of generation zero and N the first natural number for which $z_N^{(N-2)}(\Omega_0) \cap B_\delta \neq \emptyset$. We may redefine Ω_0 and N so that*

$$length\left(z_N^{(N-2)}(\Omega_0)\right) \geq \sqrt{\delta}.$$

Proof. For any natural number N_1, Proposition 2.5 yields an interval $\Omega_{N_1} = [a_{N_1}, a(\lambda)]$ such that $\xi_{N_1}(\Omega_{N_1}) \supset U_\lambda$ and $\xi_j(\Omega_{N_1}) \cap U_\lambda = \emptyset$ for $1 \leq j \leq N_1-1$. Let $\tilde{a} \in \Omega_{N_1}$ be such that $\xi_{N_1}(\tilde{a}) = c_\lambda + \delta$ and define $\Omega_0^1 = [a_{N_1}, \tilde{a}]$. Let N_2 be the first return situation of Ω_0^1. From Lemma 2.17, N_2 is an escape situation of Ω_0^1. Hence, there exists $\Omega_0^2 \subset \Omega_0^1$ such that $|\xi_{N_2}(\Omega_0^2)| \geq \sqrt{\delta}$. If $|\Omega_0^1| \leq \frac{1}{160}\lambda^4 F(\lambda, \tilde{a})$, then the proposition follows by taking $\Omega_0 = \Omega_0^1$. Otherwise, we repeat the arguments replacing Ω_0^1 by Ω_0^2. From Proposition 2.5 we get, at worst, a sequence of intervals $\{\Omega_0^i\}_i$ such that the first return situation of each Ω_0^i is an escape situation and, moreover, $\lim_{i\to\infty}|\Omega_0^i| = 0$. Thus, for some i, $|\Omega_0^i| \leq \frac{1}{160}\lambda^4 F(\lambda, \tilde{a}) < \frac{1}{160}\lambda^4 F(\lambda, \sup \Omega_0^i)$. So (4.6), and therefore Proposition 4.11, holds. Furthermore, the first return situation of Ω_0^i is $N = N_{i+1}$, which is an escape situation. These arguments are easily extended to the bidimensional case by taking b small enough with respect to N. $\square$

Now, let us assume that $n \geq N$. The inductive process used in the construction of E_k is described in the following inductive hypothesis:

I.H.7. *Assume that for $k \leq n - 1$, a set $E_k \subset \Omega_0$ has already been constructed. Furthermore, for each $a_0 \in E_k$ and for every $z_0^{(k-1)} \in C_k(a_0)$, there exist an interval $\omega \subset \Omega_0$ and an iterate $\nu \in \left[\frac{1}{2}(k+1), k+1\right]$ satisfying the following properties:*

(a) If $\zeta_0^{(\nu-2)}(a_0) \in C_{\nu-1}(a_0)$ is the critical approximation of order $\nu - 2$ to which $z_0^{(k-1)}$ is bound, then $\zeta_0^{(\nu-2)}$ has an analytic continuation to ω such that

$$\left\| D_a\left(\zeta_0^{(\nu-2)}(a)\right)\right\| \leq \sum_{i=1}^{g} b^{\frac{1}{30}i} + \sum_{j=1}^{\nu-1} b^{\frac{1}{3}j},$$

where g is the generation of $\zeta_0^{(\nu-2)}(a_0)$.

(b) For each $a \in \omega$, $\zeta_0^{(\nu-2)}(a)$ satisfies (BA) and (FA) in $[1, \nu - 1]$. Moreover, $\zeta_0^{(\nu-2)}(a)$ satisfies, in $[1, \nu - 1]$, all the estimates stated for binding periods and folding

periods in the last three chapters. Consequently, $\zeta_1^{(\nu-2)}(a)$ is e^c-expanding up to time $\nu - 1$.

(c) $K^{-\frac{3}{2}\nu} \le |\omega| \le e^{-\frac{2}{3}c\nu}$.

(d) ν belongs to an escape period of $\zeta_0^{(\nu-2)} : \omega \to \mathbf{R}^2$.

On each ω we could construct, as in the unidimensional case, a sequence $\omega = E_{\nu-1}(\zeta_0^{(\nu-2)}) \supset E_\nu(\widetilde{\zeta}_0^{(\nu-1)}) \supset \dots \supset E_k(\widetilde{\zeta}_0^{(k-1)})$ such that $\widetilde{\zeta}_1^{(j-1)}(a)$ is e^c-expanding up to time j for each $a \in E_j(\widetilde{\zeta}_0^{(j-1)})$, where $\widetilde{\zeta}_0^{(j-1)}(a)$ is the critical approximation obtained from $\zeta_0^{(\nu-2)}(a)$ by successively applying algorithm A. However, $z_0^{(k-1)}(a_0)$ may be different from $\widetilde{\zeta}_0^{(k-1)}(a_0)$ and so, we cannot achieve a set of parameters for which $z_0^{(k-1)}$ exists and is e^c-expanding up to time k. This means that $z_0^{(k-1)}(a_0)$ cannot be constructed from $\zeta_0^{(\nu-2)}(a_0)$ by only using algorithm A. In Figure 8.1, we display how the sequence of algorithms used in the construction of critical approximations can depend on the parameter.

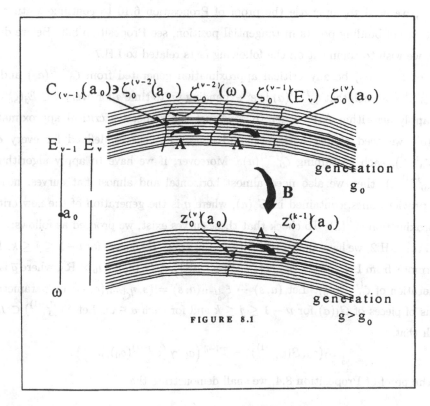

FIGURE 8.1

This situation will be avoided by a more restrictive use of algorithm B. In this sense, we go back to the definition of the sets C_j'' given in Chapter 5, see Remark 5.4, to complete it in the following way:

Let $a_0 \in E_k$ and $\zeta_0^{(k-1)} \in C_k(a_0)$. Let ω be the interval of parameters associated to $\zeta_0^{(k-1)}$ given by I.H.7. Assume that, by means of the definition given in Chapter 5, we may construct from $\zeta_0^{(k-1)}(a_0)$ a critical approximation $z_0^{(k)}(a_0)$ with generation $g \in (\theta k, \theta(k+1)]$. Now, we take $z_0^{(k)} \in C_{k+1}''$ if and only if:

(a) For every $a \in E_k(\zeta_0^{(k-1)})$, $z_0^{(k)}(a)$ can be constructed from $\zeta_0^{(k)}(a)$ by using algorithm B, where $\zeta_0^{(k)}(a)$ is the critical approximation constructed from $\zeta_0^{(k-1)}(a)$ by applying algorithm A.

(b) For every $a \in E_k(\zeta_0^{(k-1)})$, the following condition holds:

$$\left\| z_0^{(k)}(a) - \zeta_0^{(k)}(a) \right\| \leq b^{\frac{1}{10}g}. \tag{8.4}$$

Now, since many critical approximations are ruled out in the new construction of C_j''', we need to supervise the proof of Proposition 6.10 to continue stating the existence of binding points in tangential position, see Proposition 8.3. Before doing this, we wish to comment on the following facts related to I.H.7:

Let $\zeta_0^{(j-1)}(a_0)$ be any critical approximation generated from $\zeta_0^{(\nu-2)}(a_0)$ and let $E_j(\zeta_0^{(j-1)})$ be such that $\zeta_0^{(j-1)}(a)$ is e^c-expanding up to time j for every $a \in E_j(\zeta_0^{(j-1)})$. To apply algorithm A in order to construct the respective critical approximation $\zeta_0^{(j)}(a)$, we need almost horizontal and almost flat curves defined for every $a \in E_j(\zeta_0^{(j-1)})$, each containing $\zeta_0^{(j-1)}(a)$. Moreover, if we have to apply algorithm B to $\zeta_0^{(j-1)}(a)$, then we also need almost horizontal and almost flat curves, near to the previous ones, contained in $G_g(a)$, where g is the generation of the new critical approximation $z_0^{(j)}(a)$. To check that these curves exist, we proceed as follows:

From I.H.2, we have that $\gamma\left(\zeta_0^{(j-1)}(a_0), \rho_0^{\theta j}\right)$ are $C^2(b)$ curves for $\nu-1 \leq j \leq k$. Furthermore, from I.H.3, $T^{1-\bar{g}}\left(a_0, \gamma\left(\zeta_0^{(j-1)}(a_0), \rho_0^{\theta j}\right)\right) \subset G_1(a_0) \cap (I_0 \times \mathbf{R})$, where $\bar{g}$ is the generation of $\zeta_0^{(j-1)}(a_0)$. Let $(a,s) \to \xi_{\zeta_0^{(j-1)}}(a,s) = (s, \eta_{\zeta_0^{(j-1)}}(a,s))$ be parametrizations of pieces of $G_1(a)$ for $\nu - 1 \leq j \leq k$ and for each $a \in \omega$. Let $S(\zeta_0^{(j-1)}) \subset I_0$ be such that

$$\xi_{\zeta_0^{(j-1)}}(a_0, S(\zeta_0^{(j-1)})) = T^{1-\bar{g}}\left(a_0, \gamma\left(\zeta_0^{(j-1)}(a_0), \rho_0^{\theta j}\right)\right).$$

In the proof of Proposition 8.4, we shall demonstrate that

$$R_{\zeta_0^{(j-1)}}(a, S(\zeta_0^{(j-1)})) = T^{\bar{g}-1}\left(a, \xi_{\zeta_0^{(j-1)}}(a, S(\zeta_0^{(j-1)}))\right)$$

are almost horizontal and almost flat curves for every $a \in \omega$ and for every $j \in [\nu - 1, k]$. We may apply algorithms A and B on these curves.

To prove that $R_{\zeta_0^{(j-1)}}(a, S(\zeta_0^{(j-1)}))$ is close to $R_{\zeta_0^{(j-1)}}(a_0, S(\zeta_0^{(j-1)}))$ notice that, from $\left\| \xi_{\zeta_0^{(j-1)}} \right\|_{C^3(a,s)} \leq K_1$, it follows that $\left\| R_{\zeta_0^{(j-1)}} \right\|_{C^3(a,s)} \leq K_2^{\bar{g}}$. Then, I.H.7 leads to

$$\left\| R_{\zeta_0^{(j-1)}}(a_0, s) - R_{\zeta_0^{(j-1)}}(a, s) \right\| \leq K_2^{\bar{g}} |\omega| \leq K_2^{\bar{g}} e^{-\frac{1}{3} cj}$$

for every $a \in \omega$ and for every $s \in S(\zeta_0^{(j-1)})$. Furthermore, since $\bar{g}$ is the generation of $\zeta_0^{(j-1)}(a_0)$, it follows that $\bar{g} \leq \theta j$. Then $K_2^{\bar{g}} \leq \left(e^{\frac{1}{8} c} \rho_0^{\frac{1}{2}\theta} \right)^j$, provided that θ is small enough. Thus,

$$\left\| R_{\zeta_0^{(j-1)}}(a_0, s) - R_{\zeta_0^{(j-1)}}(a, s) \right\| \leq \left(\rho_0^{\frac{1}{2}\theta} e^{-\frac{1}{8}c} \right)^j << \rho_0^{\theta j},$$

whenever $j \geq N$ is sufficiently large. Hence, $R_{\zeta_0^{(j-1)}}(a, S(\zeta_0^{(j-1)}))$ are defined on

$$I_{\zeta_0^{(j-1)}} = \left(x_0^{(j-1)}(a_0) - \frac{1}{2}\rho_0^{\theta j}, x_0^{(j-1)}(a_0) + \frac{1}{2}\rho_0^{\theta j} \right),$$

where $\zeta_0^{(j-1)}(a_0) = \left(x_0^{(j-1)}(a_0), y_0^{(j-1)}(a_0) \right)$. Let us set

$$R_{\zeta_0^{(j-1)}}(a, S'(\zeta_0^{(j-1)}, a)) = \left\{ (x, y) \in R_{\zeta_0^{(j-1)}}(a, S(\zeta_0^{(j-1)})) : x \in I_{\zeta_0^{(j-1)}} \right\}$$

and let $\Psi_a : I_{\zeta_0^{(j-1)}} \to \mathbf{R}$ be the map whose graph is $R_{\zeta_0^{(j-1)}}(a, S'(\zeta_0^{(j-1)}, a))$. With these definitions we may accurately state I.H.7(a) in the following way: The analytic continuation of $\zeta_0^{(\nu-2)}(a_0)$ to ω is made by applying the implicit function theorem to

$$H(a, x) = \left\langle t \left(W_{a,m}^u, T_a(\Psi_a(x)) \right), f^{(\nu-2)}(a, T_a(\Psi_a(x))) \right\rangle,$$

where $(a, x) \in \omega \times I_{\zeta_0^{(\nu-2)}}$, $t \left(W_{a,m}^u, T_a(\Psi_a(x)) \right)$ is the tangent vector to $W_{a,m}^u$ at $T_a(\Psi_a(x))$ and $f^{(\nu-2)}$ is the maximally expanding vector by $DT_a^{(\nu-2)}(T_a(\Psi_a(x)))$.

Consequently, each $\zeta_0^{(\nu-2)}(a)$ belongs to $R_{\zeta_0^{(\nu-2)}}(a, S'(\zeta_0^{(\nu-2)}, a))$. Furthermore, since I.H.7 yields

$$\left\| \zeta_0^{(\nu-2)}(a) - \zeta_0^{(\nu-2)}(a_0) \right\| \leq b^{\frac{1}{40}} |\omega| \leq b^{\frac{1}{40}} e^{-\frac{2}{3} cv} << \rho_0^{\theta(\nu-1)},$$

we have

$$\gamma \left(\zeta_0^{(\nu-2)}(a), \frac{1}{4}\rho_0^{\theta(\nu-1)} \right) \subset R_{\zeta_0^{(\nu-2)}}(a, S'(\zeta_0^{(\nu-2)}, a)).$$

When we apply algorithm A to a critical approximation $\zeta_0^{(j-1)}(a)$, with $a \in E_j(\zeta_0^{(j-1)})$, we do it on the respective $\gamma \left(\zeta_0^{(j-1)}(a), \frac{1}{4}\rho_0^{\theta j} \right)$. When we apply algorithm B to a critical approximation $\zeta_0^{(j)}(a)$, $a \in E_j(\zeta_0^{(j-1)})$, we do it on the respective $\gamma\left(\zeta_0^{(j)}(a), \frac{1}{4}\rho_0^{\theta(j+1)}\right)$ and $R_{z_0^{(j)}}(a, S'(z_0^{(j)}, a))$. Notice that, if we inductively assume that

$$\gamma\left(\zeta_0^{(j-1)}(a), \frac{1}{4}\rho_0^{\theta j}\right) \subset R_{\zeta_0^{(j-1)}}(a, S'(\zeta_0^{(j-1)}, a)),$$

then, from (4.12), we obtain

$$\left\|\zeta_0^{(j)}(a) - \zeta_0^{(j-1)}(a)\right\| << \rho_0^{\theta(j+1)} \quad \text{and} \quad \left\|\zeta_0^{(j)}(a_0) - \zeta_0^{(j-1)}(a_0)\right\| << \rho_0^{\theta(j+1)}.$$

Therefore, $\gamma\left(\zeta_0^{(j)}(a), \frac{1}{4}\rho_0^{\theta(j+1)}\right) \subset R_{\zeta_0^{(j)}}(a, S'(\zeta_0^{(j)}, a))$. Applying algorithm B and from the new definition of critical sets, it also follows that $\left\|z_0^{(j)}(a) - \zeta_0^{(j)}(a)\right\| \leq b^{\frac{1}{10}g}$ for every $a \in E_j(\zeta_0^{(j-1)})$, where $g \in (\theta j, \theta(j+1)]$ is the generation of $z_0^{(j)}(a_0)$. Thus,

$$\left\|z_0^{(j)}(a) - \zeta_0^{(j)}(a)\right\| \leq b^{\frac{1}{10}\theta j} << \rho_0^{\theta(j+1)}$$

and

$$\left\|z_0^{(j)}(a_0) - \zeta_0^{(j)}(a_0)\right\| \leq b^{\frac{1}{10}\theta j} << \rho_0^{\theta(j+1)}.$$

So,

$$\gamma\left(z_0^{(j)}(a), \frac{1}{4}\rho_0^{\theta(j+1)}\right) \subset R_{z_0^{(j)}}(a, S'(z_0^{(j)}, a)).$$

Hence, for $\nu - 1 \leq j \leq k$ and for every $a \in E_j(\zeta_0^{(j-1)})$, not only the critical approximations $\zeta_0^{(j)}(a)$ have the same generation g, but also

$$T^{1-g}\left(a, \gamma\left(\zeta_0^{(j)}(a), \frac{1}{4}\rho_0^{\theta(j+1)}\right)\right) \subset S(\zeta_0^{(j)}) \times \mathbf{R}.$$

Proposition 8.3. *(Tangential position). Let n be a free return of $z_0 \in C_n$ and $\zeta_0 \in C_n$ be the binding point associated to z_n. Then,*

$$\left\|z_n - \eta^{[k]}\right\| \leq b^{\frac{3}{10}}d_n(z_0) \quad \text{and} \quad \left|ang\left(h_{n-1}(z_1), t(\gamma^{[k]}; \eta^{[k]})\right)\right| \leq b^{\frac{3}{10}}d_n(z_0).$$

Consequently, there exists a $C^2(b)$-curve $\tilde{\gamma}$ containing z_n and ζ_0, tangent to $\gamma^{[k]}$ at ζ_0 and to $h_{n-1}(z_1)$ at z_n.

Proof. From the first stage of the proof of Proposition 6.10, we get a sequence of critical approximations $\zeta_0^{(n-2)}(a_0), ..., \zeta_0^{(\mu)}(a_0)$ of generation g_k. We have proved that if tangential position does not hold, then we may contruct, by means of algorithm B, a critical approximation $\overline{\zeta}_0^{(\mu)}(a_0) \in C''_{\mu+1}$ of generation g_{k+1}, which contradicts the definition of binding point. Next, we shall check that the same takes place when the new definition of critical sets is considered. That is, we shall prove that, for every $a \in E_\mu(\zeta_0^{(\mu-1)}(a_0))$, we may also apply algorithm B to $\zeta_0^{(\mu-1)}(a)$ so as to construct a critical approximation $\overline{\zeta}_0^{(\mu)}(a)$ of generation g_{k+1}.

Let ω and $\tilde{\nu} \in \left[\frac{1}{2}(\mu+1), \mu+1\right]$ be respectively the interval and the iterate associated to $\zeta_0^{(\mu-1)}(a_0) \in C_\mu(a_0)$ given by I.H.7. Recall that $\mu < n$ satisfies $g_k \le \theta\mu < g_{k+1} \le \theta(\mu+1)$. From I.H.7(c),

$$|\omega| \le e^{-\frac{2}{3}\tilde{c}\tilde{\nu}} \le e^{-\frac{1}{3}c(\mu+1)} \le e^{-\frac{1}{3}c\theta^{-1}g_{k+1}}. \tag{8.5}$$

We shall show that the point $y_0 = z_{n-m_{k+1}}(a_0)$, used in the proof of Proposition 6.10, is $\left(\frac{1}{2}CK^{-1}\lambda_0\right)^8$-expanding up to time m_{k+1} for every $a \in \omega$. That is,

$$\left\|DT^j(a,y_0)(1,0)\right\| \ge \left(\frac{1}{2}CK^{-1}\lambda_0\right)^{8j} \quad \text{for } 1 \le j \le m_{k+1}. \tag{8.6}$$

Indeed, from Lemma 6.4, $\left\|DT^j(a_0, z_{n-m_i}(a_0))(1,0)\right\| \ge (CK^{-1}\lambda_0)^{8j}$ for $1 \le j \le m_i$. Now, as in (3.6),

$$\left\|DT^j(a,y_0)(1,0) - DT^j(a_0,y_0)(1,0)\right\| \le K^{3j}|\omega| \le K^{3j}e^{-\frac{1}{3}c\theta^{-1}g_{k+1}}.$$

Hence, from Lemma 6.4, $\left\|DT^j(a,y_0)(1,0)\right\| \ge (CK^{-1}\lambda_0)^{8j} - K^{3j}e^{-\frac{1}{3}c\theta^{-1}g_{k+1}}$. Furthermore, if $j \le m_{k+1} < g_{k+1}$, then $K^{3j}e^{-\frac{1}{3}c\theta^{-1}g_{k+1}} < \frac{1}{2}(CK^{-1}\lambda_0)^{8j}$, provided that $\theta = \theta(b)$ is small enough. Therefore (8.6) is proved and so, for every $a \in \omega$, there exist integral curves $\Gamma^{k+1}(a)$ and $\Gamma^k(a)$ containing y_0 and $\tilde{y}_0 = z_{n-m_k}(a_0)$, respectively. Next, we shall extend the arguments used in the second stage of the proof of Proposition 6.10 to each $a \in E_\mu\left(\zeta_0^{(\mu-1)}(a_0)\right)$.

Let $\eta_0^{[k]}(a) = \Gamma^k(a) \cap G_1(a)$, $\eta_0^{[k+1]}(a) = \Gamma^{k+1}(a) \cap G_1(a)$, $\gamma_0^{[k]}(a) = \gamma\left(\eta_0^{[k]}(a), \rho_0^{m_k}\right)$ and $\gamma_0^{[k+1]}(a) = \gamma\left(\eta_0^{[k+1]}(a), \rho_0^{m_{k+1}}\right)$. First, notice that from Remark 3.11, we obtain $\left\|e^{(m_k)}(a,z) - e^{(m_k)}(a_0,z)\right\| \le const\sqrt{b}\,|\omega|$. Therefore, (8.5) and Proposition 4.13 give

$$\left\|\eta_0^{[k]}(a) - \eta_0^{[k]}(a_0)\right\| \le const\sqrt{b}\,|\omega| \le const\sqrt{b}e^{-\frac{1}{3}c\theta^{-1}g_{k+1}} \ll \rho_0^{g_k}. \tag{8.7}$$

Let S be a fixed interval such that, for every $a \in \omega$, the function $s \in S \to \xi(a,s) = (s, \eta(a,s))$ is a parametrization of $G_1(a)$ with $\xi(a_0, S) = \gamma_0^{[k]}(a_0)$. First, we shall show that the curves of generation g_k given by $\zeta(a,s) = T^{m_k}(a, \xi(a,s)) = (x(a,s), y(a,s))$ are almost horizontal and almost flat. To this end, notice that (8.5) leads to

$$\left\|DT^{m_k}(a, \xi(a,s))\frac{\partial_s\xi(a,s)}{\|\partial_s\xi(a,s)\|} - DT^{m_k}(a_0, \xi(a_0,s))\frac{\partial_s\xi(a_0,s)}{\|\partial_s\xi(a_0,s)\|}\right\| \le$$

$$\leq K^{3g_k} |\omega| < \left(K^3 e^{-\frac{1}{3}c\theta^{-1}} \right)^{g_k} . \tag{8.8}$$

Thus, from Corollary 6.7,

$$\| DT^{m_k} (a, \xi(a,s)) \partial_s \xi(a,s) \| \geq \frac{C_0}{5} \| \partial_s \xi(a,s) \| . \tag{8.9}$$

Let us write $\eta^{[k]}(a) = T^{m_k}(a, \eta_0^{[k]}(a))$ and $\eta^{[k+1]}(a) = T^{m_{k+1}}(a, \eta_0^{[k+1]}(a))$. From (8.7) and (8.9), it follows that

$$\zeta(a, S) \supset \gamma \left(\eta^{[k]}(a), \frac{1}{10} C_0 \rho_0^{m_k} \right) \supset \gamma \left(\eta^{[k]}(a), 5\rho_0^{g_k} \right) .$$

Furthermore, Lemma 6.6 gives $|\text{slope} \, (DT^{m_k}(a_0, \xi(a_0, s)) \partial_s \xi(a_0, s))| < \sqrt[4]{b}$. Therefore, by using (8.8),

$$|\text{ang} \, (DT^{m_k}(a, \xi(a,s)) \partial_s \xi(a,s), DT^{m_k}(a_0, \xi(a_0, s)) \partial_s \xi(a_0, s))| << 1.$$

Then, for every $(a, s) \in \omega \times S$, we get

$$|\text{slope} \, (DT^{m_k}(a, \xi(a,s)) \partial_s \xi(a,s))| < \frac{1}{100}. \tag{8.10}$$

Moreover, since $DT^{m_k}(a, \xi(a,s)) \partial_s \xi(a,s) = \partial_s \zeta(a,s) = (\partial_s x(a,s), \partial_s y(a,s))$, from (8.9) and (8.10) it follows that

$$|\partial_s x(a,s)| \geq \frac{C_0}{10}. \tag{8.11}$$

On the other hand, since $\| \xi \|_{C^3(a,s)} \leq K_1$, we deduce $\| \zeta \|_{C^3(a,s)} \leq K_2^{g_k}$, where $K_2 = \max \{ K, K_1 \}$. Then, denoting by

$$t(a,s) = \frac{\partial_s y(a,s)}{\partial_s x(a,s)}$$

the slope of the tangent vector to $\zeta(a,s)$, from (8.11), we obtain

$$\| t \|_{C^2(a,s)} \leq K_3^{g_k}. \tag{8.12}$$

Now, Lemma 6.6 and (8.5) imply

$$|t(a,s)| \leq \sqrt[4]{b} + K_3^{g_k} |\omega| \leq \sqrt[4]{b} + \left(K_3 e^{-\frac{1}{3}c\theta^{-1}} \right)^{g_k} < \frac{1}{100} \tag{8.13}$$

and

$$|\partial_s t(a,s)| < \frac{1}{100}. \tag{8.14}$$

Therefore, $\gamma\left(\eta^{[k]}(a), 5\rho_0^{g_k}\right)$ is an almost horizontal and almost flat curve. Clearly, the same holds for $\gamma\left(\eta^{[k+1]}(a), 5\rho_0^{g_{k+1}}\right)$. Furthermore, the bounds obtained in (8.13) and (8.14) allow us to apply algorithm B.

For every $a \in E_\mu(\zeta_0^{(\mu-1)}(a))$, let $\zeta_0^{(\mu)}(a) = (x_0(a), y_0(a))$ be the critical approximation obtained from $\zeta_0^{(\mu-1)}(a)$ by means of algorithm A. We shall demonstrate that $\gamma\left(\eta^{[k]}(a), 5\rho_0^{g_k}\right)$ and $\gamma\left(\eta^{[k+1]}(a), 5\rho_0^{g_{k+1}}\right)$ are defined on $\left(x_0(a) - \rho_0^{g_{k+1}}, x_0(a) + \rho_0^{g_{k+1}}\right)$ for every $a \in E_\mu(\zeta_0^{(\mu-1)})$. To this end, notice that I.H.7 implies $\left\| D_a\left(\zeta_0^{(\tilde\nu-2)}(a)\right)\right\| \le 2b^{\frac{1}{30}}$ for every $a \in \omega$. Furthermore, from (8.4) and (4.12),

$$\left\|\zeta_0^{(\mu)}(a) - \zeta_0^{(\tilde\nu-2)}(a)\right\| \le 2\left((Kb)^{\tilde\nu-2} + b^{\frac{1}{10}\theta(\tilde\nu-1)}\right).$$

Hence,

$$\left\|\zeta_0^{(\mu)}(a) - \zeta_0^{(\mu)}(a_0)\right\| \le 4\left((Kb)^{\frac{1}{4}\theta^{-1}g_{k+1}} + b^{\frac{1}{40}g_{k+1}}\right) + 2b^{\frac{1}{30}}e^{-\frac{1}{3}c\theta^{-1}g_{k+1}}.$$

Moreover, from Proposition 6.10, it follows that $\left\|\zeta_0^{(\mu)}(a_0) - \eta^{[k]}(a_0)\right\| \le 3b^{\frac{1}{20}g_k}$ and, from (8.7),

$$\left\|\eta^{[k]}(a_0) - \eta^{[k]}(a)\right\| \le const\left(Ke^{-\frac{1}{3}c\theta^{-1}}\right)^{g_{k+1}}.$$

Consequently,

$$\left|x_0(a) - \eta_x^{[k]}(a)\right| \le \left\|\zeta_0^{(\mu)}(a) - \eta^{[k]}(a)\right\| << \frac{1}{5}\rho_0^{g_{k+1}}$$

and, in the same way,

$$\left|\eta_x^{[k]}(a) - \eta_x^{[k+1]}(a)\right| \le \left\|\eta^{[k]}(a) - \eta^{[k+1]}(a)\right\| << \frac{1}{5}\rho_0^{g_{k+1}}.$$

Notice also that, from

$$\left\|\eta^{[k]}(a) - \zeta_0^{(\mu)}(a_0)\right\| \le 3b^{\frac{1}{20}g_k} + const\left(Ke^{-\frac{1}{3}c\theta^{-1}}\right)^{g_{k+1}} << \frac{1}{4}\rho_0^{\theta(\mu+1)},$$

it follows that $\zeta_0^{(\mu)}(a) \in \gamma(\eta^{[k]}(a), 5\rho_0^{g_k})$. Hence, $\eta^{[k]}(a) \in R_{\zeta_0^{(\mu)}}(a, S'(\zeta_0^{(\mu)}, a))$, where $R_{\zeta_0^{(\mu)}}(a, S'(\zeta_0^{(\mu)}, a))$ is the family of curves defined just before this proposition. Furthermore, we already know that $\zeta_0^{(\mu)}(a) \in R_{\zeta_0^{(\mu)}}(a, S'(\zeta_0^{(\mu)}, a))$.

Therefore, let $z_0(a) = \gamma\left(\eta^{[k+1]}(a), \rho_0^{g_{k+1}}\right) \cap \{x = x_0(a)\}$ for every $a \in E_\mu(\zeta_0^{(\mu-1)})$. In order to check, as in (6.14), that $d(a) = \left\|\zeta_0^{(\mu)}(a) - z_0(a)\right\| < const\sqrt{b}\,(const\,b)^{m_k}$, we repeat the argument used in Proposition 6.10. On the one hand,

$$length\left(T^{m_{k+1}-m_k}\left(a, \gamma_0^{[k+1]}(a)\right)\right) << \frac{1}{10}\rho_0^{m_k}$$

remains true for every $a \in \omega$. On the other, setting

$$\eta_{m_{k+1}-m_k}^{[k+1]}(a) = T^{m_{k+1}-m_k}(a, \eta_0^{[k+1]}(a)),$$

we also obtain

$$dist\left(\eta_{m_{k+1}-m_k}^{[k+1]}(a), \Gamma^k(a)\right) \leq$$

$$\leq \left\|\eta_{m_{k+1}-m_k}^{[k+1]}(a) - \eta_{m_{k+1}-m_k}^{[k+1]}(a_0)\right\| + dist\left(\eta_{m_{k+1}-m_k}^{[k+1]}(a_0), \Gamma^k(a)\right).$$

Now, from (8.7), $\left\|\eta_{m_{k+1}-m_k}^{[k+1]}(a) - \eta_{m_{k+1}-m_k}^{[k+1]}(a_0)\right\| << \frac{1}{10}\rho_0^{m_k}$ and from Proposition 6.10, $dist\left(\eta_{m_{k+1}-m_k}^{[k+1]}(a_0), \Gamma^k(a)\right) << \frac{1}{10}\rho_0^{m_k}$.

Thus, setting $z_{0,0}(a) = T^{-m_k}(a, z_0(a)) \in T^{m_{k+1}-m_k}(a, \gamma_0^{[k+1]}(a))$, then we may compute the respective integral curve $\Gamma_0^{[k]}(a)$ which intercepts $G_1(a)$ at a point $\overline{\zeta}_{0,0}(a)$. By using Gronwall's Lemma, we get $\left\|\eta_0^{[k]}(a) - \overline{\zeta}_{0,0}(a)\right\| < \frac{1}{4}\rho_0^{m_k}$. So, $T^{m_k}(a, \overline{\zeta}_{0,0}(a)) \in \zeta(a, S)$ and $\left\|T^{m_k}(a, \overline{\zeta}_{0,0}(a)) - z_0(a)\right\| \leq const\sqrt{b}\,(const\ b)^{m_k}$. Since $|slope\ \zeta(a, S)| \leq \frac{1}{100}$, we conclude that $d(a) < const\sqrt{b}\,(const\ b)^{m_k}$.

Finally, we may repeat, in a straighforward way, the arguments used in Proposition 6.10 to find, for each $a \in E_\mu(\zeta_0^{(\mu-1)})$, a critical approximation $\overline{\zeta}_0^{(\mu)}(a) \in \gamma\left(z_0(a), \rho_0^{g_{k+1}}\right)$ such that $\left\|\overline{\zeta}_0^{(\mu)}(a) - \zeta_0^{(\mu)}(a)\right\| \leq b^{\frac{1}{10}g_{k+1}}$. Then the proposition is proved. $\square$

In I.H.7 we assumed that E_k was constructed for every $1 \leq k \leq n-1$. Next, we shall show how E_n is obtained from E_{n-1}.

Let $a_0 \in E_{n-1}$ and $z_0^{(n-1)} \in C_n(a_0)$. Let $\zeta_0^{(n-2)} \in C_{n-1}(a_0)$ be the unique critical approximation of order $n-2$ related to $z_0^{(n-1)}$ by means of either algorithm A or B. Let $\nu \in \left[\frac{1}{2}n, n\right]$ and $\omega \subset \Omega_0$ be respectively the iterate and the interval associated to $\zeta_0^{(n-2)}$ given by I.H.7. Recall that $z_0^{(n-1)}$ is defined for every $a \in E_{n-1}(\zeta_0^{(n-2)})$.

Proposition 8.4. *If $\widetilde{g}$ is the generation of $z_0^{(n-1)}$, then*

$$\left\|D_a\left(z_0^{(n-1)}(a)\right)\right\| \leq \sum_{i=1}^{\widetilde{g}} b^{\frac{1}{30}i} + \sum_{j=1}^{n} b^{\frac{1}{3}j}$$

for every $a \in E_{n-1}(\zeta_0^{(n-2)})$.

Proof. We shall prove the result by recurrence. For each $a \in \omega$, let $\widetilde{\zeta}_0^{(\nu-1)}(a)$ be the critical approximation obtained from $\zeta_0^{(\nu-2)}(a)$ by applying algorithm A. Let $z_0^{(\nu-1)}(a)$ be the possible critical approximation obtained, for each $a \in \omega$, from $\zeta_0^{(\nu-2)}(a)$ by applying algorithm B. From (4.12), it follows that

$$\left\|\widetilde{\zeta}_0^{(\nu-1)}(a) - \zeta_0^{(\nu-2)}(a)\right\| \leq (Kb)^{\nu-2}$$

and, from the definition of C_ν'',

$$\left\| z_0^{(\nu-1)}(a) - \widetilde{\zeta}_0^{(\nu-1)}(a) \right\| \le b^{\frac{1}{10}g'},$$

where g' is the generation of $z_0^{(\nu-1)}(a)$. In accordance with these bounds, we shall check that $\left\| D_a\left(z_0^{(\nu-1)}(a)\right) \right\| \le \sum_{i=1}^{g'} b^{\frac{1}{30}i} + \sum_{j=1}^{\nu} b^{\frac{1}{3}j}$ and $\left\| D_a\left(\widetilde{\zeta}_0^{(\nu-1)}(a)\right) \right\| \le \sum_{i=1}^{\overline{g}} b^{\frac{1}{30}i} + \sum_{j=1}^{\nu} b^{\frac{1}{3}j}$ where $\overline{g}$ is the generation of $\zeta_0^{(\nu-2)}$ and hence, the generation of $\widetilde{\zeta}_0^{(\nu-1)}$. To this end, we shall demonstrate that, for every $a \in \omega$,

$$\left\| D_a^2(\widetilde{\zeta}_0^{(\nu-1)}(a)) - D_a^2(\zeta_0^{(\nu-2)}(a)) \right\| \le const^{\overline{g}}$$

$$\left\| D_a^2(z_0^{(\nu-1)}(a)) - D_a^2(\widetilde{\zeta}_0^{(\nu-1)}(a)) \right\| \le const^{g'}.$$

(8.15)

Then, the proposition follows from the following result:

Lemma 8.5. *(Hadamard's Lemma). Let $f : [a_1, a_2] \subset \mathbf{R} \to \mathbf{R}$ be a C^2 map such that $|f(x)| \le M_0$ and $|f''(x)| \le M_2$ for every $x \in [a_1, a_2]$. If $a_2 - a_1 \ge 2\sqrt{M_0}$, then*

$$|f'(x)| \le \sqrt{M_0}\,(1 + M_2)$$

for every $x \in [a_1, a_2]$.

First, we shall show that Lemma 8.5 implies the proposition. We consider the two possible cases depending on whether algorithm A or B is used:

1. In accordance with $\left\| \widetilde{\zeta}_0^{(\nu-1)}(a) - \zeta_0^{(\nu-2)}(a) \right\| \le (Kb)^{\nu-2}$, the first inequality in (8.15) and the inequalities $|\omega| \ge K^{-\frac{3}{2}\nu} \ge 2\,(Kb)^{\frac{1}{2}(\nu-2)}$ arising from I.H.7(c), we may apply Lemma 8.5 to conclude

$$\left\| D_a(\widetilde{\zeta}_0^{(\nu-1)}(a)) - D_a(\zeta_0^{(\nu-2)}(a)) \right\| \le$$

$$\le (Kb)^{\frac{1}{2}(\nu-2)}\left(1 + const^{\overline{g}}\right) \le (Kb)^{\frac{1}{2}(\nu-2)}\left(1 + const^{\theta\nu}\right) \le b^{\frac{1}{3}\nu}.$$

Finally, I.H.7(a) gives $\left\| D_a(\widetilde{\zeta}_0^{(\nu-1)}(a)) \right\| \le \sum_{i=1}^{\overline{g}} b^{\frac{1}{30}i} + \sum_{j=1}^{\nu} b^{\frac{1}{3}j}$.

2. Since g' and $\overline{g}$ are the generations of $z_0^{(\nu-1)}$ and $\zeta_0^{(\nu-2)}$, respectively, we get $\overline{g} \le \theta(\nu - 1) < g' \le \theta\nu$. From $\left\| z_0^{(\nu-1)}(a) - \widetilde{\zeta}_0^{(\nu-1)}(a) \right\| \le b^{\frac{1}{10}g'}$, the second inequality in (8.15) and $|\omega| \ge K^{-\frac{3}{2}\nu} \ge K^{-2(\nu-1)} \ge K^{-2\theta^{-1}g'} > 2b^{\frac{1}{20}g'}$, we obtain

$$\left\| D_a(z_0^{(\nu-1)}(a)) - D_a(\widetilde{\zeta}_0^{(\nu-1)}(a)) \right\| \le \left(b^{\frac{1}{20}}(1 + const) \right)^{g'} < b^{\frac{1}{30}g'}.$$

Thus, from the previous case, $\left\| D_a(z_0^{(\nu-1)}(a)) \right\| \le \sum_{i=1}^{g'} b^{\frac{1}{30}i} + \sum_{j=1}^{\nu} b^{\frac{1}{3}j}$.

Therefore, the proof of the proposition ends by checking (8.15). In fact, the following estimates do not depend on the choice of the critical approximation we are considering. Denote by $\zeta_0^{(\nu-1)}$ either $\tilde{\zeta}_0^{(\nu-1)}$ or $z_0^{(\nu-1)}$ and let g be its generation. Since $\zeta_0^{(\nu-1)}(a_0) \in C_\nu(a_0)$, I.H.3 allows us to claim that $T^{1-g}\left(a_0, \gamma(\zeta_0^{(\nu-1)}(a_0), \rho_0^{\theta\nu})\right)$ is a C^2-curve contained in $G_1(a_0) \cap (I_0 \times \mathbf{R})$. Moreover, its tangent vectors are $\frac{1}{4}C_0$ expanded by $DT_{a_0}^{g-1}$. Let $S \subset I_0$ be a real interval and $\xi(a, s) = (s, \eta(a, s))$ be a parametrization of $G_1(a)$ such that $\xi(a_0, S) = T^{1-g}\left(a_0, \gamma(\zeta_0^{(\nu-1)}(a_0), \rho_0^{\theta\nu})\right)$. Notice also that

$$|\omega| \le e^{-\frac{2}{3}c\nu} \le e^{-\frac{1}{3}cn} \le e^{-\frac{1}{3}c\theta^{-1}\tilde{g}} \le e^{-\frac{1}{3}c\theta^{-1}g}. \tag{8.16}$$

Therefore, setting $\zeta(a, s) = T^{g-1}(a, \xi(a, s)) = (x(a, s), y(a, s))$ and repeating the arguments used in Proposition 8.3, we obtain $|\text{slope}\,(DT^{g-1}(a, \xi(a, s))\partial_s\xi(a, s))| \le \frac{1}{100}$ and $\|DT^{g-1}(a, \xi(a, s))\partial_s\xi(a, s)\| \ge \frac{1}{5}C_0$. Hence,

$$|\partial_s x(a, s)| \ge \frac{C_0}{10}. \tag{8.17}$$

In particular, $\zeta(a, s)$ are almost horizontal and we may parametrize them by $\zeta(a, s) = (x(a, s), y(a, x(a, s)))$. Furthermore, $\|\zeta\|_{C^3(a, s)} \le K_2^g$ and (8.17) lead to

$$\|y\|_{C^3(a, x)} \le K_3^g. \tag{8.18}$$

On the other hand, $\zeta_0^{(\nu-1)}(a) = (x_0(a), y(a, x_0(a)))$ is the solution of

$$H(a, x) = \left\langle t\left(a, T(a, x, y(a, x))\right), f^{(\nu-1)}\left(a, T(a, x, y(a, x))\right)\right\rangle = 0. \tag{8.19}$$

Next, we shall prove that

$$|x_0''(a)| \le const^g \tag{8.20}$$

for every $a \in \omega$. To this end, we derive (8.19). Since $z_0^{(n-1)}(a_0) \in C_n(a_0)$, then from I.H.2 we obtain $|\partial_x y(a_0, x)| < \sqrt[4]{b}$ and $|\partial_x^2 y(a_0, x)| < \sqrt[4]{b}$. Therefore, (8.18) and (8.16) yield

$$|\partial_x y(a, x)| \le \sqrt[4]{b} + K_3^g\,|\omega| \le \sqrt[4]{b} + \left(K_3 e^{-\frac{1}{3}c\theta^{-1}}\right)^g < \frac{1}{100} \text{ and } \left|\partial_x^2 y(a, x)\right| \le \frac{1}{100} \tag{8.21}$$

for every $a \in \omega$. Furthermore, from Proposition 1.3, we have

$$\frac{5}{2} \le \|D_x t(a, T(a, x, y(a, x)))\| \le const \tag{8.22}$$

and, from (8.18),

$$\|D_a t(a, T(a, x, y(a, x)))\| \le const^g, \quad \left\|D_x^2 t(a, T(a, x, y(a, x)))\right\| \le const^g,$$

$$\left\|D^2_{a,x}t(a,T(a,x,y(a,x)))\right\| \le const^g, \quad \left\|D^2_a t(a,T(a,x,y(a,x)))\right\| \le const^g. \quad (8.23)$$

Setting $\tilde{z} = \tilde{z}(a,x) = (a,T(a,x,y(a,x)))$, Propositions 3.10 and 3.13 give

$$\left\|D_{\tilde{z}}f^{(\nu-1)}(\tilde{z})\right\| \le const\sqrt{b} \text{ and } \left\|D^2_{\tilde{z}}f^{(\nu-1)}(\tilde{z})\right\| \le const\sqrt{b}.$$

Moreover, from (8.21), we get $\|D_x\tilde{z}(a,x)\| \le const$ and from (8.18)

$$\|D_a\tilde{z}(a,x)\| \le const^g, \ \left\|D^2_x\tilde{z}(a,x)\right\| \le const^g,$$

$$\left\|D^2_{a,x}\tilde{z}(a,x)\right\| \le const^g, \ \left\|D^2_a\tilde{z}(a,x)\right\| \le const^g.$$

Therefore, $\left\|D_x f^{(\nu-1)}(\tilde{z}(a,x))\right\| \le const\sqrt{b}$ and

$$\left\|D_a f^{(\nu-1)}(\tilde{z}(a,x))\right\| \le const^g, \ \left\|D^2_x f^{(\nu-1)}(\tilde{z}(a,x))\right\| \le const^g,$$

$$\left\|D^2_{a,x} f^{(\nu-1)}(\tilde{z}(a,x))\right\| \le const^g, \ \left\|D^2_a f^{(\nu-1)}(\tilde{z}(a,x))\right\| \le const^g.$$

So, from (8.22) and the fact that $D_x t$ and $f^{(\nu-1)}$ are nearly horizontal, we conclude that $2 \le \|\partial_x H(a,x)\| \le const$. Furthermore, the previous estimates and (8.23) imply

$$\|\partial_a H(a,x)\| \le const^g, \ \left\|\partial^2_x H(a,x)\right\| \le const^g,$$

$$\left\|\partial^2_{a,x} H(a,x)\right\| \le const^g, \ \left\|\partial^2_a H(a,x)\right\| \le const^g.$$

Therefore, (8.20) is proved.

Notice also that $|x'_0(a)| \le const^g$. Thus, (8.18) leads to $\left\|D^2_a(\zeta_0^{(\nu-1)}(a))\right\| \le const^g$. In the same way we get $\left\|D^2_a(\zeta_0^{(\nu-2)}(a))\right\| \le const^g$. Then (8.15) holds for every $a \in \omega$.

The proof of the proposition ends by recurrence, bearing in mind that, from Lemma 8.12, each connected component ω_{n-1} of $E_{n-1}(\zeta_0^{(n-2)})$ satisfies

$$K^{-\frac{3}{2}n} \le |\omega_{n-1}| \le e^{-\frac{2}{3}cn}.$$

Thus, we may apply (8.16) and Hadamard's Lemma for each ω_{n-1}. $\square$

Hence, given $a_0 \in E_{n-1}$ and $z_0^{(n-1)} \in C_n(a_0)$, we may consider, for each $a \in E_{n-1}(\zeta_0^{(n-2)})$, the critical approximation $z_0^{(n-1)}(a)$ defined from $\zeta_0^{(n-2)}(a)$ by using either algorithm A or B. We shall define a set of parameter values $E_n(z_0^{(n-1)}) \subset E_{n-1}(\zeta_0^{(n-2)})$ for which $z_1^{(n-1)}$ is e^c-expanding up to time n. $E_n(z_0^{(n-1)})$ will be defined in the same way as Ω_n was in Chapter 2. We need to define partitions P_j of $E_j(\zeta_0^{(j-1)})$ for $\nu - 1 \le j \le n$. Notice that $\zeta_0^{(j-1)}(a)$ is the critical approximation obtained from

$\zeta_0^{(\nu-2)}(a)$ by successively applying algorithms A or B. The choice of the algorithm in each stage is determined by the constructive process for getting $\zeta_0^{(n-2)}(a_0)$ from $\zeta_0^{(\nu-2)}(a_0)$.

For $j = \nu - 1$, we take $E_{\nu-1}(\zeta_0^{(\nu-2)}) = \omega$ and $P_{\nu-1} = \{\omega\}$. Next, we proceed by induction to obtain $E_j(\zeta_0^{(j-1)})$ from $E_{j-1}(\zeta_0^{(j-2)})$ in the following way:

Let $\overline{\omega} \subset E_{j-1}(\zeta_0^{(j-2)})$, $\overline{\omega} \in P_{j-1}$. We say that j is a return situation of $\overline{\omega}$ if there exists $a \in \overline{\omega}$ such that j is a return of $\zeta_0^{(j-1)}(a)$. If j does not belong to the binding period associated to any previous return of $\overline{\omega}$, then j is called a free return situation. Clearly, we must assume that the binding periods do not depend on the chosen parameter. To this end, we shall slightly modify the definition of binding periods later on.

We distinguish between two types of free return situations. Let us assume that j is a free return situation of $\overline{\omega} \subset E_{j-1}(\zeta_0^{(j-2)})$, $\overline{\omega} \in P_{j-1}$ and let $I_{m,k}$ be the intervals introduced in Chapter 2. Let $\overline{a} \in \overline{\omega}$ be such that $\zeta_0^{(j-1)}(\overline{a})$ has a (free) return at time j and let $\tilde{z}_0(\overline{a})$ be the binding point associated to $\zeta_j^{(j-1)}(\overline{a})$ as in Chapter 6. Let $\Delta(\overline{\omega}) = \left\{ \zeta_{j,x}^{(j-1)}(a) - \tilde{z}_{0,x}(\overline{a}) : a \in \overline{\omega} \right\}$, where $\left(\zeta_{j,x}^{(j-1)}(a), \zeta_{j,y}^{(j-1)}(a) \right) = \zeta_j^{(j-1)}(a)$ and $(\tilde{z}_{0,x}(\overline{a}), \tilde{z}_{0,y}(\overline{a})) = \tilde{z}_0(\overline{a})$. We shall prove that $\zeta_j^{(j-1)}(\overline{\omega}) = \bigcup_{a \in \overline{\omega}} \zeta_j^{(j-1)}(a)$ is an almost horizontal and almost flat curve. Hence, the map $a \in \overline{\omega} \to \Delta(a) = \zeta_{j,x}^{(j-1)}(a) - \tilde{z}_{0,x}(\overline{a})$ is a homeomorphism and consequently, $\Delta(\overline{\omega})$ is an interval. We say that j is an essential return situation if and only if $\Delta(\overline{\omega})$ contains an interval $I_{r,k}$. Otherwise, j is called an inessential return situation.

Now, if j is not an essential return situation, we take $\overline{\omega} \subset E_j(\zeta_0^{(j-1)})$. Otherwise, we split $\overline{\omega}$ into the disjoint union of two sets, ω' and ω'', in such a way that $a \in \omega''$ if and only if $|\Delta(a)| > \delta$.

Therefore, ω'' has at most two connected components. We also split ω' into intervals $\omega_{r,k}$ such that if $a \in \omega_{r,k}$, then $\Delta(a) \in I_{r,k}$. As in dimension one, $I_{r,k}$ must be completely contained in $\Delta(\omega_{r,k})$. Thus, we join $\omega_{r,k}$ to either $\omega_{r,k-1}$ or $\omega_{r,k+1}$, if necessary.

From (BA), we remove those $\omega_{r,k}$ such that $|r| \geq \alpha j$ and denote by $E_j'(\zeta_0^{(j-1)})$ the union of the remaining elements of $E_{j-1}(\zeta_0^{(j-2)})$. We shall prove that there exists a positive constant ϵ such that

$$m\left(E_{j-1}(\zeta_0^{(j-2)}) \setminus E_j'(\zeta_0^{(j-1)}) \right) \leq e^{-\frac{1}{3}\epsilon\alpha j} m\left(E_{j-1}(\zeta_0^{(j-2)}) \right). \qquad (8.24)$$

Let $P_j' = \left\{ \overline{\omega} \in P_{j-1} : \overline{\omega} \subset E_j'(\zeta_0^{(j-1)}) \right\}$. Now, we remove those $\tilde{\omega}' \in P_j'$ which do not satisfy (FA), that is, those $\tilde{\omega}'$ such that

$$F_j(\tilde{\omega}') < (1 - \alpha)j. \tag{8.25}$$

To this end, as in Chapter 2, we introduce the following concepts: Let j be an essential return situation of $\bar{\omega}$. If $length\left(\zeta_j^{(j-1)}(\bar{\omega})\right) > \sqrt{\delta}$, then j is said to be an escape situation, in which case, at least one of the connected components $\tilde{\omega}$ of ω'' satisfies $length\left(\zeta_j^{(j-1)}(\tilde{\omega})\right) > \frac{1}{3}\sqrt{\delta}$. We say that $\tilde{\omega}$ is an escape component. Escape situations will be crucial to prove that

$$m\left(E_j'(\zeta_0^{(j-1)}) \setminus E_j(\zeta_0^{(j-1)})\right) \le e^{-\frac{1}{100}c\alpha j}|\Omega_0|. \tag{8.26}$$

Once the elements $\tilde{\omega}' \in P_j'$ which do not satisfy (FA) are excluded, the remaining $\omega_{r,k}$ belong to P_j and we say that j is an essential return of $\omega_{r,k}$. Let $p(a)$ be the length of the binding period associated to the return j of $\zeta_0^{(j-1)}(a)$. If we define

$$p(\omega_{r,k}) = \min_{a \in \omega_{r,k}} p(a),$$

then $p(\omega_{r,k})$ also verifies all the estimates proved for binding periods. To check this, notice that there exist constants A_1 and A_2, which only depend on K, c and β, such that $A_1 \log d_j\left(\zeta_0^{(j-1)}(a)\right)^{-1} \le p(a) \le A_2 \log d_j\left(\zeta_0^{(j-1)}(a)\right)^{-1}$. Furthermore, if E_{j-1} denotes the set of parameters for which every critical approximation of order $j-2$ is e^c-expanding up to time $j-1$, then, for each $a \in (\cup \omega_{r,k}) \cap E_{j-1}$, the binding point associated to $\zeta_j^{(j-1)}(a)$ is $\tilde{z}_0(a)$, where $\tilde{z}_0(a)$ is the analytic continuation of the critical approximation $\tilde{z}_0(\bar{a})$ used for constructing the sets $\omega_{r,k}$. Therefore, from I.H.7,

$$\|\tilde{z}_0(\bar{a}) - \tilde{z}_0(a)\| \le b^{\frac{1}{40}}|\omega| \le b^{\frac{1}{40}}e^{-\frac{2}{3}c\nu} \le b^{\frac{1}{40}}e^{-\frac{1}{3}cj} << e^{-\alpha j}.$$

In particular, the choice of the parameter $\bar{a}$ does not play an important role in the above construction. Furthermore, using also the tangential position, we obtain

$$\log d_j(\zeta_0^{(j-1)}(a))^{-1} = \log \left\|\zeta_j^{(j-1)}(a) - \tilde{z}_0(a)\right\|^{-1} \approx |r|.$$

Finally, $p(a) \approx p(a')$ for every $a, a' \in \omega_{r,k}$ and thus $p(a) \approx p(\omega_{r,k})$.

It will also be necessary to define the folding period as constant on $\omega_{r,k}$. Once again, since the folding periods essentially depend on $\log d_j(\zeta_0^{(j-1)}(a))^{-1}$, all the estimates stated in Chapter 5 for folding periods still hold.

Now, to check (8.24) and (8.26) we need to prove a distortion result between the behaviour of $\|\omega_{j-1}(a)\| = \left\|DT^{j-1}(a, \zeta_1^{(j-1)}(a))(1,0)\right\|$ and $\|\omega_{j-1}(a')\|$ for every

$a, a' \in \overline{\omega}$ and for every $\overline{\omega} \in P_{j-1}$, $\overline{\omega} \subset E_{j-1}(\zeta_0^{(j-2)})$.

Proposition 8.6. *There exists a constant $C_3 = C_3(\lambda, K, \alpha, \beta) > 0$ such that if j is a free return situation of $\overline{\omega} \in P_{j-1}$, $\overline{\omega} \subset E_{j-1}(\zeta_0^{(j-2)})$ and $\zeta_j^{(j-1)}(\overline{\omega}) \subset B_{\sqrt[4]{\delta}}$, then*

$$\frac{\|\omega_{j-1}(a)\|}{\|\omega_{j-1}(a')\|} \leq C_3 \quad and \quad |ang(\omega_{j-1}(a), \omega_{j-1}(a'))| \leq \sqrt[4]{b}$$

for every $a, a' \in \overline{\omega}$.

First, we need a bound distortion result between $\left\|\zeta_j'(a)\right\| = \left\|D_a(\zeta_j^{(j-1)}(a))\right\|$ and $\|\omega_{j-1}(a)\|$, that is, we need to state an equivalent bidimensional result to Proposition 2.3:

Proposition 8.7. *There exists a constant $A > 1$ such that if $\Omega_0 = \Omega_0(\lambda)$ is sufficiently close to $a(\lambda)$, then*

$$\frac{1}{A} \leq \frac{\left\|\zeta_j'(a)\right\|}{\|\omega_{j-1}(a)\|} \leq A$$

for every $a \in \overline{\omega} \in P_{j-1}$, $\overline{\omega} \subset E_{j-1}(\zeta_0^{(j-2)})$. Moreover, if j is a free iterate, then

$$\left|ang\left(\zeta_j'(a), \omega_{j-1}(a)\right)\right| < \sqrt[3]{b}.$$

In the proof of Proposition 8.7 we use the following result:

Lemma 8.8. *There exist positive constants C'' and c'' such that if $j \leq n$ is a free iterate of $z_0 \in C_n$, then*

$$\left\|DT^{j-m}(z_m)\right\| \leq C'' e^{-c''(m-1)} \|\omega_{j-1}(z_1)\|$$

for every $m \leq j$.

Finally, we need the following statement to prove Lemma 8.8:

Lemma 8.9. *Let $j \leq n$ be a free iterate of $z_0 \in C_n$. Let $R = \bigcup_\mu (\mu - 100l_\mu, \mu + 100l_\mu)$ and $S = \bigcup_\mu (\mu - l_\mu, \mu + l_\mu)$, where the unions are extended to all the returns $\mu < j$ of z_0 and l_μ is the length of the respective folding period. Let R' and S' be the complementaries of R and S in $(0, j)$. Then:*

(a) Let $m \in R$ and let $I_\mu = (\mu - 100l_\mu, \mu + 100l_\mu)$ be the largest interval which contains m. Then, there exists $k \in [200l_\mu, 300l_\mu]$ such that $m + k \in R'$.

(b) Let $m \in R'$ and $i > 0$. Then, there exists $k \in [i, 4i]$ such that $m + k \in S'$.

Proof. To prove (a), let p_μ be the length of the binding period associated to μ and let $m' = m + 200l_\mu$. If $m' \in R'$, then the statement is proved. If $m' \in R$, take $I_{\mu'} = (\mu' - 100l_{\mu'}, \mu' + 100l_{\mu'})$ the largest interval which contains m'. From the maximality of I_μ it follows that $\mu' > \mu$. First, we prove that $\mu' \in [\mu, \mu + p_\mu]$. To this end, let us suppose that $\mu' > \mu + p_\mu$. Then, since $\mu' - 100l_{\mu'} \leq \mu + 300l_\mu$, we get $\mu' - \mu < p_{\mu'} - p_\mu$. To check this last inequality notice that, from $p_{\mu'} >> l_{\mu'}$, we obtain

$$\frac{1}{4}p_{\mu'} \geq \mu' - \mu - 300l_\mu > p_\mu - 300l_\mu > \frac{1}{2}p_\mu$$

and consequently,

$$p_\mu + \mu' - \mu \leq \frac{1}{2}p_{\mu'} + 100l_{\mu'} + 300l_\mu < p_{\mu'}.$$

Now, from (FA),

$$(1 - \alpha)(2\mu' - \mu + p_\mu) \leq (1 - \alpha)(\mu' + p_{\mu'}) \leq F_{\mu' + p_{\mu'}}(z_0) \leq \mu' - p_\mu$$

and thus, $(1 - 2\alpha)\mu' + (2 - \alpha)p_\mu \leq (1 - \alpha)\mu$. Hence, since $p_\mu \leq 2c^{-1}\alpha\mu$, we get $(1 - 2\alpha)\mu' \leq (1 - \alpha - 4c^{-1}\alpha + 2c^{-1}\alpha^2)\mu$ and this contradicts $\mu' > \mu$.

Therefore, $\mu' \in [\mu, \mu + p_\mu]$ and so $\mu' + p_{\mu'} \leq \mu + p_\mu$. Furthermore, once again from (FA),

$$(1 - \alpha)(\mu' + p_{\mu'} - \mu) \leq F_{\mu' + p_{\mu'} - \mu}(\zeta_0) \leq \mu' - \mu.$$

Thus, $p_{\mu'} \leq \mu' - \mu$. We claim that $\mu' \leq \mu + 400l_\mu$. In fact, if we suppose that $\mu' - \mu > 400l_\mu$, then $\mu' - 100l_{\mu'} \leq m' \leq \mu + 300l_\mu$ implies $l_\mu < l_{\mu'}$. Hence, $\mu < \mu' - p_{\mu'} < \mu' - 500l_{\mu'} < \mu' - 500l_\mu$ and so $\mu' - \mu > 500l_\mu$. Repeating the above argument, we obtain $2l_\mu < l_{\mu'}$ and $\mu' - \mu > 600l_\mu$. Thus, $3l_\mu < l_{\mu'}$ and so on.

In summary, if some $I_{\mu'}$ contains m', then $\mu' < \mu + 400l_\mu$. From (BA),

$$d_{\mu' - \mu}(\zeta_0) \geq e^{-\alpha(\mu' - \mu)} \geq e^{-400\alpha l_\mu}$$

and from Proposition 5.23, $l_\mu > 400\alpha l_\mu \geq \log (2\delta)^{-1} >> 4$. Hence,

$$l_{\mu'} \leq \frac{const}{\log b^{-1}}l_\mu + 4 << l_\mu.$$

Now, we repeat the arguments. Let us define $m'' = m' + 200l_{\mu'} << m + 300l_\mu$. Since the number of returns between μ and $\mu + 600l_\mu$ is finite, some $m + 200l_\mu + 200l_{\mu'} + \ldots \in [m + 200l_\mu, m + 300l_\mu]$ has to belong to R'. Statement (a) is proved.

Next, we shall prove (b). Let us assume that $m \in R'$ and let $i > 0$. If $m + i \in S'$, then the statement is proved. Otherwise, let $\tilde{I}_\mu = (\mu - l_\mu, \mu + l_\mu)$ be the largest interval of S which contains $m + i$. Notice that $l_\mu < i$. Let $m' = m + i + 2l_\mu < m + 3i$. If $m' \in S'$, then the statement is proved. If not, there exists a return $\mu' > \mu$ such that $m' \in (\mu' - l_{\mu'}, \mu' + l_{\mu'})$. Then, we argue as in (a) so as to find $k \in [i, 4i]$ such that $m + k \in S'$. $\square$

Proof of Lemma 8.8. First, let us assume that $m \in R'$. We shall prove that

$$\frac{\omega_{m-1}}{\|\omega_{m-1}\|}$$

is $b^{\frac{1}{8}}$-expanding up to time $j - m$. Let $i \in (0, j - m]$ be arbitrary and take $k \in [i, 4i]$ such that $m + k \in S'$. Notice that $h_{m+k-1}(z_1) = \omega_{m+k-1}(z_1)$ and $\|\omega_{m+i-1}(z_1)\| \geq K^{-(k-i)} \|\omega_{m+k-1}(z_1)\|$. If $m + i \in S'$, then we may take $k = i$. Otherwise, there exists a return μ', with $l_{\mu'}$ maximum, such that $m + i \in [\mu' - l_{\mu'}, \mu' + l_{\mu'}]$. From the proof of Lemma 8.9(b), we get $k = i + \tau l_{\mu'}$, with $\tau \in [2, 3]$. So, $k - i \leq 3l_{\mu'}$.

If there are no returns in $[m, m + k]$, then the proof follows from Proposition 5.15. Let $\mu \in [m, m + k]$ be the return of z_0 where $d_t(z_0)$ reaches the minimum. Then, from Proposition 5.10,

$$l_{\mu'} \leq \frac{20 \log K \log d_\mu(z_0)^{-1}}{c \log b^{-1}} + 4 \leq 6l_\mu + 4 \leq 7l_\mu. \tag{8.27}$$

Furthermore, since $m \in R'$, we have $100l_\mu \leq k \leq 4i$. Therefore, from (8.27), we get

$$\log d_\mu(z_0)^{-1} \leq \frac{3ci}{250 \log K} \log \frac{1}{b} < \frac{i}{10} \log \frac{1}{b}$$

and so, $d_\mu(z_0) \geq b^{\frac{1}{10}i}$. Then, from Proposition 5.15, it follows that

$$\frac{\|\omega_{m+k-1}(z_1)\|}{\|\omega_{m-1}(z_1)\|} = \frac{\|h_{m+k-1}(z_1)\|}{\|h_{m-1}(z_1)\|} \geq C^k d_\mu(z_0) \geq C^k b^{\frac{i}{10}}.$$

So, from (8.27),

$$\|\omega_{m+i-1}(z_1)\| \geq K^{-3l_{\mu'}} C^k b^{\frac{1}{10}i} \|\omega_{m-1}(z_1)\| \geq b^{\frac{1}{8}i} \|\omega_{m-1}(z_1)\|.$$

Finally, from Proposition 3.4,

$$\left\| DT^{j-m}(z_m) \right\| \leq 2 \frac{\|\omega_{j-1}(z_1)\|}{\|\omega_{m-1}(z_1)\|} \leq 2e^{-c(m-1)} \|\omega_{j-1}(z_1)\|$$

and the lemma is proved.

Now, let us assume that $m \in R$. Let $(\mu - 100l_\mu, \mu + 100l_\mu)$ be the largest interval which contains m. Then, from Lemma 8.9, there exists $k \le 300l_\mu$ such that $m' = m + k \in R'$. According to the previous case, $\|DT^{j-m}(z_m)\| \le 2e^{-c(m+k-1)}K^k \|\omega_{j-1}(z_1)\|$. Furthermore, from (BA), it follows that

$$k \le 300l_\mu \le \frac{const}{\log b^{-1}}m + 1200.$$

Therefore,

$$e^{-ck}K^k \le const \left(e^{-c}K\right)^{\frac{const}{\log b^{-1}}m} \le const \, e^{\frac{1}{2}cm},$$

provided that b is small enough. Hence,

$$\left\|DT^{j-m}(z_m)\right\| \le const \, e^{-c(m-1)}e^{\frac{1}{2}cm} \|\omega_{j-1}(z_1)\| \le C''e^{-c''(m-1)} \|\omega_{j-1}(z_1)\|. \quad \square$$

Proof of Proposition 8.7. Let $\overline{\omega} \subset E_{j-1}(\zeta_0^{(j-2)})$ and

$$\gamma_i = \bigcup_{a \in \overline{\omega}} \zeta_i^{(j-1)}(a) = \bigcup_{a \in \overline{\omega}} (x_i(a), y_i(a)).$$

Let $\zeta_i'(a) = (x_i'(a), y_i'(a))$ be the tangent vector to γ_i at $\zeta_i(a)$. Notice that

$$\begin{cases} x_{i+1} = \lambda^{-1} \log a + x_i + \lambda^{-1} \log \cos x_i + \lambda^{-1} \log \left(1 + \sqrt{b}y_i\right) \\ y_{i+1} = \sqrt{b}(1 + \sqrt{b}y_i)e^{\lambda x_i} \sin x_i \end{cases}$$

Therefore, $\zeta_{i+1}' = DT(\zeta_i)\zeta_i' + \left((\lambda a)^{-1}, 0\right) = M_i\zeta_i' + V$ and, by recurrence,

$$\zeta_j' = M_{j-1}...M_1\zeta_1' + \sum_{i=1}^{j-2} (M_{j-1}...M_{i+1}) V + V =$$

$$= DT^{j-1}(\zeta_1)\zeta_1' + \sum_{i=1}^{j-2} DT^{j-i-1}(\zeta_{i+1})V + V. \tag{8.28}$$

First, let us assume that j is a free return. Writing $\omega_{i,j-1} = DT^i(\zeta_1^{(j-1)})(1,0)$, Proposition 5.18 leads to

$$\left\|DT^{j-i-1}(\zeta_{i+1})\frac{\omega_{i,j-1}}{\|\omega_{i,j-1}\|}\right\| \ge K^{-5}e^{\widetilde{c}(j-i-1)}.$$

Therefore, if $f^{(j-i-1)}$ and $e^{(j-i-1)}$ respectively denote the maximally expanding and the maximally contracting vector by $DT^{j-i-1}(\zeta_{i+1})$, then Proposition 1.3 gives

$$K^{-5}e^{\widetilde{c}(j-i-1)} \left\|e_{j-i-1}^{(j-i-1)}\right\| \le \left\|f_{j-i-1}^{(j-i-1)}\right\| \left\|e_{j-i-1}^{(j-i-1)}\right\| \le (Kb)^{j-i-1}$$

and consequently,

$$\left\|e_{j-i-1}^{(j-i-1)}\right\| \le \left(\frac{K^6 b}{e^{\tilde{c}}}\right)^{j-i-1}. \tag{8.29}$$

Let $J = \left\{i \in [1, j-2] : \|DT^{j-i-1}(\zeta_{i+1})(1,0)\| \ge \left(\sqrt{b}\right)^{j-i-1}\right\}$. If $i \notin J$, the respective term in (8.28) will not play an important role to bound

$$\frac{\left\|\zeta_j'(a)\right\|}{\|\omega_{j-1}(a)\|}.$$

On the other hand, if v is a unit vector such that $\|DT^{j-i-1}(\zeta_{i+1})v\| \ge \left(\sqrt{b}\right)^{j-i-1}$, then (8.29) implies

$$\left|\sin\left(\text{ang}\left(DT^{j-i-1}(\zeta_{i+1})v, f_{j-i-1}^{(j-i-1)}\right)\right)\right| \le \frac{\left\|e_{j-i-1}^{(j-i-1)}\right\|}{\|DT^{j-i-1}(\zeta_{i+1})v\|} \le \left(const\sqrt{b}\right)^{j-i-1}$$

and consequently,

$$\left|\text{ang}\left(DT^{j-i-1}(\zeta_{i+1})v, f_{j-i-1}^{(j-i-1)}\right)\right| \le \left(const\sqrt{b}\right)^{j-i-1}.$$

Then, if $i \in J$, we obtain

$$\left|\text{ang}\left(DT^{j-i-1}(\zeta_{i+1})(1,0), \omega_{j-1}\right)\right| \le \left(const\sqrt{b}\right)^{j-i-1}.$$

Hence, we split

$$DT^{j-i-1}(\zeta_{i+1})(1,0) = \alpha_i \frac{\omega_{j-1}}{\|\omega_{i,j-1}\|} + \epsilon_i,$$

where ϵ_i is a vector on the direction of $e_{j-i-1}^{(j-i-1)}$ with $\|\epsilon_i\| \le \left(const\sqrt{b}\right)^{j-i-1}$. Furthermore, if we define $\Lambda_i = \alpha_i \|\omega_{i,j-1}\|^{-1}$, then, from Lemma 8.8,

$$|\Lambda_i| \le 2\frac{\|DT^{j-i-1}(\zeta_{i+1})\|}{\|\omega_{j-1}\|} \le const\, e^{-c''i}.$$

Now, let us write $(1,0) = \Lambda_{j-1}\omega_{j-1} + \epsilon_{j-1}$, where ϵ_{j-1} is a vertical vector. Since j is a fold-free iterate, Proposition 5.13 implies $\|\epsilon_{j-1}\| = |\text{slope}(\omega_{j-1})| \le const\sqrt{b}$ and so $|\Lambda_{j-1}| \le const\, e^{-c(j-1)}$. Then, from (8.28),

$$\zeta_j' = M_{j-1}...M_1\zeta_1' + \frac{1}{\lambda a}\left\{\sum_{i \in J}\Lambda_i + \Lambda_{j-1}\right\}\omega_{j-1} + \frac{1}{\lambda a}\epsilon_j, \tag{8.30}$$

where $\epsilon_j = \sum_{i \notin J} DT^{j-i-1}(\zeta_{i+1})(1,0) + \sum_{i \in J}\epsilon_i + \epsilon_{j-1}$. Therefore,

$$\|\epsilon_j\| \le \sum_{i \notin J}\left\|DT^{j-i-1}(\zeta_{i+1})(1,0)\right\| + \sum_{i \in J}\left(const\sqrt{b}\right)^{j-i-1} + const\sqrt{b} \le const\sqrt{b}.$$

Since $\zeta_1' = M_0\zeta_0' + V$ and as it follows from I.H.7 that $\|\zeta_0'(a)\| \leq b^{\frac{1}{40}}$, we obtain $\zeta_1' = \tilde{e}^0 + (\lambda a)^{-1}(1,0)$ with $\|\tilde{e}^0\| \leq b^{\frac{1}{40}}$. Hence, $|\mathrm{slope}(\zeta_1')| < \frac{1}{10}$ and, from Remark 3.5, $\|M_{j-1}...M_1\zeta_1'\| \geq (2\lambda a)^{-1} e^{c(j-1)}$. In the same way, we may split $M_{j-1}...M_1\zeta_1' = \Lambda_0\omega_{j-1} + \epsilon_0$, where ϵ_0 is a vector on the direction of $e_{j-1}^{(j-1)}$ with $\|\epsilon_0\| \leq (const\sqrt{b})^{j-1}$. In summary, we may write $\zeta_j' = \Lambda^j\omega_{j-1} + \epsilon^j$, where

$$\Lambda^j = \Lambda_0 + \frac{1}{\lambda a}\left\{\sum_{i \in J}\Lambda_i + \Lambda_{j-1}\right\} \quad \text{and} \quad \epsilon^j = \frac{1}{\lambda a}\epsilon_j + \epsilon_0.$$

From the previous estimates, we get $|\Lambda^j| < const$ and $\|\epsilon^j\| \leq const\sqrt{b}$. Therefore,

$$\frac{\|\zeta_j'\|}{\|\omega_{j-1}\|} \leq const.$$

In order to obtain a lower bound for

$$\frac{\|\zeta_j'\|}{\|\omega_{j-1}\|},$$

we fix $N_0 \in \mathbf{N}$ sufficiently large so that $\sum_{i=N_0-1}^{\infty}|\Lambda_i| \leq const \sum_{i=N_0-1}^{\infty}e^{-c''i} < \chi$, where $\chi > 0$ is a small constant to be fixed later. From (8.28) we may write

$$\zeta_j' = M_{j-1}...M_{N_0}\left\{M_{N_0-1}...M_1\zeta_1' + \sum_{i=1}^{N_0-2}DT^{N_0-i-1}(\zeta_{i+1})V\right\} +$$

$$+ \sum_{i=N_0-1}^{j-2}DT^{j-i-1}(\zeta_{i+1})V + V = M_{j-1}...M_{N_0}W + \sum_{i=N_0-1}^{j-2}DT^{j-i-1}(\zeta_{i+1})V + V. \quad (8.31)$$

Let us choose $\Omega_0(\lambda)$ close enough to $a(\lambda)$ to get, as in the proof of Proposition 2.3, $\left|f'_{\lambda,a}(\zeta_{1,x})\right| > \frac{3}{2}$ and $f'_{\lambda,a}\left(f^i_{\lambda,a}(\zeta_{1,x})\right) \geq 1 + \sqrt{5}$ for $1 \leq i \leq N_0$. Then,

$$\frac{\sum_{i=1}^{N_0-2}\left\{\left(f^{N_0-i-1}_{\lambda,a}\right)'\left(f^i_{\lambda,a}(\zeta_{1,x})\right)\right\}}{\left|\left(f^{N_0-1}_{\lambda,a}\right)'(\zeta_{1,x})\right|} \leq \frac{\sum_{i=1}^{N_0-2}(r_{N_0-2}...r_i)}{\frac{3}{2}r_1...r_{N_0-2}},$$

where $r_i = f'_{\lambda,a}\left(f^i_{\lambda,a}(\zeta_{1,x})\right) > 3$. Hence,

$$\frac{\sum_{i=1}^{N_0-2}\left\{\left(f^{N_0-i-1}_{\lambda,a}\right)'\left(f^i_{\lambda,a}(\zeta_{1,x})\right)\right\}}{\left|\left(f^{N_0-1}_{\lambda,a}\right)'(\zeta_{1,x})\right|} \leq 1 - 8\chi,$$

provided that N_0 is large enough. Now, taking b sufficiently small, we get

$$\frac{\|W\|}{\|\omega_{N_0-1,j-1}\|} \geq \frac{4\chi}{\lambda a}.$$

Furthermore, since ζ_1' and V are nearly horizontal and ζ_i does not belong to B_δ for $1 \leq i \leq N_0$, we may apply Lemma 4.1 to claim that W is also a nearly horizontal vector.

In the proof of Lemma 8.8 we proved that

$$\frac{\omega_{N_0-1,j-1}}{\|\omega_{N_0-1,j-1}\|}$$

is $b^{\frac{1}{8}}$-expanding up to time $j - N_0$. Then, from Proposition 3.4, $e^{(j-N_0)}$ is nearly vertical and, since $\omega_{N_0-1,j-1}$ is nearly horizontal, we may write

$$\frac{W}{\|W\|} = \alpha \frac{\omega_{N_0-1,j-1}}{\|\omega_{N_0-1,j-1}\|} + \beta e^{(j-N_0)},$$

with $|\alpha| > \frac{3}{4}$ and $|\beta| < \frac{1}{3}$. Thus, (8.29) and Proposition 5.18 lead to

$$\frac{\|M_{j-1}...M_{N_0}W\|}{\|W\|} \geq \frac{1}{2} \frac{\|\omega_{j-1}\|}{\|\omega_{N_0-1,j-1}\|}.$$

Consequently,

$$\frac{\|M_{j-1}...M_{N_0}W\|}{\|\omega_{j-1}\|} \geq \frac{1}{2} \frac{\|W\|}{\|\omega_{N_0-1,j-1}\|} \geq \frac{2\chi}{\lambda a}.$$

Hence, from (8.31),

$$\frac{\|\zeta_j'\|}{\|\omega_{j-1}\|} \geq \frac{2\chi}{\lambda a} - \frac{1}{\lambda a} \left(|\Lambda_{j-1}| + \sum_{\substack{i=N_0-1 \\ i \in J}}^{j-2} |\Lambda_i| + \|\epsilon_j\| \right) \geq$$

$$\geq \frac{2\chi}{\lambda a} - \frac{1}{\lambda a} \left(const \sum_{i=N_0-1}^{\infty} e^{-c''i} + const\sqrt{b} \right) \geq \frac{\chi}{2\lambda a} > \frac{\chi}{4\lambda}.$$

Moreover, since we have written $\zeta_j' = \Lambda^j \omega_{j-1} + \epsilon^j$, with

$$\|\omega_{j-1}\| \geq e^{c(j-1)}, \quad \|\epsilon^j\| \leq const\sqrt{b} \quad \text{and} \quad |\Lambda^j| > \frac{\|\zeta_j'\|}{\|\omega_{j-1}\|} - \frac{\|\epsilon^j\|}{\|\omega_{j-1}\|} \geq \frac{\chi}{5\lambda},$$

we conclude that $\left| ang \left(\zeta_j', \omega_{j-1} \right) \right| \leq \sqrt[3]{b}$.

If j is not a return but is a free iterate, from the proof of Proposition 5.18 we obtain $\|\omega_{j-1}\| \geq \delta K^{-5} e^{\tilde{c}(j-i-1)} \|\omega_{i,j-1}\|$ for every $i > 1$. Then, we have

$$\left\| e_{j-i-1}^{(j-i-1)} \right\| \leq \left(\frac{K^6 b}{\delta e^{\tilde{c}}} \right)^{j-i-1}$$

instead of (8.29) and the proposition follows as above.

Finally, let us assume that j belongs to a binding period $[\mu + 1, \mu + p]$. Take μ minimum. Thus, $\mu \geq N_0$ is a free return and, repeating the previous process, we get $\zeta'_\mu = \Lambda^\mu \omega_{\mu-1,j-1} + \epsilon^\mu$. Then, we may write

$$\zeta'_j = \Lambda^\mu \omega_{j-1} + DT^{j-\mu}(\zeta_\mu)\epsilon^\mu + \frac{1}{\lambda a}\left(\sum_{i=\mu}^{j-2} DT^{j-i-1}(\zeta_{i+1})(1,0) + (1,0)\right).$$

Therefore, taking α sufficiently small, $\mu > N$ large enough and bearing in mind that $\zeta_1^{(j-1)}$ is e^c-expanding up to time $j - 1$, we obtain

$$\left\|\frac{1}{\lambda a}\left(\sum_{i=\mu}^{j-2} DT^{j-i-1}(\zeta_{i+1})(1,0) + (1,0)\right) + DT^{j-\mu}(\zeta_\mu)\epsilon^\mu\right\| \leq$$

$$\leq \frac{2}{\lambda c}\alpha\mu K^{2c^{-1}\alpha\mu} \leq \sqrt{\alpha}e^{c\mu} \leq \sqrt{\alpha}\|\omega_{j-1}\|.$$

Hence, we conclude that

$$0 < const \leq \frac{\chi}{5\lambda} - \sqrt{\alpha} \leq \frac{\|\zeta'_j\|}{\|\omega_{j-1}\|} \leq const + \sqrt{\alpha} \leq const$$

and the proposition is proved for $j > N_0$. On the other hand, if $j \leq N_0$, then $\left\|\zeta'_j(a)\right\| \geq const\left|1 + \sum_{i=0}^{j-2}(r_i...r_{j-2})\right|$. Furthermore, since $r_0 < -\frac{3}{2}$ and $r_i > 1 + \sqrt{5}$ for $1 \leq i \leq j - 2$, we get $\left\|\zeta'_j(a)\right\| \geq const > 0$ and this completes the proof. □

As an immediate consequence of Proposition 8.7 we obtain that if $\overline{\omega} \in P_{j-1}$, $\overline{\omega} \subset E_{j-1}(\zeta_0^{(j-2)})$ and k is a free iterate of $\overline{\omega}$, then $\zeta_k(t) = (x'_k(t), y'_k(t)) = D_a(\zeta_k^{(j-1)}(a))\mid_{a=t}$ is a nearly horizontal vector for every $t \in \overline{\omega}$. Hence, if $1 \leq i_1 \leq i_2 \leq j$ are two free iterates of $\overline{\omega}$, then we may define the homeomorphism $\Psi : x_{i_1}^{(j-1)}(t) \longrightarrow x_{i_2}^{(j-1)}(t)$, where $\left(x_{i_k}^{(j-1)}(t), y_{i_k}^{(j-1)}(t)\right) = \zeta_{i_k}^{(j-1)}(t)$. Therefore, in the same way as Proposition 2.4, we obtain the following result:

Proposition 8.10. *Let $\overline{\omega} \subset P_{j-1}$, $\overline{\omega} \subset E_{j-1}(\zeta_0^{(j-2)})$. Let i_1 and i_2 be free iterates of $\overline{\omega}$ with $1 \leq i_1 \leq i_2 \leq j$ and*

$$v_{i_2-i_1}(t) = DT^{i_2-i_1}(\zeta_{i_1}^{(j-1)}(t))\frac{\omega_{i_1-1,j-1}(t)}{\|\omega_{i_1-1,j-1}(t)\|}.$$

Then:

(a) For every $a_1, a_2 \in \overline{\omega}$, there exists $t \in \overline{\omega}$ such that

$$\frac{1}{16A^2}\|v_{i_2-i_1}(t)\| \leq \frac{\left\|\zeta_{i_2}^{(j-1)}(a_1) - \zeta_{i_2}^{(j-1)}(a_2)\right\|}{\left\|\zeta_{i_1}^{(j-1)}(a_1) - \zeta_{i_1}^{(j-1)}(a_2)\right\|} \leq 16A^2\|v_{i_2-i_1}(t)\|.$$

(b) There exists $t \in \overline{\omega}$ such that

$$\frac{1}{16A^2} \left\| v_{i_2 - i_1}(t) \right\| \leq \frac{length\left(\zeta_{i_2}^{(j-1)}(\omega) \right)}{length\left(\zeta_{i_1}^{(j-1)}(\omega) \right)} \leq 16A^2 \left\| v_{i_2 - i_1}(t) \right\|.$$

Corollary 8.11. *Let $\overline{\omega} \in P_{\nu_0}$ be an escape component in the construction of $P_{n-1}(z_0)$. If j is the next return situation of $\overline{\omega}$, then $length\left(\zeta_j^{(j-1)}(\overline{\omega}) \right) \geq \sqrt{\delta}$ and consequently, j is an escape situation of $\overline{\omega}$.*

Proof. If $\overline{\omega}$ is an escape component, then ν_0 is a free iterate and thus $\zeta_{\nu_0}^{(\nu_0-1)}(\overline{\omega})$ is an almost horizontal curve. Furthermore, $length\left(\zeta_{\nu_0}^{(\nu_0-1)}(\overline{\omega}) \right) \geq \frac{1}{3}\sqrt{\delta}$. Let j be the next return situation of $\overline{\omega}$. Then, $\overline{\omega} \subset E_{j-1}(\zeta_0^{(j-2)})$ and $\zeta_0^{(j-1)}(a)$ is defined on every $a \in \overline{\omega}$. Moreover, $\zeta_0^{(j-1)}(a)$ is bound to $\zeta_0^{(\nu_0-1)}(a)$ up to time ν_0. Hence,

$$\left\| \zeta_{\nu_0}^{(\nu_0-1)}(a) - \zeta_{\nu_0}^{(j-1)}(a) \right\| \leq e^{-\beta\nu_0} << \delta,$$

provided that $\nu_0 \geq N$ is large enough. Therefore,

$$length\left(\zeta_{\nu_0}^{(j-1)}(\overline{\omega}) \right) \geq \frac{1}{4}\sqrt{\delta}.$$

Now, the proof follows from Propositions 8.10 and 4.4 as does Lemma 2.17. $\quad\square$

Before proving Proposition 8.6, we shall demonstrate a result which will allows us to get I.H.7(c) at time n.

Lemma 8.12. *For each $j \in [\nu, n]$ and for every $\overline{\omega} \in P_j$, $\omega \subset E_{j-1}(\zeta_0^{(j-2)})$, we have*

$$K^{-\frac{3}{2}j} \leq |\overline{\omega}| \leq e^{-\frac{2}{3}cj}.$$

Proof. First, notice that there always exists a free iterate $\mu \in \left[\frac{9}{10}j, j \right]$. Then, from Proposition 8.7, it follows that $\zeta_\mu^{(j-1)}(\overline{\omega})$ is an almost horizontal curve. So, there exists $\tilde{a} \in \overline{\omega}$ such that

$$\frac{1}{4} length\left(\zeta_\mu^{(j-1)}(\overline{\omega}) \right) \leq \left\| \zeta_\mu'(\tilde{a}) \right\| |\overline{\omega}| \leq 4 length\left(\zeta_\mu^{(j-1)}(\overline{\omega}) \right).$$

Furthermore, Proposition 8.7 implies

$$\frac{1}{A} \leq \frac{\left\| \zeta_\mu'(\tilde{a}) \right\|}{\left\| \omega_{\mu-1, j-1}(\tilde{a}) \right\|} \leq A.$$

Hence, $\frac{1}{4}A^{-1}e^{c(\mu-1)}|\varpi| \leq \frac{1}{4}\left\|\zeta'_\mu(\tilde{a})\right\||\varpi| \leq 2$. Therefore, $|\varpi| \leq e^{-\frac{2}{3}cj}$ whenever $j \geq N$ is sufficiently large.

On the other hand, let $i \in [\nu, j]$ be the iterate at which ϖ was created. Then, since i is an essential return, it follows that $\zeta_i^{(i-1)}(\varpi)$ is an almost horizontal curve such that

$$length\left(\zeta_i^{(i-1)}(\varpi)\right) \geq \frac{e^{-\alpha i}}{2(\alpha i)^2}.$$

Furthermore, since $\varpi \subset E_{i-1}(\zeta_0^{(i-2)})$, there exists $a' \in \varpi$ such that

$$length\left(\zeta_i^{(i-1)}(\varpi)\right) \leq 4A\|\omega_{i-1}(a')\||\varpi| \leq 4AK^{i-1}|\varpi|.$$

Hence, $|\varpi| \geq constK^{-j}e^{-\alpha i}(\alpha i)^{-2}$. Thus,

$$const\, e^{-\alpha i}(\alpha i)^{-2} > e^{-\frac{3}{2}\alpha i} > K^{-\frac{1}{2}j},$$

provided that $i \geq N$ is large enough. Therefore, $|\varpi| \geq K^{-\frac{3}{2}j}$. $\square$

Proof of Proposition 8.6. Let us fix $a, a' \in \varpi$ and proceed as in the proof of Proposition 7.2. Then, we deduce that for every $\eta_0(a)$ and $\eta_0(a')$ bound to $\zeta_0^{(j-1)}(a)$ and $\zeta_0^{(j-1)}(a')$, respectively, and for

$$V_k = \sum_{t=1}^{k}\left(\sqrt[4]{b}\right)^{k-t}\left(|a - a'| + \|\eta_t(a) - \eta_t(a')\|\right),$$

it follows that

$$(1) \qquad \frac{\|h_i(\eta_1(a'))\|}{\|h_i(\eta_1(a))\|} \leq \exp\left(4K\sum_{r=0}^{i}\left(\sqrt[4]{b}\right)^r \sum_{k=1}^{i}\frac{V_k}{d_k(\eta_0(a))}\right)$$

$$(2) \qquad |\text{ang}\left(h_i(\eta_1(a)), h_i(\eta_1(a'))\right)| \leq \sum_{r=1}^{i}\left(\sqrt[4]{b}\right)^r V_i$$

for every $1 \leq i \leq j - 1$. Therefore, to prove the first statement of the proposition it suffices to bound

$$\sum_{k=1}^{j-1}V_k\left(d_k\left(\zeta_0^{(j-1)}(a)\right)\right)^{-1}.$$

Let $\nu_1 < \nu_2 < ... < \nu_{s-1}$ be all the free returns of ϖ in $(1, j)$. First, we shall prove that there exists $\tau = \tau(\lambda, K, \alpha, \beta)$ such that

$$\sum_{k=1}^{\nu_s-1+p_s-1}V_k\left(d_k\left(\zeta_0^{(j-1)}(a)\right)\right)^{-1} < \tau.$$

Since from Proposition 5.19 and Lemma 8.12 we get

$$\sum_{k=1}^{\nu_{s-1}+p_{s-1}} \left(d_k \left(\zeta_0^{(j-1)}(a) \right) \right)^{-1} \sum_{t=1}^{k} \left(\sqrt[4]{b} \right)^{k-t} |a - a'| \leq const(\lambda, \alpha),$$

it is sufficient to prove

$$\sum_{k=1}^{\nu_{s-1}+p_{s-1}} \theta_k \left(d_k \left(\zeta_0^{(j-1)}(a) \right) \right)^{-1} \leq \tau' = \tau'(\lambda, K, \alpha, \beta),$$

where

$$\theta_k = \theta_k(a, a') = \sum_{t=1}^{k} \left(\sqrt[4]{b} \right)^{k-t} \left\| \zeta_t^{(j-1)}(a) - \zeta_t^{(j-1)}(a') \right\|.$$

First, let us assume that $t \leq \nu_1$. According to Propositions 8.10 and 4.4(b) we have

$$\left\| \zeta_t^{(j-1)}(a) - \zeta_t^{(j-1)}(a') \right\| \leq 16 C_0^{-1} A^2 e^{c_0(t-\nu_1)} \left\| \zeta_{\nu_1}^{(j-1)}(a) - \zeta_{\nu_1}^{(j-1)}(a') \right\|.$$

Therefore,

$$\theta_k \leq 20 C_0^{-1} A^2 \left\| \zeta_{\nu_1}^{(j-1)}(a) - \zeta_{\nu_1}^{(j-1)}(a') \right\| e^{c_0(k-\nu_1)}$$

for every $k \leq \nu_1$. In particular,

$$\theta_{\nu_1} \leq 20 C_0^{-1} A^2 \left\| \zeta_{\nu_1}^{(j-1)}(a) - \zeta_{\nu_1}^{(j-1)}(a') \right\|.$$

We shall inductively prove that

$$\theta_{\nu_i} \leq 20 C_0^{-1} A^2 K^5 \left\| \zeta_{\nu_i}^{(j-1)}(a) - \zeta_{\nu_i}^{(j-1)}(a') \right\| \tag{8.32}$$

for every $1 \leq i \leq s-1$. Indeed, let us assume that (8.32) holds up to the return ν_{i-1} and let p_{i-1} be the length of its binding period. From Proposition 8.7,

$$\frac{\left\| D_a \left(\zeta_t^{(j-1)}(a) \right) |_{a=a_0} \right\|}{\left\| D_a \left(\zeta_{\nu_{i-1}}^{(j-1)}(a) \right) |_{a=a_0} \right\|} = \frac{\left\| \zeta_t'(a_0) \right\|}{\left\| \zeta_{\nu_{i-1}}'(a_0) \right\|} \leq A^2 \frac{\left\| \omega_{t-1,j-1}(a_0) \right\|}{\left\| \omega_{\nu_{i-1}-1,j-1}(a_0) \right\|}$$

for $\nu_{i-1} + 1 \leq t \leq \nu_{i-1} + p_{i-1}$ and for every $a_0 \in \bar{\omega}$. Let $e = (q, 1)$ be a vector on the direction of $e^{(p_{i-1})}$ and let us split

$$\frac{\omega_{\nu_{i-1},j-1}(a_0)}{\left\| \omega_{\nu_{i-1},j-1}(a_0) \right\|} = \hat{\alpha}_{\nu_{i-1}}(a_0)e(a_0) + \hat{\beta}_{\nu_{i-1}}(a_0)(1,0).$$

Now, as in Proposition 6.11, we get

$$\left| \hat{\alpha}_{\nu_{i-1}}(a_0) \right| < 1 \quad \text{and} \quad \left| \hat{\beta}_{\nu_{i-1}}(a_0) \right| \leq \frac{2}{\lambda} d_{\nu_{i-1}} \left(\zeta_0^{(j-1)}(a_0) \right).$$

Thus, repeating the arguments in the proof of Proposition 7.5(a), we get

$$\left|\hat{\beta}_{\nu_{i-1}}(a_0)\right| \left\|\omega_{t-\nu_{i-1}-1}(\zeta_{\nu_{i-1}+1}^{(j-1)}(a_0))\right\| \le \frac{4}{\lambda}e^{-\beta(t-\nu_{i-1}-1)}\left(d_{\nu_{i-1}}\left(\zeta_0^{(j-1)}(a_0)\right)\right)^{-1}.$$

Since Proposition 3.3 leads to $\left\|DT^{t-\nu_{i-1}-1}(\zeta_{\nu_{i-1}+1}^{(j-1)}(a_0))e(a_0)\right\| \le (K_3 b)^{t-\nu_{i-1}-1}$, we obtain

$$\frac{\|\zeta_t'(a_0)\|}{\left\|\zeta_{\nu_{i-1}}'(a_0)\right\|} \le \frac{8}{\lambda}A^2 e^{-\beta(t-\nu_{i-1}-1)}\left(d_{\nu_{i-1}}\left(\zeta_0^{(j-1)}(a_0)\right)\right)^{-1}.$$

Furthermore, according to Proposition 8.7, $\zeta_{\nu_{i-1}}(\varpi)$ is an almost horizontal curve and, from the mean value theorem, there exists $a_0 \in \varpi$ such that

$$\left\|\zeta_t^{(j-1)}(a) - \zeta_t^{(j-1)}(a')\right\| \le 2\frac{\|\zeta_t'(a_0)\|}{\left\|\zeta_{\nu_{i-1}}'(a_0)\right\|}\left\|\zeta_{\nu_{i-1}}^{(j-1)}(a) - \zeta_{\nu_{i-1}}^{(j-1)}(a')\right\|.$$

Therefore,

$$\left\|\zeta_t^{(j-1)}(a) - \zeta_t^{(j-1)}(a')\right\| \le \frac{32}{\lambda}A^2 e^{-\beta(t-\nu_{i-1}-1)}\frac{\left\|\zeta_{\nu_{i-1}}^{(j-1)}(a) - \zeta_{\nu_{i-1}}^{(j-1)}(a')\right\|}{d_{\nu_{i-1}}(\zeta_0^{(j-1)}(a))}.$$

So, if $k \in [\nu_{i-1}+1, \nu_{i-1}+p_{i-1}]$, then we obtain

$$\theta_k \le \left(\sqrt[4]{b}\right)^{k-\nu_{i-1}}\theta_{\nu_{i-1}} + \sum_{t=\nu_{i-1}+1}^{k}\left(\sqrt[4]{b}\right)^{k-t}\left\|\zeta_t^{(j-1)}(a) - \zeta_t^{(j-1)}(a')\right\| \le$$

$$\le \frac{100A^2}{\lambda}e^{-\beta(k-\nu_{i-1})}\frac{\left\|\zeta_{\nu_{i-1}}^{(j-1)}(a) - \zeta_{\nu_{i-1}}^{(j-1)}(a')\right\|}{d_{\nu_{i-1}}(\zeta_0^{(j-1)}(a))}$$

and Propositions 5.19 and 5.21 yield

$$\frac{\theta_k}{d_k(\zeta_0^{(j-1)}(a))} \le \frac{200A^2}{\tilde{\Lambda}\lambda}e^{(\alpha-\beta)(k-\nu_{i-1})}\frac{\left\|\zeta_{\nu_{i-1}}^{(j-1)}(a) - \zeta_{\nu_{i-1}}^{(j-1)}(a')\right\|}{d_{\nu_{i-1}}(\zeta_0^{(j-1)}(a))}. \tag{8.33}$$

Now, if $\nu_{i-1}+p_{i-1} < k \le \nu_i$, then, repeating the above arguments, we prove that there exists $a_t \in \varpi$ such that

$$\left\|\zeta_t^{(j-1)}(a) - \zeta_t^{(j-1)}(a')\right\| \le 2\frac{\|\zeta_t'(a_t)\|}{\|\zeta_{\nu_i}'(a_t)\|}\left\|\zeta_{\nu_i}^{(j-1)}(a) - \zeta_{\nu_i}^{(j-1)}(a')\right\|$$

for every $t \le \nu_i$. Finally, from Propositions 8.7 and 5.18, it follows that

$$\left\|\zeta_{\nu_i}^{(j-1)}(a) - \zeta_{\nu_i}^{(j-1)}(a')\right\| \ge \frac{K^{-5}}{2A^2}e^{\tilde{c}(\nu_i-t)}\left\|\zeta_t^{(j-1)}(a) - \zeta_t^{(j-1)}(a')\right\|. \tag{8.34}$$

Notice that, since $\zeta_j^{(j-1)}(\varpi) \subset B_{\sqrt[4]{\delta}}$, (8.34) also holds for $\nu_s = j$.

In summary, if $k \in (\nu_{i-1} + p_{i-1}, \nu_i]$, then

$$\theta_k \leq 20C_0^{-1} A^2 K^5 e^{-\widetilde{c}(\nu_i - k)} \left\| \zeta_{\nu_i}^{(j-1)}(a) - \zeta_{\nu_i}^{(j-1)}(a') \right\|.$$

In particular, $\theta_{\nu_i} \leq 20C_0^{-1} A^2 K^5 \left\| \zeta_{\nu_i}^{(j-1)}(a) - \zeta_{\nu_i}^{(j-1)}(a') \right\|$ and (8.32) is proved at time ν_i. Furthermore, if $k \in (\nu_{i-1} + p_{i-1}, \nu_i)$, then Proposition 5.23 implies

$$\frac{\theta_k}{d_k(\zeta_0^{(j-1)}(a))} \leq 80C_0^{-1} A^2 K^5 e^{-\widetilde{c}(\nu_i - k)} \frac{\left\| \zeta_{\nu_i}^{(j-1)}(a) - \zeta_{\nu_i}^{(j-1)}(a') \right\|}{d_{\nu_i}(\zeta_0^{(j-1)}(a))},$$

which, together with (8.33), leads to

$$\sum_{k=1}^{\nu_s - 1 + p_s - 1} \frac{\theta_k}{d_k(\zeta_0^{(j-1)}(a))} \leq const \sum_{i=1}^{s} \frac{\left\| \zeta_{\nu_i}^{(j-1)}(a) - \zeta_{\nu_i}^{(j-1)}(a') \right\|}{d_{\nu_i}(\zeta_0^{(j-1)}(a))},$$

where the constant depends on λ, K, α and β. On the other hand, taking δ sufficiently small and Ω_0 close enough to $a(\lambda)$, (8.34) implies

$$\left\| \zeta_{\nu_i}^{(j-1)}(a) - \zeta_{\nu_i}^{(j-1)}(a') \right\| \geq 2 \left\| \zeta_{\nu_{i-1}}^{(j-1)}(a) - \zeta_{\nu_{i-1}}^{(j-1)}(a') \right\|. \tag{8.35}$$

Now, once $r \in \mathbf{N}$ is fixed, we define $N_r = \{i \in \{1, ..., s\} : \Delta_{\nu_i}(\overline{\omega}) \subset I_r^+\}$ and $\mu_r = \max N_r$. From Remark 5.22 and (8.35),

$$\sum_{i \in N_r} \frac{\left\| \zeta_{\nu_i}^{(j-1)}(a) - \zeta_{\nu_i}^{(j-1)}(a') \right\|}{d_{\nu_i} \left(\zeta_0^{(j-1)}(a) \right)} \leq const \frac{\left\| \zeta_{\nu_{\mu_r}}^{(j-1)}(a) - \zeta_{\nu_{\mu_r}}^{(j-1)}(a') \right\|}{d_{\nu_{\mu_r}} \left(\zeta_0^{(\nu_{\mu_r} - 1)}(a) \right)}.$$

Furthermore, since $\Delta_{\nu_{\mu_r}}(\overline{\omega}) \subset I_r^+$, from the tangential position it follows that

$$d_{\nu_{\mu_r}} \left(\zeta_0^{(\nu_{\mu_r} - 1)}(a) \right) > \frac{1}{2} e^{-(|r|+2)}$$

and Definition 5.5 yields

$$\left\| \zeta_{\nu_{\mu_r}}^{(j-1)}(a) - \zeta_{\nu_{\mu_r}}^{(j-1)}(a') \right\| \leq 2e^{-\beta \nu_{\mu_r}} + 5e^{-|r|} r^{-2}.$$

But, from (BA), we have $2e^{-\beta \nu_{\mu_r}} < 2e^{-2\alpha \nu_{\mu_r}} < 2e^{-2|r|}$, provided that $2\alpha < \beta$. Since $|r| > \log \delta^{-1}$ is arbitrarily large, we eventually get

$$\left\| \zeta_{\nu_{\mu_r}}^{(j-1)}(a) - \zeta_{\nu_{\mu_r}}^{(j-1)}(a') \right\| \leq 6e^{-|r|} r^{-2}$$

and consequently,

$$\sum_{i \in N_r} \left\| \zeta_{\nu_i}^{(j-1)}(a) - \zeta_{\nu_i}^{(j-1)}(a') \right\| \left(d_{\nu_i}(\zeta_0^{(j-1)}(a)) \right)^{-1} \leq const \; r^{-2}.$$

Therefore,

$$\sum_{k=1}^{\nu_{s-1}+p_{s-1}} \theta_k \left(d_k(\zeta_0^{(j-1)}(a))\right)^{-1} \leq const \sum_{|r|\geq \log \delta^{-1}} r^{-2} < const.$$

Finally, to bound

$$\sum_{k=\nu_{s-1}+p_{s-1}+1}^{j-1} \theta_k \left(d_k(\zeta_0^{(j-1)}(a))\right)^{-1},$$

notice that if $j-1$ is not a free iterate, then $j-1 = \nu_{s-1}+p_{s-1}$. So, we assume that $j-1$ is a free iterate and distinguish between two cases:

If $length\left(\zeta_{j-1}^{(j-1)}(\varpi)\right) \leq 2\delta^2$ then, from (8.34) and Proposition 4.4(a), it follows that

$$\left\|\zeta_{j-1}^{(j-1)}(a) - \zeta_{j-1}^{(j-1)}(a')\right\| \geq \frac{C_0\delta}{4A^2} e^{c_0(j-t-1)} \left\|\zeta_t^{(j-1)}(a) - \zeta_t^{(j-1)}(a')\right\|$$

for every $t \in (\nu_{s-1}+p_{s-1}, j-1)$. Therefore

$$\theta_k \leq const\, \delta \left(\sqrt[8]{b}\right)^{k-\nu_{s-1}-p_{s-1}} + 12C_0^{-1}A^2\delta e^{c_0(k+1-j)}$$

for every $k \in (\nu_{s-1}+p_{s-1}, j-1)$. Furthermore, since every $k \in (\nu_{s-1}+p_{s-1}+1, j)$ is a free iterate, we get $d_k(\zeta_0^{(j-1)}(a)) > \frac{1}{2}\delta$ and consequently,

$$\sum_{k=\nu_{s-1}+p_{s-1}+1}^{j-1} \theta_k \left(d_k(\zeta_0^{(j-1)}(a))\right)^{-1} \leq const.$$

Let us assume that $length\left(\zeta_{j-1}^{(j-1)}(\varpi)\right) > 2\delta^2$. Let $k_0 \geq \nu_{s-1}+p_{s-1}+1$ be the first natural number for which $length\left(\zeta_{k_0}^{(j-1)}(\varpi)\right) > 2\delta^2$. From the previous case,

$$\sum_{k=1}^{k_0-1} \theta_k \left(d_k(\zeta_0^{(j-1)}(a))\right)^{-1} < const$$

and therefore,

$$\frac{\|\omega_{k_0-1,j-1}(a)\|}{\|\omega_{k_0-1,j-1}(a')\|} < const(\lambda, K, \alpha, \beta).$$

Furthermore, from Lemma 4.2, we get

$$\frac{\|\omega_{j-1}(a')\|}{\|\omega_{k_0-1,j-1}(a')\|} \geq \left(1 - \sqrt[6]{b}\right)^{j-k_0} \left|\left(f_{\lambda,a'}^{j-k_0}\right)'(\zeta_{x,k_0}^{(j-1)}(a'))\right|,$$

whenever b is taken small enough with respect to $j - k_0$. Likewise, we also have

$$\frac{\|\omega_{j-1}(a)\|}{\|\omega_{k_0-1,j-1}(a)\|} \leq \left(1 + \sqrt[6]{b}\right)^{j-k_0} \left|\left(f_{\lambda,a}^{j-k_0}\right)'(\zeta_{x,k_0}^{(j-1)}(a))\right|.$$

Therefore,

$$\frac{\|\omega_{j-1}(a)\|}{\|\omega_{j-1}(a')\|} \le const \left(\frac{1+\sqrt[6]{b}}{1-\sqrt[6]{b}}\right)^{j-k_0} \frac{\left|\left(f_{\lambda,a}^{j-k_0}\right)'\left(\zeta_{x,k_0}^{(j-1)}(a)\right)\right|}{\left|\left(f_{\lambda,a'}^{j-k_0}\right)'\left(\zeta_{x,k_0}^{(j-1)}(a')\right)\right|}.$$

Moreover, since k_0 is a free iterate, $\zeta_{k_0}^{(j-1)}(\overline{\omega})$ is an almost horizontal curve. So, $\zeta_{x,k_0}^{(j-1)}(\overline{\omega})$ is an interval of length greater than δ^2. Hence, repeating the latter arguments of the proof of Proposition 2.20, there exists $n_0 = n_0(\delta)$ such that $j - k_0 \le n_0$ for every $\overline{\omega}$ such that $length\left(\zeta_{j-1}^{(j-1)}(\overline{\omega})\right) > 2\delta^2$. Therefore, the proof ends by taking b sufficiently small so that

$$\left(\frac{1+\sqrt[6]{b}}{1-\sqrt[6]{b}}\right)^{n_0} \le 2$$

and bearing in mind that, as in dimension one,

$$\frac{\left|\left(f_{\lambda,a}^{j-k_0}\right)'\left(\zeta_{x,k_0}^{(j-1)}(a)\right)\right|}{\left|\left(f_{\lambda,a'}^{j-k_0}\right)'\left(\zeta_{x,k_0}^{(j-1)}(a')\right)\right|} \le const,$$

whenever Ω_0 is close enough to $a(\lambda)$.

To prove that $|ang\left(\omega_{j-1}(a), \omega_{j-1}(a')\right)| < \sqrt[4]{b}$, notice that, from $\zeta_j^{(j-1)}(\overline{\omega}) \subset B_{\sqrt[4]{\delta}}$, we obtain

$$\theta_{j-1} \le 20 C_0^{-1} A^2 K^5 e^{-\tilde{c}} \left\|\zeta_j^{(j-1)}(a) - \zeta_j^{(j-1)}(a')\right\| \le 50 C_0^{-1} A^2 K^5 e^{-\tilde{c}} \sqrt[4]{\delta}.$$

Hence, $V_{j-1} \le \frac{1}{2}$ and thus

$$|ang\left(\omega_{j-1}(a), \omega_{j-1}(a')\right)| < \frac{1}{2} \sum_{r=1}^{i} \left(\sqrt[4]{b}\right)^r < \sqrt[4]{b}.$$

This completes the proof. □

From Propositions 8.6 and 8.7 we obtain the following result:

Corollary 8.13. *There exists a constant $C_4 = C_4(\lambda, K, \alpha, \beta)$ such that if j is a free return situation of $\overline{\omega} \in P_{j-1}$, $\overline{\omega} \subset E_{j-1}(\zeta_0^{(j-2)})$ and $\zeta_j^{(j-1)}(\overline{\omega}) \subset B_{\sqrt[4]{\delta}}$, then*

$$\frac{\|\zeta_j'(a)\|}{\|\zeta_j'(a')\|} \le C_4 \quad and \quad \left|ang\left(\zeta_j'(a), \zeta_j'(a')\right)\right| \le 3\sqrt[4]{b}$$

for every $a, a' \in \overline{\omega}$.

Since the distortion constant in Proposition 8.6 depends on λ, K, α and β (just as in the unidimensional case), the remaining arguments follow in the same way as the latter ones in Chapter 2. We are now going to show how they adapt to the bidimensional case.

As in Lemma 2.19, we shall prove that if j is a return situation of $\varpi \in P_{j-1}$, $\varpi \subset E_{j-1}(\zeta_0^{(j-2)})$, then there exists $\epsilon > 0$ such that

$$length\left(\zeta_j^{(j-1)}(\varpi)\right) \geq e^{-(1-\frac{1}{2}\epsilon)\alpha j}. \tag{8.36}$$

To this end, let $\mu \leq j-1$ be the first natural number for which $\omega_\mu = \varpi$. If ω_μ is an escape component, then j is an escape situation and (8.36) follows from Corollary 8.11 by taking $j \geq N$ large enough. If ω_μ is not an escape component, then there exists a host interval $I_{r,k}^+$, associated to μ, with $|r| \geq \Delta = \log \delta^{-1}$. Then, as in the proof of Lemma 2.18, we get

$$length\left(\zeta_j^{(j-1)}(\varpi)\right) \geq e^{(-1+\epsilon)|r|-\Delta},$$

where $\epsilon = \epsilon(K, c)$ is a small positive constant. Now, since μ and $\mu + p + 1$ are free iterates, from Propositions 8.10, 6.13 and 7.7 we obtain

$$\frac{length\left(\zeta_{\mu+p+1}^{(j-1)}(\varpi)\right)}{length\left(\zeta_\mu^{(j-1)}(\varpi)\right)} \geq \frac{1}{16A^2} \frac{\|h_{\mu+p,j-1}(t)\|}{\|h_{\mu,j-1}(t)\|} \frac{\|h_{\mu,j-1}(t)\|}{\|h_{\mu-1,j-1}(t)\|} \geq$$

$$\geq \frac{\tau_2}{32\lambda A^2} \exp\left(\frac{1}{3}c(p+1)\right) \geq const \exp\left(\frac{c|r|}{15 \log K}\right). \tag{8.37}$$

Let μ' be the next return to μ and $q = \mu' - (\mu + p + 1)$. Propositions 4.4(b) and 8.10 lead to

$$\frac{length\left(\zeta_{\mu'}^{(j-1)}(\varpi)\right)}{length\left(\zeta_{\mu+p+1}^{(j-1)}(\varpi)\right)} \geq \frac{1}{16A^2} \frac{\|\omega_{\mu'-1,j-1}(t')\|}{\|\omega_{\mu+p,j-1}(t')\|} \geq const \, e^{coq}. \tag{8.38}$$

Therefore,

$$\frac{length\left(\zeta_{\mu'}^{(j-1)}(\varpi)\right)}{length\left(\zeta_\mu^{(j-1)}(\varpi)\right)} \geq e^{coq} \exp\left(\frac{c|r|}{30 \log K}\right).$$

If there are no returns between μ and j, then, from Proposition 4.4(a),

$$\frac{length\left(\zeta_j^{(j-1)}(\varpi)\right)}{length\left(\zeta_\mu^{(j-1)}(\varpi)\right)} \geq e^{coq-\Delta} \exp\left(\frac{c|r|}{30 \log K}\right).$$

Otherwise, from the previous arguments

$$\frac{length\left(\zeta_j^{(j-1)}(\overline{\omega})\right)}{length\left(\zeta_\mu^{(j-1)}(\overline{\omega})\right)} \geq e^{-\Delta} \exp\left(\frac{c\,|r|}{30\log K}\right).$$

In either case, since μ is an essential return situation, it follows that

$$length\left(\zeta_j^{(j-1)}(\overline{\omega})\right) \geq \exp\left(\left(-1+\frac{c}{40\log K}\right)|r|-\Delta\right).$$

Finally, (BA) yields $|r| < \alpha\mu < \alpha j$ and consequently (8.36) is proved.

Next, we shall extend Proposition 2.21 to the bidimensional case to check (8.24). Let $\overline{\omega} \subset E_{j-1}(\zeta_0^{(j-2)})$ and $\omega_{exc} = \left\{a \in \overline{\omega} : a \notin E_j'(\zeta_0^{(j-1)})\right\}$. If j is not a return situation, then $\omega_{exc} = \emptyset$. Furthermore, from the arguments given in the proof of Proposition 2.16, that is, by using (8.36), we also have $\omega_{exc} = \emptyset$ when j is an inessential return situation. Therefore, assume that j is an essential return situation and let $\tilde{\omega} = \zeta_j^{-1}(B_{\sqrt[4]{\delta}})$. Then, since $\zeta_j(\tilde{\omega})$ is almost horizontal, there exist $a_1, a_2 \in \tilde{\omega}$ such that $length\left(\zeta_j(\omega_{exc})\right) \geq \frac{1}{4}\left\|\zeta_j'(a_1)\right\| |\omega_{exc}|$ and $length\left(\zeta_j(\tilde{\omega})\right) \leq 4\left\|\zeta_j'(a_2)\right\| |\tilde{\omega}|$. Hence,

$$\frac{|\omega_{exc}|}{|\overline{\omega}|} \leq \frac{|\omega_{exc}|}{|\tilde{\omega}|} \leq 16\frac{\left\|\zeta_j'(a_2)\right\|}{\left\|\zeta_j'(a_1)\right\|}\frac{length\left(\zeta_j(\omega_{exc})\right)}{length\left(\zeta_j(\tilde{\omega})\right)}.$$

Furthermore, by definition, $length\left(\zeta_j(\omega_{exc})\right) \leq const\, e^{-\alpha j}$.

On the other hand, if $\tilde{\omega} = \overline{\omega}$, then (8.36) implies $length\left(\zeta_j(\tilde{\omega})\right) \geq e^{-(1-\frac{1}{2}\epsilon)\alpha j}$. If $\tilde{\omega} \neq \overline{\omega}$, then $length\left(\zeta_j(\tilde{\omega})\right) \geq \sqrt[4]{\delta} - \delta \geq e^{-(1-\frac{1}{2}\epsilon)\alpha j}$, provided that $j \geq N$ is large enough. So, Corollary 8.13 yields

$$\frac{|\omega_{exc}|}{|\overline{\omega}|} \leq const\, e^{-\frac{1}{2}\epsilon\alpha j} < e^{-\frac{1}{3}\epsilon\alpha j}$$

and consequently, (8.24) is proved.

To prove (8.26), let us consider, for each $\overline{\omega} \subset E_j'(\zeta_0^{(j-1)})$, the following sequences:

$$1 = e_0 < e_1 < ... < e_{k-1} = e_k = ... = e_\nu < ... < e_s < e_{s+1} = j,$$
$$N = \hat{\mu}_1 < \hat{\mu}_2 < ... < \hat{\mu}_k = \hat{\mu}_{k+1} = ... = \hat{\mu}_{\nu+1} < ... < \hat{\mu}_s < \hat{\mu}_{s+1} \leq j,$$
$$\Omega_0 = \omega^0 \supset \omega^1 \supset ... \supset \omega^{k-1} = \omega^k = ... = \omega^\nu = \omega \supset ... \supset \omega^s \supset \omega^{s+1} = \overline{\omega},$$

where, for $i = 1, ..., s$, e_i are the escape situations of $\overline{\omega}$, ω^i the respective escape components and $\hat{\mu}_i$ the first return situation of ω^{i-1} after e_{i-1}. Notice that, for $1 \leq i \leq \nu$, ω^{i-1}, e_{i-1} and $\hat{\mu}_i$ do not depend on $\overline{\omega} \subset E_j'(\zeta_0^{(j-1)})$.

Denote by $\tilde{E}_i\left(\zeta_0^{(j-1)}\right)$ the union of all the escape components ω^i of every $\overline{\omega} \subset E_j'\left(\zeta_0^{(j-1)}\right)$. Let $\langle\omega^{i-1}\rangle = \omega^{i-1} \cap \tilde{E}_i\left(\zeta_0^{(j-1)}\right)$ for each escape component ω^{i-1} of

$\bar{\omega}$. Then, $\tilde{E}_{i-1}\left(\varsigma_0^{(j-1)}\right) = \omega^{i-1}$ for $1 \leq i \leq \nu$ and $\tilde{E}_\nu\left(\varsigma_0^{(j-1)}\right) = \omega$. Therefore, $\tilde{E}_{\nu+1}\left(\varsigma_0^{(j-1)}\right) = \langle\omega\rangle$ and $\tilde{E}_i\left(\varsigma_0^{(j-1)}\right) = \omega^i \subset \tilde{E}_i\left(\varsigma_0^{(\nu-1)}\right)$ for every $1 \leq i \leq \nu$.

For each $\bar{\omega} \subset E'_j\left(\varsigma_0^{(j-1)}\right)$ and for every $1 \leq i \leq j$, we set $S_i(\bar{\omega}) = 0$ if either $k \leq i \leq \nu$ or $s+2 \leq i \leq j$; otherwise, $S_i(\bar{\omega}) = e_i - \hat{\mu}_i$. We also define $T_i(a) = S_1(a) + \ldots + S_i(a)$ for each $a \in \bar{\omega}$. We shall check that

$$m\left(\left\{a \in E'_j\left(\varsigma_0^{(j-1)}\right) : T_j(a) \geq \alpha j\right\}\right) \leq e^{-\frac{1}{100}c\alpha j} |\Omega_0|.$$

Indeed, since $\omega \subset \tilde{E}_\nu\left(\varsigma_0^{(\nu-1)}\right)$, from the proof of Proposition 2.25, we obtain

$$\int_\omega e^{\frac{1}{40}cT_\nu(a)} da \leq e^{\frac{1}{100}c\alpha\nu} |\Omega_0|.$$

Now, since $T_{\nu+1}(a) = T_\nu(a) + S_{\nu+1}(a)$ and $T_\nu(a)$ is constant on ω, we have

$$\int_{\tilde{E}_{\nu+1}\left(\varsigma_0^{(j-1)}\right)} e^{\frac{1}{40}cT_{\nu+1}(a)} da = e^{\frac{1}{40}cT_\nu(\omega)} \int_{\langle\omega\rangle} e^{\frac{1}{40}cS_{\nu+1}(a)} da.$$

Furthermore, as in Proposition 2.24, since $\hat{\mu}_{\nu+1}$ is an escape situation of ω, it follows that

$$\int_{\langle\omega\rangle} e^{\frac{1}{40}cS_{\nu+1}(a)} da \leq e^{\frac{1}{100}c\alpha} |\omega|.$$

Hence,

$$\int_{\tilde{E}_{\nu+1}\left(\varsigma_0^{(j-1)}\right)} e^{\frac{1}{40}cT_{\nu+1}(a)} da \leq e^{\frac{1}{100}c\alpha(\nu+1)} |\Omega_0|.$$

Then, we inductively prove that

$$\int_{\tilde{E}_j\left(\varsigma_0^{(j-1)}\right)} e^{\frac{1}{40}cT_j(a)} da \leq e^{\frac{1}{100}c\alpha j} |\Omega_0|.$$

Finally, since $E'_j\left(\varsigma_0^{(j-1)}\right) = \tilde{E}_j\left(\varsigma_0^{(j-1)}\right)$, from

$$e^{\frac{1}{40}c\alpha j} m\left(\left\{a \in E'_j\left(\varsigma_0^{(j-1)}\right) : T_j(a) \geq \alpha j\right\}\right) \leq e^{\frac{1}{100}c\alpha j} |\Omega_0|,$$

we conclude that

$$m\left(\left\{a \in E'_j\left(\varsigma_0^{(j-1)}\right) : T_j(a) \geq \alpha j\right\}\right) \leq e^{-\frac{1}{100}c\alpha j} |\Omega_0|.$$

Clearly, this implies (8.26).

Actually, Lemma 2.22 is the key to prove the last inequality. To obtain an equivalent bidimensional result, let $\mu_1, \ldots, \mu_s$ be all the inessential returns between two essential returns ν_0 and ν_1 of $\bar{\omega} \in P_{\nu_0}$. Then, from (8.37) and (8.38), we deduce

$$\frac{length\left(\varsigma_{\mu_{i+1}}(\bar{\omega})\right)}{length\left(\varsigma_{\mu_i}(\bar{\omega})\right)} \geq const \ e^{\frac{1}{3}c(\mu_{i+1}-\mu_i)},$$

where p_i is the length of the binding period associated to μ_i and $q_i = \mu_{i+1} - \mu_i - p_i - 1$. Therefore,

$$length\left(\zeta_{\nu_1}(\varpi)\right) \geq (const)^{s+1} \delta e^{\frac{1}{3}c(\nu_1 - \nu_0)} length\left(\zeta_{\nu_0}(\varpi)\right).$$

Now, since $length\left(\zeta_{\nu_1}(\varpi)\right) < e$ and

$$length\left(\zeta_{\nu_0}(\varpi)\right) \geq \frac{e^{-|r_0|}}{2r_0^2},$$

we obtain $1 \geq (s+1)\log const + \frac{1}{3}c(\nu_1 - \nu_0) - |r_0| - 3\log|r_0| - \Delta$, where $const < 1$ and $\Delta = \log\delta^{-1}$. Furthermore, since

$$p_i \geq \frac{|r_i|}{5\log K} \geq \frac{\Delta}{5\log K},$$

it follows that $(s+1) \leq (\nu_1 - \nu_0)\, 5\Delta^{-1}\log K$. Therefore, taking δ small enough, we have $(s+1)\log const + \frac{1}{3}c(\nu_1 - \nu_0) \geq \frac{1}{4}c(\nu_1 - \nu_0)$ and consequently, $\nu_1 - \nu_0 \leq 10c^{-1}|r_0|$.

In summary, from (8.24) and (8.26), we get

$$m\left(E_{j-1}(\zeta_0^{(j-2)}) \setminus E_j(\zeta_0^{(j-1)})\right) \leq e^{-const\, j}|\Omega_0|$$

for every $j = \nu, ..., n$. Thus, since $2\nu \geq n$, it follows that

$$m\left(\omega \setminus E_n(z_0^{(n-1)})\right) \leq |\Omega_0| \sum_{j=\nu}^{n} e^{-const\, j} \leq const\, e^{-const\, n}|\Omega_0|.$$

Let us define $E_n = E_{n-1} \setminus \left(\bigcup_{z_0 \in C_n}\left(\omega \setminus E_n(z_0^{(n-1)})\right)\right)$. From (5.1),

$$m\left(E_{n-1} \setminus E_n\right) \leq const\left(\frac{K}{\rho_0}\right)^{\theta n} e^{-const\, n}|\Omega_0|$$

and, for every sufficiently small b, $m\left(E_{n-1} \setminus E_n\right) \leq const\, e^{-const\, n}|\Omega_0|$ where the constants do not depend on b.

Now, let us check that I.H.7 holds at time n. Notice that this immediately holds if $\nu \geq \frac{1}{2}(n+1)$. Otherwise, we replace ω in I.H.7 at time $n-1$ by one of the escape components which arise in $\left[\frac{1}{2}(n+1), n+1\right]$. Then I.H.7(d) holds trivially. I.H.7(a) follows from Proposition 8.4, I.H.7(b) was already stated in the three previous chapters and I.H.7(c) follows from Lemma 8.12. We may construct $E = \bigcap_{n \in \mathbb{N}} E_n$ and, since $E_n = \Omega_0$ for every $n \leq N$, we eventually achieve

$$m(E) \geq |\Omega_0|\left(1 - const \sum_{n \geq N} e^{-const\, n}\right) > 0. \tag{8.39}$$

For every $a \in E$, the critical points may be defined as the limits of the critical approximations of the same generation. These critical points are bound to their critical approximations and consequently, their images are e^c-expanding for every $a \in E$. Finally, in order to conclude that $\Lambda_{a,m} = cl\left(W^u_{a,m}\right)$ is a strange attractor for a positive Lebesgue measure set of parameters a, we claim that the orbit of the critical point of generation zero is dense in $W^u_{a,m}$ for almost every $a \in E$. This claim is a consequence of the following facts:

1. The backward orbit $O^-(P_{a(\lambda)})$ of the fixed point $P_{a(\lambda)}$ of $f_{a(\lambda)}$ is dense in $W^u(P_{a(\lambda)})$. Indeed, let $J\left(f_{a(\lambda)}\right)$ be the Julia set of $f_{a(\lambda)}$, which coincides with the α-limit set, $\alpha(c_\lambda)$, of the critical point. Since $P_{a(\lambda)} \in \alpha(c_\lambda)$, it follows that $cl\left(O^-(P_{a(\lambda)})\right) \subset J\left(f_{a(\lambda)}\right)$. Furthermore, since $c_\lambda \in cl\left(O^-(P_{a(\lambda)})\right)$ we have $J(f_{a(\lambda)}) = cl\left(O^-(P_{a(\lambda)})\right)$. On the other hand, from Singer's theorem, [26], it follows that the complementary of $J\left(f_{a(\lambda)}\right)$, which is called the Fatou set, is empty. See [16].

2. From the continuity of $a \to f_a$, it follows that, for every $\delta > 0$, there exists $a_0 = a_0(\delta) \leq a(\lambda)$ such that if $a \in [a_0(\delta), a(\lambda)]$, then $O^-(P_a)$ is δ-dense in $W^u(a)$, that is, any interval with length 2δ contains a point $x \in O^-(P_a)$. Therefore, for any disk $B \subset W^u(P_{a,\infty}) \times \mathbf{R}$ with radius δ, we have $W^s(P_{a,\infty}) \cap B \neq \emptyset$, where $P_{a,\infty} = (P_a, 0)$ is a fixed point of $\Psi_a(x,y) = (f_a(x), 0)$. Now, let U_m be any domain given in (1.6) with m large enough. Then, again by continuity, for any disk $B \subset U_m$ with radius δ, we also have $W^s(P_{a,m}) \cap B \neq \emptyset$.

3. Let $\{G_j\}$ be a countable basis of the topology of U_m. Let us assume that there exists a set of values of the parameter $E_0 \subset E$, with $m(E_0) > 0$, such that for every $a \in E_0$, the orbit of $z_0(a)$ is not dense in $\Lambda_{a,m}$. This means that, for each $a \in E_0$, there exists $G_j = G_j(a)$ such that the orbit of $z_0(a)$ does not intercept G_j. Then, there exists a basic open G_j and a set $E_0' \subset E_0$ with $m(E_0') > 0$, such that, for every $a \in E_0'$, the orbit of $z_0(a)$ does not intercept G_j. Indeed, let $A(G_j) = \{a \in E_0 : G_j = G_j(a)\}$. Then, $E_0 = \cup_{j \in \mathbf{N}} A(G_j)$ and, since $m(E_0) > 0$, there exists $A(G_j)$ such that $m(A(G_j)) > 0$. Let $E_0' = A(G_j)$.

4. Let us assume that there exists E_0' as above and let a_0 be a density point of E_0'. Let $\{\omega_n\}_{n \in \mathbf{N}}$ be the family of intervals constructed in this chapter such that $a_0 \in \omega_n \subset E_n(z_0^{(n-1)})$. From the results related to escape periods we obtain an infinite set of natural numbers n satisfying that $z_n^{(n-1)}(\omega_n)$ is an almost horizontal curve of length greater than $\sqrt{\delta}$. Therefore, $z_n^{(n-1)}(\omega_n)$ intercepts $W^s(P_{a,m}) = W^s_{a,m}$

for every $a \in \omega_n$. Hence, there exists $\tilde{a} \in \omega_n$ such that $z_n^{(n-1)}(\tilde{a}) \in W_{\tilde{a},m}^s$ and, by compactness, $z_{n+k}^{(n-1)}(\tilde{a}) \in \left(W_{\tilde{a},m}^s\right)_{loc}$ for some $k \leq k_0 = k_0(\delta)$. Thus, since $T_{\tilde{a}}^{-\tilde{k}}(G_j)$ is close enough to $\left(W_{\tilde{a},m}^s\right)_{loc}$ for a sufficiently large $\tilde{k}$, it follows that $z_{n+k+\tilde{k}}^{(n-1)}(\omega_n)$ intercepts G_j with $length\left(z_{n+k+\tilde{k}}^{(n-1)}(\omega_n) \cap G_j\right) \geq d$, where d is a constant which does not depend on n. Now, let $L_n = \left\{a \in \omega_n : z_{n+k+\tilde{k}}^{(n-1)}(a) \in G_j'\right\}$ be such that $cl\left(G_j'\right) \subset G_j$ and $length(z_n^{(n-1)}(L_n)) > const$. So, we get $L_n \cap E_0' = \emptyset$ and, from Corollary 8.13,

$$\frac{length(L_n)}{length(\omega_n)} \geq const \frac{length(z_n^{(n-1)}(L_n))}{length(z_n^{(n-1)}(\omega_n))} \geq const.$$

This contradicts the fact that $a_0 \in \omega_n$ is a density point of E_0'.

We have just proved that, for every positive $\lambda < 0.3319$ and for every $m \geq m_0$ large enough, there exists $E_m \subset [a_0(\lambda), a(\lambda)]$ with $m(E_m) > \epsilon > 0$, such that $T_{\lambda,a,b}$ has a strange attractor in U_m for every $a \in E_m$. Finally, let $F = \{A_i\}_{i \in \mathbb{N}}$ be such that $A_i = E_{m_0+i}$ and let $k \in \mathbb{N}$ such that $k > \frac{5}{2}\epsilon^{-1}(a(\lambda) - a_0(\lambda))$. Then, there exist $i, j \in \{1, ..., k\}$ such that

$$m(A_i \cap A_j) > \frac{1}{k^2}(a(\lambda) - a_0(\lambda)) = \tilde{\epsilon}.$$

Let us define $\tilde{A}_1 = A_i \cap A_j$. So, we may obtain a new family $\tilde{F} = \left\{\tilde{A}_i\right\}_{i \in \mathbb{N}}$ such that $m(\tilde{A}_i) \geq \tilde{\epsilon}$. Therefore, by repeating this process, we achieve, for each $j \in \mathbb{N}$, 2^j sets in F whose intersection has positive Lebesgue measure. Since A_i can be assumed to be compact sets, we also obtain an infinite number of A_i whose intersection is non-null. The main theorem is proved.

Appendix A

NUMERICAL EXPERIMENTS

Finally, we display some pictures which show numerical experiments related to the map

$$T_{\lambda,a}(\eta,\xi) = \left(1 + \eta e^{\lambda\xi}\sin\xi, \xi + \frac{1}{\lambda}\left(\log a + \log\eta + \log\cos\xi\right)\right)$$

given in (1.3). Namely, in the domain

$$\left\{(\eta,\xi) \in \mathbf{R}^2 : 1.015 \le \eta \le 1.03, -13.5 \le \xi \le -11\right\} \subset C_2$$

we represent the first 30000 iterates of the $T_{\lambda,a}$-orbit of the initial point $(1.02, -11.45)$ for $\lambda = 0.3$ and $a = 1.3$:

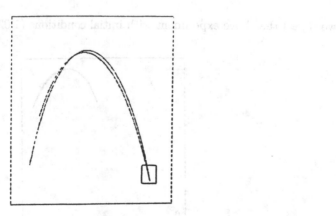

The following pictures are successive amplifications of the previous one. We show the same experiment in

$$\left\{(\eta,\xi) \in \mathbf{R}^2 : 1.0275 \le \eta \le 1.02825, -13.25 \le \xi \le -12.75\right\}$$

and

$$\left\{(\eta,\xi) \in \mathbf{R}^2 : 1.02803 \le \eta \le 1.028045, -13.21 \le \xi \le -13.19\right\},$$

respectively:

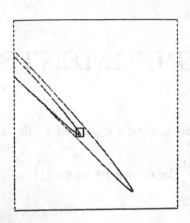

At last, in the domain

$$\left\{(\eta,\xi) \in \mathbf{R}^2 : 1.002 \le \eta \le 1.03, -19.5 \le \xi \le -11\right\}$$

we repeat the above experiment with initial conditions $(1.02, -11.45)$ and $(1, -18)$.

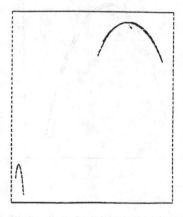

Bibliography

[1] A. Arneodo, P. Coullet, E. Spiegel and C. Tresser. *Asymptotic chaos.* Physica, **14 D**, 327-347, *(1985)*.

[2] M. Benedicks and L. Carleson. *On iterations of* $1 - ax^2$ *on (-1,1).* Ann. Math., **122**, 1-25, *(1985)*.

[3] M. Benedicks and L. Carleson. *The dynamics of the Hénon map.* Ann. Math., **133**, 73-169, *(1991)*.

[4] G. D. Birkhoff. *Nouvelles recherches sur les systèmes dynamiques.* Mem. Pont. Acad. Sci. Novi. Lyncaei **1**, 85-216, *(1935)*.

[5] M. Bosch and C. Simó. *Attractors in a Sil'nikov-Hopf scenario and a related one-dimensional map.* Physica-D **62**, 217-229, *(1993)*.

[6] P. Hartman. *Ordinary differential equations.* Wiley, *(1964)*.

[7] M. Hénon. *A two-dimensional mapping with a strange attractor.* Comm. Math. Phys. **50**, 69-77, *(1976)*.

[8] M. Hirsch, C. Pugh and M. Shub. *Invariant manifolds.* Lecture Notes in Math. **583**, Springer, *(1977)*.

[9] S. Ibáñez and J. A. Rodríguez. *Sil'nikov bifurcations in generic 4-unfolding of a codimension-4 singularity.* Jour. Diff. Eq. **120**, 411-428, *(1995)*.

[10] M. V. Jacobson. *Absolutely continuous invariant measures for one-parameter families of one-dimensional maps.* Comm. Math. Phys. **81**, 39-88, *(1981)*.

[11] E. N. Lorenz. *Deterministic non-periodic flow.* J. Atmos. Sci. **20**, 130-141, *(1963)*.

[12] **M. Yu. Lyubich.** *Non-existence of wandering intervals and structure of topological attractors of one dimensional dynamical systems I. The case of negative Schwarzian derivative.* Ergod. Th & Dynam. Sys. **9**, 737-750, *(1989)*.

[13] **W. de Melo** and **S. Van Strien.** *One-dimensional dynamics.* Springer-Verlag. Berlin, *(1993)*.

[14] **J. Milnor.** *On the concept of attractor.* Comm. Math. Phys. **99**, 177-195, *(1985)*.

[15] **L. Mora** and **M. Viana.** *Abundance of strange attractors.* Acta Math. **171**, 1-71, *(1993)*.

[16] **F. J. S. Moreira.** *Chaotic dynamics of quadratic maps.* Universidad do Porto, *(1992)*.

[17] **J. Palis** and **F. Takens.** *Hyperbolicity and sensitive chaotic dynamics at homoclinic bifurcations.* Cambridge University Press, *(1993)*.

[18] **H. Poincaré.** *Sur le problème des trois corps et les équations de la dynamique.* Acta Math. **13**, 1-270, *(1890)*.

[19] **J. A. Rodríguez.** *Bifurcations to homoclinic connections of the focus-saddle type.* Arch. Rat. Mech. and Anal. **93**, 81-90, *(1986)*.

[20] **D. Ruelle** and **F. Takens.** *On the nature of turbulence.* Comm. Math. Phys. **20**, 167-192 (1971) / **23**, 343-344, *(1971)*.

[21] **S. Smale.** *Differentiable dynamical systems.* Bull. Amer. Math. Soc. **73**, 747-817, *(1967)*.

[22] **L. P. Sil'nikov.** *A case of the existence of a denumerable set of periodic motions.* Sov. Math. Dokl. **6**, 163-166, *(1965)*.

[23] **D. Singer.** *Stable orbits and bifurcations of maps of the interval.* SIAM. J. Appl. Math. **35**, 260-267, *(1978)*.

[24] **C. Tresser.** *Modèles simples de transitions vers la turbulence.* Thèse d'Etat. Universitè de Nice, *(1981)*.

[25] **M. Viana.** *Strange attractors in higher dimensions.* Thesis. IMPA, *(1991)*.

Index

algorithm A, 85
algorithm B, 87
analytic continuation, 155

basic assumption, 23, 105
binding period, 23, 27, 34, 96, 150
binding point, 95
bound return, 96
bounded distortion, 29, 136, 170

contracting directions, 54
contractive fields, 87
critical approximation, 79
critical point, 73
critical set, 85, 89, 158

escape component, 35, 169
escape period, 45, 157
escape situation, 35, 169
essential return, 35, 168
expanding directions, 54
expanding points, 53
exponential growth, 22

favourable position, 119
fold-free iterates, 103
folding period, 100
free assumption, 23, 106
free period, 23, 34
free return, 96

generation, 79

homoclinic orbit, 11
homoclinic point, 19
homoclinic tangency, 78
host interval, 33, 168

inessential return, 35, 168

Liapunov exponent, 1

Misiurewicz map, 22

return, 23, 33, 95
return situation, 34, 168

semifavourable position, 119
splitting algorithm, 100
stair structure, 97, 100
strange attractor, 13

tangential position, 119

Lecture Notes in Mathematics

For information about Vols. 1–1469
please contact your bookseller or Springer-Verlag

Vol. 1470: E. Odell, H. Rosenthal (Eds.), Functional Analysis. Proceedings, 1987-89. VII, 199 pages. 1991.

Vol. 1471: A. A. Panchishkin, Non-Archimedean L-Functions of Siegel and Hilbert Modular Forms. VII, 157 pages. 1991.

Vol. 1472: T. T. Nielsen, Bose Algebras: The Complex and Real Wave Representations. V, 132 pages. 1991.

Vol. 1473: Y. Hino, S. Murakami, T. Naito, Functional Differential Equations with Infinite Delay. X, 317 pages. 1991.

Vol. 1474: S. Jackowski, B. Oliver, K. Pawałowski (Eds.), Algebraic Topology, Poznań 1989. Proceedings. VIII, 397 pages. 1991.

Vol. 1475: S. Busenberg, M. Martelli (Eds.), Delay Differential Equations and Dynamical Systems. Proceedings, 1990. VIII, 249 pages. 1991.

Vol. 1476: M. Bekkali, Topics in Set Theory. VII, 120 pages. 1991.

Vol. 1477: R. Jajte, Strong Limit Theorems in Noncommutative L_2-Spaces. X, 113 pages. 1991.

Vol. 1478: M.-P. Malliavin (Ed.), Topics in Invariant Theory. Seminar 1989-1990. VI, 272 pages. 1991.

Vol. 1479: S. Bloch, I. Dolgachev, W. Fulton (Eds.), Algebraic Geometry. Proceedings, 1989. VII, 300 pages. 1991.

Vol. 1480: F. Dumortier, R. Roussarie, J. Sotomayor, H. Żoładek, Bifurcations of Planar Vector Fields: Nilpotent Singularities and Abelian Integrals. VIII, 226 pages. 1991.

Vol. 1481: D. Ferus, U. Pinkall, U. Simon, B. Wegner (Eds.), Global Differential Geometry and Global Analysis. Proceedings, 1991. VIII, 283 pages. 1991.

Vol. 1482: J. Chabrowski, The Dirichlet Problem with L^2-Boundary Data for Elliptic Linear Equations. VI, 173 pages. 1991.

Vol. 1483: E. Reithmeier, Periodic Solutions of Nonlinear Dynamical Systems. VI, 171 pages. 1991.

Vol. 1484: H. Delfs, Homology of Locally Semialgebraic Spaces. IX, 136 pages. 1991.

Vol. 1485: J. Azéma, P. A. Meyer, M. Yor (Eds.), Séminaire de Probabilités XXV. VIII, 440 pages. 1991.

Vol. 1486: L. Arnold, H. Crauel, J.-P. Eckmann (Eds.), Lyapunov Exponents. Proceedings, 1990. VIII, 365 pages. 1991.

Vol. 1487: E. Freitag, Singular Modular Forms and Theta Relations. VI, 172 pages. 1991.

Vol. 1488: A. Carboni, M. C. Pedicchio, G. Rosolini (Eds.), Category Theory. Proceedings, 1990. VII, 494 pages. 1991.

Vol. 1489: A. Mielke, Hamiltonian and Lagrangian Flows on Center Manifolds. X, 140 pages. 1991.

Vol. 1490: K. Metsch, Linear Spaces with Few Lines. XIII, 196 pages. 1991.

Vol. 1491: E. Lluis-Puebla, J.-L. Loday, H. Gillet, C. Soulé, V. Snaith, Higher Algebraic K-Theory: an overview. IX, 164 pages. 1992.

Vol. 1492: K. R. Wicks, Fractals and Hyperspaces. VIII, 168 pages. 1991.

Vol. 1493: E. Benoît (Ed.), Dynamic Bifurcations. Proceedings, Luminy 1990. VII, 219 pages. 1991.

Vol. 1494: M.-T. Cheng, X.-W. Zhou, D.-G. Deng (Eds.), Harmonic Analysis. Proceedings, 1988. IX, 226 pages. 1991.

Vol. 1495: J. M. Bony, G. Grubb, L. Hörmander, H. Komatsu, J. Sjöstrand, Microlocal Analysis and Applications. Montecatini Terme, 1989. Editors: L. Cattabriga, L. Rodino. VII, 349 pages. 1991.

Vol. 1496: C. Foias, B. Francis, J. W. Helton, H. Kwakernaak, J. B. Pearson, H_∞-Control Theory. Como, 1990. Editors: E. Mosca, L. Pandolfi. VII, 336 pages. 1991.

Vol. 1497: G. T. Herman, A. K. Louis, F. Natterer (Eds.), Mathematical Methods in Tomography. Proceedings 1990. X, 268 pages. 1991.

Vol. 1498: R. Lang, Spectral Theory of Random Schrödinger Operators. X, 125 pages. 1991.

Vol. 1499: K. Taira, Boundary Value Problems and Markov Processes. IX, 132 pages. 1991.

Vol. 1500: J.-P. Serre, Lie Algebras and Lie Groups. VII, 168 pages. 1992.

Vol. 1501: A. De Masi, E. Presutti, Mathematical Methods for Hydrodynamic Limits. IX, 196 pages. 1991.

Vol. 1502: C. Simpson, Asymptotic Behavior of Monodromy. V, 139 pages. 1991.

Vol. 1503: S. Shokranian, The Selberg-Arthur Trace Formula (Lectures by J. Arthur). VII, 97 pages. 1991.

Vol. 1504: J. Cheeger, M. Gromov, C. Okonek, P. Pansu, Geometric Topology: Recent Developments. Editors: P. de Bartolomeis, F. Tricerri. VII, 197 pages. 1991.

Vol. 1505: K. Kajitani, T. Nishitani, The Hyperbolic Cauchy Problem. VII, 168 pages. 1991.

Vol. 1506: A. Buium, Differential Algebraic Groups of Finite Dimension. XV, 145 pages. 1992.

Vol. 1507: K. Hulek, T. Peternell, M. Schneider, F.-O. Schreyer (Eds.), Complex Algebraic Varieties. Proceedings, 1990. VII, 179 pages. 1992.

Vol. 1508: M. Vuorinen (Ed.), Quasiconformal Space Mappings. A Collection of Surveys 1960-1990. IX, 148 pages. 1992.

Vol. 1509: J. Aguadé, M. Castellet, F. R. Cohen (Eds.), Algebraic Topology - Homotopy and Group Cohomology. Proceedings, 1990. X, 330 pages. 1992.

Vol. 1510: P. P. Kulish (Ed.), Quantum Groups. Proceedings, 1990. XII, 398 pages. 1992.

Vol. 1511: B. S. Yadav, D. Singh (Eds.), Functional Analysis and Operator Theory. Proceedings, 1990. VIII, 223 pages. 1992.

Vol. 1512: L. M. Adleman, M.-D. A. Huang, Primality Testing and Abelian Varieties Over Finite Fields. VII, 142 pages. 1992.

Vol. 1513: L. S. Block, W. A. Coppel, Dynamics in One Dimension. VIII, 249 pages. 1992.

Vol. 1514: U. Krengel, K. Richter, V. Warstat (Eds.), Ergodic Theory and Related Topics III, Proceedings, 1990. VIII, 236 pages. 1992.

Vol. 1515: E. Ballico, F. Catanese, C. Ciliberto (Eds.), Classification of Irregular Varieties. Proceedings, 1990. VII, 149 pages. 1992.

Vol. 1516: R. A. Lorentz, Multivariate Birkhoff Interpolation. IX, 192 pages. 1992.

Vol. 1517: K. Keimel, W. Roth, Ordered Cones and Approximation. VI, 134 pages. 1992.

Vol. 1518: H. Stichtenoth, M. A. Tsfasman (Eds.), Coding Theory and Algebraic Geometry. Proceedings, 1991. VIII, 223 pages. 1992.

Vol. 1519: M. W. Short, The Primitive Soluble Permutation Groups of Degree less than 256. IX, 145 pages. 1992.

Vol. 1520: Yu. G. Borisovich, Yu. E. Gliklikh (Eds.), Global Analysis – Studies and Applications V. VII, 284 pages. 1992.

Vol. 1521: S. Busenberg, B. Forte, H. K. Kuiken, Mathematical Modelling of Industrial Process. Bari, 1990. Editors: V. Capasso, A. Fasano. VII, 162 pages. 1992.

Vol. 1522: J.-M. Delort, F. B. I. Transformation. VII, 101 pages. 1992.

Vol. 1523: W. Xue, Rings with Morita Duality. X, 168 pages. 1992.

Vol. 1524: M. Coste, L. Mahé, M.-F. Roy (Eds.), Real Algebraic Geometry. Proceedings, 1991. VIII, 418 pages. 1992.

Vol. 1525: C. Casacuberta, M. Castellet (Eds.), Mathematical Research Today and Tomorrow. VII, 112 pages. 1992.

Vol. 1526: J. Azéma, P. A. Meyer, M. Yor (Eds.), Séminaire de Probabilités XXVI. X, 633 pages. 1992.

Vol. 1527: M. I. Freidlin, J.-F. Le Gall, Ecole d'Eté de Probabilités de Saint-Flour XX – 1990. Editor: P. L. Hennequin. VIII, 244 pages. 1992.

Vol. 1528: G. Isac, Complementarity Problems. VI, 297 pages. 1992.

Vol. 1529: J. van Neerven, The Adjoint of a Semigroup of Linear Operators. X, 195 pages. 1992.

Vol. 1530: J. G. Heywood, K. Masuda, R. Rautmann, S. A. Solonnikov (Eds.), The Navier-Stokes Equations II – Theory and Numerical Methods. IX, 322 pages. 1992.

Vol. 1531: M. Stoer, Design of Survivable Networks. IV, 206 pages. 1992.

Vol. 1532: J. F. Colombeau, Multiplication of Distributions. X, 184 pages. 1992.

Vol. 1533: P. Jipsen, H. Rose, Varieties of Lattices. X, 162 pages. 1992.

Vol. 1534: C. Greither, Cyclic Galois Extensions of Commutative Rings. X, 145 pages. 1992.

Vol. 1535: A. B. Evans, Orthomorphism Graphs of Groups. VIII, 114 pages. 1992.

Vol. 1536: M. K. Kwong, A. Zettl, Norm Inequalities for Derivatives and Differences. VII, 150 pages. 1992.

Vol. 1537: P. Fitzpatrick, M. Martelli, J. Mawhin, R. Nussbaum, Topological Methods for Ordinary Differential Equations. Montecatini Terme, 1991. Editors: M. Furi, P. Zecca. VII, 218 pages. 1993.

Vol. 1538: P.-A. Meyer, Quantum Probability for Probabilists. X, 287 pages. 1993.

Vol. 1539: M. Coornaert, A. Papadopoulos, Symbolic Dynamics and Hyperbolic Groups. VIII, 138 pages. 1993.

Vol. 1540: H. Komatsu (Ed.), Functional Analysis and Related Topics, 1991. Proceedings. XXI, 413 pages. 1993.

Vol. 1541: D. A. Dawson, B. Maisonneuve, J. Spencer, Ecole d´ Eté de Probabilités de Saint-Flour XXI - 1991. Editor: P. L. Hennequin. VIII, 356 pages. 1993.

Vol. 1542: J.Fröhlich, Th.Kerler, Quantum Groups, Quantum Categories and Quantum Field Theory. VII, 431 pages. 1993.

Vol. 1543: A. L. Dontchev, T. Zolezzi, Well-Posed Optimization Problems. XII, 421 pages. 1993.

Vol. 1544: M.Schürmann, White Noise on Bialgebras. VII, 146 pages. 1993.

Vol. 1545: J. Morgan, K. O'Grady, Differential Topology of Complex Surfaces. VIII, 224 pages. 1993.

Vol. 1546: V. V. Kalashnikov, V. M. Zolotarev (Eds.), Stability Problems for Stochastic Models. Proceedings, 1991. VIII, 229 pages. 1993.

Vol. 1547: P. Harmand, D. Werner, W. Werner, M-ideals in Banach Spaces and Banach Algebras. VIII, 387 pages. 1993.

Vol. 1548: T. Urabe, Dynkin Graphs and Quadrilateral Singularities. VI, 233 pages. 1993.

Vol. 1549: G. Vainikko, Multidimensional Weakly Singular Integral Equations. XI, 159 pages. 1993.

Vol. 1550: A. A. Gonchar, E. B. Saff (Eds.), Methods of Approximation Theory in Complex Analysis and Mathematical Physics IV, 222 pages, 1993.

Vol. 1551: L. Arkeryd, P. L. Lions, P.A. Markowich, S.R. S. Varadhan. Nonequilibrium Problems in Many-Particle Systems. Montecatini, 1992. Editors: C. Cercignani, M. Pulvirenti. VII, 158 pages 1993.

Vol. 1552: J. Hilgert, K.-H. Neeb, Lie Semigroups and their Applications. XII, 315 pages. 1993.

Vol. 1553: J.-L- Colliot-Thélène, J. Kato, P. Vojta. Arithmetic Algebraic Geometry. Trento, 1991. Editor: E. Ballico. VII, 223 pages. 1993.

Vol. 1554: A. K. Lenstra, H. W. Lenstra, Jr. (Eds.), The Development of the Number Field Sieve. VIII, 131 pages. 1993.

Vol. 1555: O. Liess, Conical Refraction and Higher Microlocalization. X, 389 pages. 1993.

Vol. 1556: S. B. Kuksin, Nearly Integrable Infinite-Dimensional Hamiltonian Systems. XXVII, 101 pages. 1993.

Vol. 1557: J. Azéma, P. A. Meyer, M. Yor (Eds.), Séminaire de Probabilités XXVII. VI, 327 pages. 1993.

Vol. 1558: T. J. Bridges, J. E. Furter, Singularity Theory and Equivariant Symplectic Maps. VI, 226 pages. 1993.

Vol. 1559: V. G. Sprindžuk, Classical Diophantine Equations. XII, 228 pages. 1993.

Vol. 1560: T. Bartsch, Topological Methods for Variational Problems with Symmetries. X, 152 pages. 1993.

Vol. 1561: I. S. Molchanov, Limit Theorems for Unions of Random Closed Sets. X, 157 pages. 1993.

Vol. 1562: G. Harder, Eisensteinkohomologie und die Konstruktion gemischter Motive. XX, 184 pages. 1993.

Vol. 1563: E. Fabes, M. Fukushima, L. Gross, C. Kenig, M. Röckner, D. W. Stroock, Dirichlet Forms. Varenna, 1992. Editors: G. Dell'Antonio, U. Mosco. VII, 245 pages. 1993.

Vol. 1564: J. Jorgenson, S. Lang, Basic Analysis of Regularized Series and Products. IX, 122 pages. 1993.

Vol. 1565: L. Boutet de Monvel, C. De Concini, C. Procesi, P. Schapira, M. Vergne. D-modules, Representation Theory, and Quantum Groups. Venezia, 1992. Editors: G. Zampieri, A. D'Agnolo. VII, 217 pages. 1993.

Vol. 1566: B. Edixhoven, J.-H. Evertse (Eds.), Diophantine Approximation and Abelian Varieties. XIII, 127 pages. 1993.

Vol. 1567: R. L. Dobrushin, S. Kusuoka, Statistical Mechanics and Fractals. VII, 98 pages. 1993.

Vol. 1568: F. Weisz, Martingale Hardy Spaces and their Application in Fourier Analysis. VIII, 217 pages. 1994.

Vol. 1569: V. Totik, Weighted Approximation with Varying Weight. VI, 117 pages. 1994.

Vol. 1570: R. deLaubenfels, Existence Families, Functional Calculi and Evolution Equations. XV, 234 pages. 1994.

Vol. 1571: S. Yu. Pilyugin, The Space of Dynamical Systems with the C^0-Topology. X, 188 pages. 1994.

Vol. 1572: L. Göttsche, Hilbert Schemes of Zero-Dimensional Subschemes of Smooth Varieties. IX, 196 pages. 1994.

Vol. 1573: V. P. Havin, N. K. Nikolski (Eds.), Linear and Complex Analysis – Problem Book 3 – Part I. XXII, 489 pages. 1994.

Vol. 1574: V. P. Havin, N. K. Nikolski (Eds.), Linear and Complex Analysis – Problem Book 3 – Part II. XXII, 507 pages. 1994.

Vol. 1575: M. Mitrea, Clifford Wavelets, Singular Integrals, and Hardy Spaces. XI, 116 pages. 1994.

Vol. 1576: K. Kitahara, Spaces of Approximating Functions with Haar-Like Conditions. X, 110 pages. 1994.

Vol. 1577: N. Obata, White Noise Calculus and Fock Space. X, 183 pages. 1994.

Vol. 1578: J. Bernstein, V. Lunts, Equivariant Sheaves and Functors. V, 139 pages. 1994.

Vol. 1579: N. Kazamaki, Continuous Exponential Martingales and BMO. VII, 91 pages. 1994.

Vol. 1580: M. Milman, Extrapolation and Optimal Decompositions with Applications to Analysis. XI, 161 pages. 1994.

Vol. 1581: D. Bakry, R. D. Gill, S. A. Molchanov, Lectures on Probability Theory. Editor: P. Bernard. VIII, 420 pages. 1994.

Vol. 1582: W. Balser, From Divergent Power Series to Analytic Functions. X, 108 pages. 1994.

Vol. 1583: J. Azéma, P. A. Meyer, M. Yor (Eds.), Séminaire de Probabilités XXVIII. VI, 334 pages. 1994.

Vol. 1584: M. Brokate, N. Kenmochi, I. Müller, J. F. Rodriguez, C. Verdi, Phase Transitions and Hysteresis. Montecatini Terme, 1993. Editor: A. Visintin. VII. 291 pages. 1994.

Vol. 1585: G. Frey (Ed.), On Artin's Conjecture for Odd 2-dimensional Representations. VIII, 148 pages. 1994.

Vol. 1586: R. Nillsen, Difference Spaces and Invariant Linear Forms. XII, 186 pages. 1994.

Vol. 1587: N. Xi, Representations of Affine Hecke Algebras. VIII, 137 pages. 1994.

Vol. 1588: C. Scheiderer, Real and Étale Cohomology. XXIV, 273 pages. 1994.

Vol. 1589: J. Bellissard, M. Degli Esposti, G. Forni, S. Graffi, S. Isola, J. N. Mather, Transition to Chaos in Classical and Quantum Mechanics. Montecatini Terme, 1991. Editor: S. Graffi. VII, 192 pages. 1994.

Vol. 1590: P. M. Soardi, Potential Theory on Infinite Networks. VIII, 187 pages. 1994.

Vol. 1591: M. Abate, G. Patrizio, Finsler Metrics – A Global Approach. IX, 180 pages. 1994.

Vol. 1592: K. W. Breitung, Asymptotic Approximations for Probability Integrals. IX, 146 pages. 1994.

Vol. 1593: J. Jorgenson & S. Lang, D. Goldfeld, Explicit Formulas for Regularized Products and Series. VIII, 154 pages. 1994.

Vol. 1594: M. Green, J. Murre, C. Voisin, Algebraic Cycles and Hodge Theory. Torino, 1993. Editors: A. Albano, F. Bardelli. VII, 275 pages. 1994.

Vol. 1595: R.D.M. Accola, Topics in the Theory of Riemann Surfaces. IX, 105 pages. 1994.

Vol. 1596: L. Heindorf, L. B. Shapiro, Nearly Projective Boolean Algebras. X, 202 pages. 1994.

Vol. 1597: B. Herzog, Kodaira-Spencer Maps in Local Algebra. XVII, 176 pages. 1994.

Vol. 1598: J. Berndt, F. Tricerri, L. Vanhecke, Generalized Heisenberg Groups and Damek-Ricci Harmonic Spaces. VIII, 125 pages. 1995.

Vol. 1599: K. Johannson, Topology and Combinatorics of 3-Manifolds. XVIII, 446 pages. 1995.

Vol. 1600: W. Narkiewicz, Polynomial Mappings. VII, 130 pages. 1995.

Vol. 1601: A. Pott, Finite Geometry and Character Theory. VII, 181 pages. 1995.

Vol. 1602: J. Winkelmann, The Classification of Three-dimensional Homogeneous Complex Manifolds. XI, 230 pages. 1995.

Vol. 1603: V. Ene, Real Functions – Current Topics. XIII, 310 pages. 1995.

Vol. 1604: A. Huber, Mixed Motives and their Realization in Derived Categories. XV, 207 pages. 1995.

Vol. 1605: L. B. Wahlbin, Superconvergence in Galerkin Finite Element Methods. XI, 166 pages. 1995.

Vol. 1606: P.-D. Liu, M. Qian, Smooth Ergodic Theory of Random Dynamical Systems. XI, 221 pages. 1995.

Vol. 1607: G. Schwarz, Hodge Decomposition – A Method for Solving Boundary Value Problems. VII, 155 pages. 1995.

Vol. 1608: P. Biane, R. Durrett, Lectures on Probability Theory. Editor: P. Bernard. VII, 210 pages. 1995.

Vol. 1609: L. Arnold, C. Jones, K. Mischaikow, G. Raugel, Dynamical Systems. Montecatini Terme, 1994. Editor: R. Johnson. VIII, 329 pages. 1995.

Vol. 1610: A. S. Üstünel, An Introduction to Analysis on Wiener Space. X, 95 pages. 1995.

Vol. 1611: N. Knarr, Translation Planes. VI, 112 pages. 1995.

Vol. 1612: W. Kühnel, Tight Polyhedral Submanifolds and Tight Triangulations. VII, 122 pages. 1995.

Vol. 1613: J. Azéma, M. Emery, P. A. Meyer, M. Yor (Eds.), Séminaire de Probabilités XXIX. VI, 326 pages. 1995.

Vol. 1614: A. Koshelev, Regularity Problem for Quasilinear Elliptic and Parabolic Systems. XXI, 255 pages. 1995.

Vol. 1615: D. B. Massey, Lê Cycles and Hypersurface Singularities. XI, 131 pages. 1995.

Vol. 1616: I. Moerdijk, Classifying Spaces and Classifying Topoi. VII, 94 pages. 1995.

Vol. 1617: V. Yurinsky, Sums and Gaussian Vectors. XI, 305 pages. 1995.

Vol. 1618: G. Pisier, Similarity Problems and Completely Bounded Maps. VII, 156 pages. 1996.

Vol. 1619: E. Landvogt, A Compactification of the Bruhat-Tits Building. VII, 152 pages. 1996.

Vol. 1620: R. Donagi, B. Dubrovin, E. Frenkel, E. Previato, Integrable Systems and Quantum Groups. Montecatini Terme, 1993. Editors: M. Francaviglia, S. Greco. VIII, 488 pages. 1996.

Vol. 1621: H. Bass, M. V. Otero-Espinar, D. N. Rockmore, C. P. L. Tresser, Cyclic Renormalization and Auto-morphism Groups of Rooted Trees. XXI, 136 pages. 1996.

Vol. 1622: E. D. Farjoun, Cellular Spaces, Null Spaces and Homotopy Localization. XIV, 199 pages. 1996.

Vol. 1623: H.P. Yap, Total Colourings of Graphs. VIII, 131 pages. 1996.

Vol. 1624: V. Brînzănescu, Holomorphic Vector Bundles over Compact Complex Surfaces. X, 170 pages. 1996.

Vol. 1625: S. Lang, Topics in Cohomology of Groups. VII, 226 pages. 1996.

Vol. 1626: J. Azéma, M. Emery, M. Yor (Eds.), Séminaire de Probabilités XXX. VIII, 382 pages. 1996.

Vol. 1627: C. Graham, Th. G. Kurtz, S. Méléard, Ph. E. Protter, M. Pulvirenti, D. Talay, Probabilistic Models for Nonlinear Partial Differential Equations. Montecatini Terme, 1995. Editors: D. Talay, L. Tubaro. X, 301 pages. 1996.

Vol. 1628: P.-H. Zieschang, An Algebraic Approach to Association Schemes. XII, 189 pages. 1996.

Vol. 1629: J. D. Moore, Lectures on Seiberg-Witten Invariants. VII, 105 pages. 1996.

Vol. 1630: D. Neuenschwander, Probabilities on the Heisenberg Group: Limit Theorems and Brownian Motion. VIII, 139 pages. 1996.

Vol. 1631: K. Nishioka, Mahler Functions and Transcendence. VIII, 185 pages. 1996.

Vol. 1632: A. Kushkuley, Z. Balanov, Geometric Methods in Degree Theory for Equivariant Maps. VII, 136 pages. 1996.

Vol. 1633: H. Aikawa, M. Essén, Potential Theory – Selected Topics. IX, 200 pages. 1996.

Vol. 1634: J. Xu, Flat Covers of Modules. IX, 161 pages. 1996.

Vol. 1635: E. Hebey, Sobolev Spaces on Riemannian Manifolds. X, 116 pages. 1996.

Vol. 1636: M. A. Marshall, Spaces of Orderings and Abstract Real Spectra. VI, 190 pages. 1996.

Vol. 1637: B. Hunt, The Geometry of some special Arithmetic Quotients. XIII, 332 pages. 1996.

Vol. 1638: P. Vanhaecke, Integrable Systems in the realm of Algebraic Geometry. VIII, 218 pages. 1996.

Vol. 1639: K. Dekimpe, Almost-Bieberbach Groups: Affine and Polynomial Structures. X, 259 pages. 1996.

Vol. 1640: G. Boillat, C. M. Dafermos, P. D. Lax, T. P. Liu, Recent Mathematical Methods in Nonlinear Wave Propagation. Montecatini Terme, 1994. Editor: T. Ruggeri. VII, 142 pages. 1996.

Vol. 1641: P. Abramenko, Twin Buildings and Applications to S-Arithmetic Groups. IX, 123 pages. 1996.

Vol. 1642: M. Puschnigg, Asymptotic Cyclic Cohomology. XXII, 138 pages. 1996.

Vol. 1643: J. Richter-Gebert, Realization Spaces of Polytopes. XI, 187 pages. 1996.

Vol. 1644: A. Adler, S. Ramanan, Moduli of Abelian Varieties. VI, 196 pages. 1996.

Vol. 1645: H. W. Broer, G. B. Huitema, M. B. Sevryuk, Quasi-Periodic Motions in Families of Dynamical Systems. XI, 195 pages. 1996.

Vol. 1646: J.-P. Demailly, T. Peternell, G. Tian, A. N. Tyurin, Transcendental Methods in Algebraic Geometry. Cetraro, 1994. Editors: F. Catanese, C. Ciliberto. VII, 257 pages. 1996.

Vol. 1647: D. Dias, P. Le Barz, Configuration Spaces over Hilbert Schemes and Applications. VII. 143 pages. 1996.

Vol. 1648: R. Dobrushin, P. Groeneboom, M. Ledoux, Lectures on Probability Theory and Statistics. Editor: P. Bernard. VIII, 300 pages. 1996.

Vol. 1649: S. Kumar, G. Laumon, U. Stuhler, Vector Bundles on Curves – New Directions. Cetraro, 1995. Editor: M. S. Narasimhan. VII, 193 pages. 1997.

Vol. 1650: J. Wildeshaus, Realizations of Polylogarithms. XI, 343 pages. 1997.

Vol. 1651: M. Drmota, R. F. Tichy, Sequences, Discrepancies and Applications. XIII, 503 pages. 1997.

Vol. 1652: S. Todorcevic, Topics in Topology. VIII, 153 pages. 1997.

Vol. 1653: R. Benedetti, C. Petronio, Branched Standard Spines of 3-manifolds. VIII, 132 pages. 1997.

Vol. 1654: R. W. Ghrist, P. J. Holmes, M. C. Sullivan, Knots and Links in Three-Dimensional Flows. X, 208 pages. 1997.

Vol. 1655: J. Azéma, M. Emery, M. Yor (Eds.), Séminaire de Probabilités XXXI. VIII, 329 pages. 1997.

Vol. 1656: B. Biais, T. Björk, J. Cvitanić, N. El Karoui, E. Jouini, J. C. Rochet, Financial Mathematics. Bressanone, 1996. Editor: W. J. Runggaldier. VII, 316 pages. 1997.

Vol. 1657: H. Reimann, The semi-simple zeta function of quaternionic Shimura varieties. IX, 143 pages. 1997.

Vol. 1658: A. Pumariño, J. A. Rodríguez, Coexistence and Persistence of Strange Attractors. VIII, 195 pages. 1997.